Pragmatism's Evolution

Pragmatism's Evolution

Organism and Environment in American Philosophy

TREVOR PEARCE

The University of Chicago Press   Chicago and London

PUBLICATION OF THIS BOOK HAS BEEN AIDED BY A GRANT FROM THE BEVINGTON FUND.

The University of Chicago Press, Chicago 60637
The University of Chicago Press, Ltd., London
© 2020 by The University of Chicago
All rights reserved. No part of this book may be used or reproduced in any manner whatsoever without written permission, except in the case of brief quotations in critical articles and reviews. For more information, contact the University of Chicago Press, 1427 E. 60th St., Chicago, IL 60637.
Published 2020
Printed in the United States of America

29 28 27 26 25 24 23 22 21 20   1 2 3 4 5

ISBN-13: 978-0-226-71988-7 (cloth)
ISBN-13: 978-0-226-71991-7 (paper)
ISBN-13: 978-0-226-72008-1 (e-book)
DOI: https://doi.org/10.7208/chicago/9780226720081.001.0001

Library of Congress Cataloging-in-Publication Data

Names: Pearce, Trevor, author.
Title: Pragmatism's evolution : organism and environment in American philosophy / Trevor Pearce.
Description: Chicago ; London : The University of Chicago Press, 2020. | Includes bibliographical references and index.
Identifiers: LCCN 2019058280 | ISBN 9780226719887 (cloth) | ISBN 9780226719917 (paperback) | ISBN 9780226720081 (ebook)
Subjects: LCSH: Pragmatism. | Evolution. | Philosophy and science—United States. | Philosophy, American—19th century. | United States—Intellectual life—19th century.
Classification: LCC B944.P72 P33 2020 | DDC 144/.3—dc23
LC record available at https://lccn.loc.gov/2019058280

Contents

List of Tables and Figures vii
Abbreviations of Manuscript Sources ix
Abbreviations of Scholarly Editions xi
Note to the Reader xiii

Introduction 1

1 The Metaphysical Club and the *Origin of Species* 23

2 Products of the Environment: Spencer's Challenge 58

   Spencerian Evolution 60
   Spencerian Psychology 67
   Spencerian Sociology 86

3 Evolution at School: Educating a New Generation 101

   Evolution in College 102
   Evolution in Graduate School 127
   Teaching Evolution 143

4 "Hegelianism Needs to Be Darwinized": Evolution and Idealism 159

   Hegel and Evolution 161
   The Organism-Environment Dialectic 168
   Evolutionary Strivings 180

5 Weismannism Comes to America: The Factors of Evolution 196

   The Reception of Weismann 201
   Peirce and Neo-Lamarckism 211
   Dewey and the Spencer-Weismann Debate 225

6 Pragmatist Ethics: Evolution, Experiment, and Social Progress 248

Fieldwork in Ethics 255
Organism and Environment in Social Reform 268
Social Science and Social Evolution 275
Eugenics and Civilization 283

7 Pragmatist Logic: Evolution, Experiment, and Inquiry 290

The "Natural History" Approach 295
Evolutionary Experimentalism 313

Conclusion 332

Acknowledgments 343   Index 347

Tables and Figures

Table 1	The first cohort of pragmatists, 1851–69 11
Table 2	The second cohort of pragmatists, 1875–98 12
Table 3	The third cohort of pragmatists, 1900–1916 14
Table 4	The fourth cohort of pragmatists, 1919–39 15
Table 5	Individuals connected to the Metaphysical Club of Cambridge, Massachusetts 24
Figure 1	*Crust of the Earth as Related to Zoology* 28
Figure 2	Structure of honeycomb 31
Figure 3	Comparison of circular and hexagonal cells 33
Figure 4	Detail of United States Coast Survey map 37
Figure 5	Amazon River in northwest Brazil 44
Figure 6	Advertisement for *Spencer's System of Philosophy* 61
Figure 7	William James's analogy between zoological and social evolution 88
Figure 8	*Heliophyllum halli*, a rugose coral fossil 109
Figure 9	Sketch of Amphioxus, with legs added by Henry Castle 111
Figure 10	Announcement of lectures on organic evolution 151
Figure 11	Marginal note on Hegel and Darwin 168
Figure 12	*The Evolution of Hull House* 287

Abbreviations of Manuscript Sources

Alexander Papers
> Samuel Alexander Papers (GB 133 ALEX), Special Collections, John Rylands Library, University of Manchester.

Bunting Papers
> Sir Percy William Bunting Papers (ICU.SPCL.BUNTING), Special Collections Research Center, University of Chicago.

Castle Papers
> Henry Northrup Castle Papers (ICU.SPCL.CASTLEHN), Special Collections Research Center, University of Chicago.

Cuddeback Letter Book
> Cuddeback Letter Book, Pike County Historical Society, Milford, PA.

Du Bois Folder
> "Du Bois, William Edward Burghardt," Box 120 VT, Student Folders (UAIII 15.88.10), Harvard University Archives.

Du Bois Papers
> W. E. B. Du Bois Papers (MS 312), Special Collections and University Archives, University of Massachusetts Amherst. Online at http://credo.library.umass.edu/view/collection/mums312. Quoted with permission of the Permissions Company, LLC, on behalf of the David Graham Du Bois Trust.

Du Bois Transcript
> "Record of W. E. B. Du Bois," Class of 1890, Box 1, Student Records, 1876–1982 (UAIII 15.75.10 mf), Harvard University Archives.

ABBREVIATIONS OF MANUSCRIPT SOURCES

Fiske Papers
: Papers of John Fiske (mssFK 1–1661), Manuscripts Department, Huntington Library, Art Collections, and Botanical Gardens, San Marino, CA.

Hull House Collection
: Hull House Collection (MSHHC_NO), Special Collections, Richard J. Daley Library, University of Illinois at Chicago.

James Papers
: William James Papers (MS Am 1092.9–1092.12), Houghton Library, Harvard University.

Library Charging Lists
: Records of the Harvard College Library: Library Charging Records, 1888–92 (UAIII 50.15.60, vols. 143–47), Harvard University Archives. Online at https://id.lib.harvard.edu/ead/hua12009/catalog.

Mead Papers
: George Herbert Mead Papers (ICU.SPCL.MEAD), Special Collections Research Center, University of Chicago.

Mead Transcript
: "Record of G. H. Mead," Class of 1888, Box 1, Student Records, 1876–1982 (UAIII 15.75.10 mf), Harvard University Archives.

Metaphysical Club Records
: Metaphysical Club Records (RG-15-040), Special Collections, Johns Hopkins University.

Norton Letters
: Letters Received by the Norton Family, 1830–1920 (MS Am 1088.1), Houghton Library, Harvard University.

Osborn Family Papers
: Osborn Family Papers (MS 474), New-York Historical Society, New York, NY.

Peirce Papers
: Charles S. Peirce Papers (MS Am 1632), Houghton Library, Harvard University.

Royce Papers
: Papers of Josiah Royce (HUG 1755), Harvard University Archives.

Wright Papers
: Chauncey Wright Papers (Mss.B.W933), American Philosophical Society, Philadelphia, PA.

Abbreviations of Scholarly Editions

Addams, *Addams Papers*
> Jane Addams. *The Jane Addams Papers*. Edited by Mary Lynn McCree Bryan. Ann Arbor, MI: University Microfilms International, 1984. Microfilm, 82 reels.

Addams, *Selected Papers*
> Jane Addams. *The Selected Papers of Jane Addams*. Edited by Mary Lynn McCree Bryan, Barbara Bair, Maree de Angury, and Ellen Skerrett. 3 vols. Urbana: University of Illinois Press, 2003–.

Castle, *Letters*
> Henry Northrup Castle. *The Collected Letters of Henry Northrup Castle*. Edited by George Herbert Mead and Helen Castle Mead. Athens: Ohio University Press, 2012.

Darwin, *Correspondence*
> Charles Darwin. *The Correspondence of Charles Darwin*. Edited by Frederic Burkhardt and Sydney Smith. 26 vols. Cambridge: Cambridge University Press, 1985–.

Dewey, *Class Lectures*
> John Dewey. *The Class Lectures of John Dewey*. Edited by Donald F. Koch. 2 vols. Charlottesville, VA: InteLex, 2010–15. Electronic edition, InteLex Past Masters.

Dewey, *Correspondence*
> John Dewey. *The Correspondence of John Dewey, 1871–2007*. Vol. 1, *1871–1918*. Edited by Larry A. Hickman. Charlottesville, VA: InteLex, 2008. Electronic edition, InteLex Past Masters.

Du Bois, *Correspondence*
    W. E. B. Du Bois. *The Correspondence of W. E. B. Du Bois*. Edited by Herbert Aptheker. 3 vols. Amherst: University of Massachusetts Press, 1973–78.

James, *Correspondence*
    William James. *The Correspondence of William James*. Edited by Ignas K. Skrupskelis and Elizabeth M. Berkeley. 12 vols. Charlottesville: University Press of Virginia, 1992–2004.

James, *Manuscript Lectures*
    William James. *Manuscript Lectures*. Works of William James. Edited by Frederick H. Burkhardt. Cambridge: Harvard University Press, 1988.

Peirce, *Collected Papers*
    Charles Sanders Peirce. *Collected Papers of Charles Sanders Peirce*. Edited by Charles Hartshorne, Paul Weiss, and Arthur W. Burks. 8 vols. Cambridge: Harvard University Press, 1931–60.

Peirce, *Essential Peirce*
    Charles Sanders Peirce. *The Essential Peirce: Selected Philosophical Writings*. Edited by Nathan Houser, Christian Kloesel, and the Peirce Edition Project. 2 vols. Bloomington: Indiana University Press, 1992–98.

Peirce, *Historical Perspectives*
    Charles Sanders Peirce. *Historical Perspectives on Peirce's Logic of Science: A History of Science*. Edited by Carolyn Eisele. Berlin: Mouton, 1985.

Peirce, *Illustrations*
    Charles Sanders Peirce. *Illustrations of the Logic of Science*. Edited by Cornelis De Waal. New York: Open Court, 2014.

Peirce, *New Elements*
    Charles Sanders Peirce. *The New Elements of Mathematics*. Edited by Carolyn Eisele. 4 vols. The Hague: Mouton, 1976.

Peirce, *Writings*
    Charles Sanders Peirce. *Writings of Charles S. Peirce: A Chronological Edition*. Edited by Edward C. Moore, Christian J. W. Kloesel, Nathan Houser, and the Peirce Edition Project. 8 vols. Bloomington: Indiana University Press, 1982–.

Wright, *Letters*
    Chauncey Wright. *Letters of Chauncey Wright, with Some Account of His Life*. Edited by James Bradley Thayer. Cambridge, MA: John Wilson & Son, 1878.

Note to the Reader

To reduce the book's length, many of its original footnotes—including most of those that took the form "for more on X, see Y"—had to be removed. For these supplemental footnotes and a full bibliography, see http://www.trevorpearce.com/books/pragmatisms-evolution/.

Introduction

William Edward Burghardt Du Bois stepped up to the podium. The man who would become America's most celebrated black intellectual asked the twelve hundred people in his New York audience whether Charles Darwin's *On the Origin of Species* undermined the view "that all are created free and equal." Rejecting the popular idea that the work of August Weismann and other evolutionists implied some "essential and inevitable inequality among men and races of men," Du Bois insisted that freedom for "social self-realization" was "the central assertion of the evolutionary theory," since what matters from the broader perspective of evolution is "not equality of present attainment but equality of opportunity for unbounded future attainment." Earlier that day, John Dewey had spoken at the first session of the same conference. The man who would become America's most famous philosopher declared that because "acquired characteristics" are not inherited and thus do not contribute directly to evolutionary progress, all individuals can "have a full, fair and free social opportunity. Each generation biologically commences over again." "In other words," he continued, "there is no 'inferior race,' and the members of a race so-called should each have the same opportunities as those of a more favored race."[1]

---

1. W. E. B. Du Bois, "Evolution of the Race Problem," in *Proceedings of the National Negro Conference, 1909: New York, May 31 and June 1* (n.p., [1909]), 149, 152, https://hdl.handle.net/2027/nyp.33433081797809; also in W. E. B. Du Bois, *John Brown* (Philadelphia: George W. Jacobs, 1909), 375, 379; John Dewey, "Address of John Dewey," in *Proceedings of the National Negro Conference*, 72; see also "Negro's

That both Dewey and Du Bois were asked to speak at this 1909 conference, the first stirring of an embryonic National Association for the Advancement of Colored People (NAACP), is not surprising. Each of them had a reputation as an expert on the topic of education, and although Dewey had written almost nothing about race, he was at least local, having moved to Columbia University in 1904. What is more surprising is that they both structured their remarks around recent developments in evolutionary biology. Why did Dewey and Du Bois turn to biology to frame their response to a sociopolitical problem? The simple answer is that social and evolutionary progress were in the air, since 1909 was the centenary of both Abraham Lincoln and Charles Darwin. The New York conference was itself the result of a letter distributed to newspapers on Lincoln's birthday, calling "upon all the believers in democracy to join in a national conference for the discussion of present evils, the voicing of protests, and the renewal of the struggle for civil and political liberty."[2] On that very same day, which was also Darwin's birthday, the paleontologist Henry Fairfield Osborn had given a lecture on the English naturalist's life and work, the first in a Columbia University series devoted to Darwin and his influence on science. Osborn ended on an optimistic note: "The conflict of opinion aroused by Darwin will subside like the evil passions of our Civil War. Surely the reverent study of nature can not lead men astray." But the Columbia series went beyond the study of nature, narrowly conceived: there were also lectures on psychology, anthropology, sociology, and even philosophy. This last topic was the responsibility of Dewey, whose lecture—"Darwinism and Modern Philosophy"—was presented two months prior to the New York conference and became the title essay of his book *The Influence of Darwin on Philosophy*.[3]

But the Darwin centenary is only a small part of a larger story. Dewey and Du Bois's interest in evolution had deeper roots, stemming from

---

Brain Discussed," *Evening Post* (New York), May 31, 1909; "Negro's Rights Discussed," *Evening Post* (New York), June 1, 1909.

2. "Conference on Negroes," *Evening Post* (New York), February 13, 1909; see also "A Lincoln Emancipation Conference," *Alexander's Magazine* 7 (1909): 230–31; Mary White Ovington, "How the National Association for the Advancement of Colored People Began," *Crisis* 8 (1914): 184–86; W. E. B. Du Bois, *Dusk of Dawn: An Essay toward an Autobiography of a Race Concept* (New York: Harcourt, 1940), 56, 95, 223–24.

3. "Darwin, 'The Dead Lion,'" *Evening Post* (New York), February 12, 1909; Henry Fairfield Osborn, "Life and Works of Darwin," *Popular Science Monthly* 74 (1909): 343; *Annual Reports of the President and Treasurer to the Trustees, with Accompanying Documents* (New York: Columbia University, 1909), 154; John Dewey, *The Influence of Darwin on Philosophy, and Other Essays in Contemporary Thought* (New York: Henry Holt, 1910), 1–19.

their shared philosophical background. In this book, I argue that pragmatism—the most famous movement in American philosophy in the early 1900s—was the outgrowth of a rich conversation between late nineteenth-century philosophers, biologists, and social scientists. For pragmatists like Du Bois and Dewey, biological ideas such as evolution, adaptation, and environment were central to debates about scientific inquiry, social reform, and moral progress.

The various thinkers associated with pragmatism were towering figures in American intellectual life: Charles Sanders Peirce, one of the most accomplished logicians in history, also made important contributions to statistics, semiotics, and a host of other subjects; William James, after opening the first American laboratory devoted to the teaching of psychology, published one of the discipline's most influential early textbooks; John Dewey, a famous champion of progressive education, founded what is now known as the Laboratory School at the University of Chicago; Jane Addams, who fought tirelessly for social and labor reform causes, won the Nobel Peace Prize in 1931; George Herbert Mead, although trained in philosophy and psychology, had a decisive influence on sociology, inspiring the theory of symbolic interactionism; and W. E. B. Du Bois, arguably the most important public intellectual of the twentieth century, founded the first American school of sociology and edited the NAACP magazine the *Crisis* for more than twenty years.[4] But despite the pragmatists' status as key figures in American intellectual history and the history of philosophy, few scholars realize the full extent of their engagement with new developments in the biological sciences at the end of the nineteenth century. Even a brief look at their life and work reveals that many of their core ideas emerged from a dialogue between philosophy and biology. These ideas are at the heart of the most influential works of pragmatism: delve into James's *Principles of Psychology* and you will discover humans and cuttlefish alike actively shaping their perceptions; browse through Du Bois's *The Souls of Black Folk* and you will encounter an evolutionary analysis of black leadership; open up Dewey's *Democracy and Education* and you will find a whole chapter on the role of the environment. If we want to understand the pragmatists

---

4. Much of this information is common knowledge, but see Stephen M. Stigler, "Mathematical Statistics in the Early States," *Annals of Statistics* 6 (1978): 246–51 (on Peirce); William James et al., "Experimental Psychology in America," *Science* 2, no. 45 (1895): 626; Daniel R. Huebner, *Becoming Mead: The Social Process of Academic Knowledge* (Chicago: University of Chicago Press, 2014), 158–72; Earl Wright II, *The First American School of Sociology: W. E. B. Du Bois and the Atlanta Sociological Laboratory* (New York: Routledge, 2016).

and their influence, we need to understand the relationship between pragmatism and biology.[5]

This may seem like well-trodden ground: Philip Paul Wiener published *Evolution and the Founders of Pragmatism* way back in 1949.[6] Moreover, although Richard Jacob Bernstein, Richard Rorty, and Hilary Putnam, who were primarily responsible for pragmatism's "intellectual renascence" in the 1980s, did not emphasize its biological aspects, others have: according to Roy Wood Sellars, "American pragmatism is strongly biological and naturalistic in its outlook"; according to Gérard Deledalle, "pragmatism is a biocentric philosophy: life lived in its evolution is the essential category of pragmatism"; and according to Joseph Margolis, pragmatists believe that "human inquiry is continuous with, and develops out of, the biological and pre-cognitive interaction between organism and environment."[7] Those inspired by Dewey's work have discussed his embrace of evolution and the organism-environment dichotomy, and scholars of Peirce have examined his evolutionary metaphysics. Intellectual historians have also highlighted the influence of Darwin on pragmatism: for example, Bruce Kuklick has argued that the *Origin of Species* (published in 1859) undermined the Scottish realism that had previously dominated American universities, ushering in "the age of pragmatism."[8] Contemporaries of the pragmatists also stressed Darwin's influence. Grace and Theodore de Laguna opened their 1910 overview

---

5. William James, *The Principles of Psychology*, 2 vols. (New York: Henry Holt, 1890), 1:288–89; W. E. B. Du Bois, *The Souls of Black Folk* (Chicago: A. C. McClurg, 1903), chap. 3; John Dewey, *Democracy and Education* (New York: Macmillan, 1916), chap. 2.

6. Philip P. Wiener, *Evolution and the Founders of Pragmatism* (Cambridge, MA: Harvard University Press, 1949).

7. Cornel West, *The American Evasion of Philosophy: A Genealogy of Pragmatism* (Madison: University of Wisconsin Press, 1989), 3 ("intellectual renascence"); Roy Wood Sellars, *Evolutionary Naturalism* (Chicago: Open Court, 1922), viii; Gérard Deledalle, *Histoire de la philosophie américaine: De la Guerre de Sécession à la Seconde Guerre Mondiale* (Paris: Presses Universitaires de France, 1954), 26; Joseph Margolis, "Skepticism, Foundationalism, and Pragmatism," *American Philosophical Quarterly* 14 (1977): 122. Organism-environment interaction was discussed briefly in Richard J. Bernstein, *Beyond Objectivism and Relativism: Science, Hermeneutics, and Praxis* (Philadelphia: University of Pennsylvania Press, 1983), 204–11.

8. Shannon Sullivan, *Living across and through Skins: Transactional Bodies, Pragmatism, and Feminism* (Bloomington: Indiana University Press, 2001), chap. 1; Jerome A. Popp, *Evolution's First Philosopher: John Dewey and the Continuity of Nature* (Albany: State University of New York Press, 2007); Carl R. Hausman, *Charles S. Peirce's Evolutionary Philosophy* (Cambridge: Cambridge University Press, 1993); Andrew Reynolds, *Peirce's Scientific Metaphysics: The Philosophy of Chance, Law, and Evolution* (Nashville, TN: Vanderbilt University Press, 2002); Bruce Kuklick, *A History of Philosophy in America, 1720–2000* (Oxford: Oxford University Press, 2001), 98–99; see also Herbert W. Schneider, *A History of American Philosophy* (New York: Columbia University Press, 1946), 319–437; Max H. Fisch, "Evolution in American Philosophy," *Philosophical Review* 56 (1947); Paul Jerome Croce, *Science and Religion in the Era of William James*, vol. 1 (Chapel Hill: University of North Carolina Press, 1995), chaps. 4–5; Melvin L. Rogers, *The Undiscovered Dewey: Religion, Morality, and the Ethos of Democracy* (New York:

of pragmatism with this passage: "No scientific hypothesis has ever exerted a more profound or far-reaching influence upon the thought of a period, than has the Darwinian theory of evolution upon that of the last half-century. Not only have the group of biological sciences been re-created, but there is scarcely one of the mental and social sciences, that has not been in large degree revolutionized." Peirce agreed, declaring in 1909 that the *Origin* had prompted "the greatest mental awakening since Newton and Leibniz."[9]

We know that pragmatism emerged in the wake of Darwin and that the pragmatists engaged with evolution. But we lack details, and details matter. Was Darwin the only important influence? How were debates about evolution in the 1860s different from those of the 1890s? Where did the vocabulary of *organism* and *environment* originate? Were the pragmatists talking to biologists and reading biology? Did the different pragmatists use biological ideas in different ways? Was biology also important for thinkers such as Addams and Du Bois, not usually included among the "classical" pragmatists?[10] In this book, I answer these questions, which have been neglected by existing scholarship. My overall thesis is that the pragmatists were deeply engaged with biology from 1860 to 1910 and that they were enthusiastic participants in a series of debates about the relationship between organism and environment. If we do not grasp the details of the pragmatists' engagement with biology, we will struggle to understand their major works, all of which deploy concepts such as organism, environment, evolution, and adaptation. Unless these concepts are placed in their original context, we will not know what they were supposed to accomplish. Such ignorance leads to scholarly mistakes: for instance, dismissing their use of "terms drawn from evolutionary biology" as belying any acquaintance with "Darwinian theory or subsequent developments in the life sciences."[11]

Because the pragmatists defended a naturalistic yet nonreductive approach to philosophy, their story is relevant not only to historians of

---

Columbia University Press, 2009), chaps. 1–2; Beth L. Eddy, *Evolutionary Pragmatism and Ethics* (Lanham, MD: Lexington, 2016).

9. Theodore de Laguna and Grace Andrus de Laguna, *Dogmatism and Evolution: Studies in Modern Philosophy* (New York: Macmillan, 1910), 117; MS 620 (1909), in Peirce, *Illustrations*, 187 (see full citation of this and other such sources in the Abbreviations of Scholarly Editions list in the front of this book).

10. On "classical" pragmatism, see David A. Hollinger, "The Problem of Pragmatism in American History," *Journal of American History* 67 (1980): 101; Christopher Hookway, "Pragmatism," in *The Stanford Encyclopedia of Philosophy*, Stanford University, 1997–, article published August 16, 2008, last modified October 7, 2013, https://plato.stanford.edu/archives/win2013/entries/pragmatism/.

11. Jennifer Welchman, *Dewey's Ethical Thought* (Ithaca, NY: Cornell University Press, 1995), 121.

philosophy but to anyone interested in the interconnections between philosophy and the life sciences. In the 1970s and '80s, just as pragmatism was being revived within mainstream philosophy, naturalism was also gaining ground: epistemologists and ethicists, along with philosophers of language and mind, began paying more attention to biology. Philosophy of biology emerged as a separate subfield at around the same time: the editors of *Biology and Philosophy*, founded in 1986, declared that they hoped "to capture, reflect, and carry forward the frothy ferment bubbling between the biological sciences, broadly construed, on the one hand, and philosophy, equally broadly construed, on the other."[12]

Philosophers working with evolutionary ideas occasionally made direct reference to the pragmatists: Thomas Goudge presented a critical summary of pragmatism's evolutionary contribution to the philosophy of mind; Donald Thomas Campbell gave William James a central role in his overview of evolutionary epistemology; and both Peter Skagestad and Christopher Hookway compared Peirce's evolutionary epistemology to more modern approaches.[13] The renewed interest in evolutionary ethics, however, was the result of two books written by popularizing biologists: Edward Osborne Wilson's *Sociobiology* and Richard Dawkins's *The Selfish Gene*, both published in the mid-1970s. These two books attempted to reduce human culture to biology, and they were widely discussed in philosophy, leading to an association between evolutionary approaches and a reductionist ethos. More nuanced views of reduction were sometimes expressed: William Church Wimsatt, for instance, argued in 1976 that "'nothing more than' talk" made little sense when applied to the biological basis of mind or culture. Nevertheless, evolutionary approaches in philosophy found it difficult to escape the association with naive biological reductionism.[14] As this book will show, the pragmatists also defended an evolutionary approach to knowledge, mind, and morality, but their approach was explicitly opposed to that of

---

12. David L. Hull et al., "Editorial," *Biology and Philosophy* 1 (1986): 1.
13. Thomas A. Goudge, "Pragmatism's Contribution to an Evolutionary View of Mind," *Monist* 57 (1973); Donald T. Campbell, "Evolutionary Epistemology," in *The Philosophy of Karl Popper*, ed. Paul Arthur Schilpp, vol. 1 (La Salle, IL: Open Court, 1974), 429–30, 438; Peter Skagestad, "Taking Evolution Seriously: Critical Comments on D. T. Campbell's Evolutionary Epistemology," *Monist* 61 (1978); Christopher Hookway, "Naturalism, Fallibilism and Evolutionary Epistemology," in *Minds, Machines and Evolution*, ed. Christopher Hookway (Cambridge: Cambridge University Press, 1984).
14. Edward O. Wilson, *Sociobiology: The New Synthesis* (Cambridge, MA: Belknap Press of Harvard University Press, 1975); Richard Dawkins, *The Selfish Gene* (Oxford: Oxford University Press, 1976); William C. Wimsatt, "Reductionism, Levels of Organization, and the Mind-Body Problem," in *Consciousness and the Brain: A Scientific and Philosophical Inquiry*, ed. Gordon G. Globus, Grover Maxwell, and Irwin Savodnik (New York: Plenum, 1976), 223.

Herbert Spencer, who was linked at the time (somewhat unfairly) with a more reductive viewpoint.[15] The pragmatists thus offer us a model of how biological ideas, suitably reframed, can ground a nonreductionist evolutionary account of mental and moral life. In other words, they urge us to embrace a historical and ecological approach to philosophy.

Inspired by this approach, my historical method in this book is unabashedly contextualist: I focus on how the context of pragmatism illumines its content rather than on reconstructing arguments. I have limited my explorations to the pragmatists' more immediate intellectual context: what they learned in college, works they referenced, debates with which they would have been familiar, and so on. For the most part, I do not discuss broader political movements, economic changes, or social and religious transformations, which of course had their own connections to biological ideas. I am not suggesting that argument reconstruction or the broader social context is historically irrelevant—on the contrary, I think they are both essential for a complete understanding of pragmatism. But there must to some extent be a division of labor among historians of philosophy, with each book providing only a thread of the tapestry: the biological context is not the whole story, but it is an important story.

Following Sarah Hutton and others, my approach also centers on conversations, institutions, and venues.[16] Paying attention to where and how philosophy was practiced and published yields important insights. For example, I have argued elsewhere that Peirce's strange interest in protoplasm (the fluid contents of living cells) makes more sense when he is placed in conversation with other contributors to the *Monist* and its sister journal *Open Court*.[17] Hutton's "conversation model" proposes that we examine not only professional debates but also correspondence, lecture notes, minutes from formal and informal clubs, newspaper articles, and any other sources at our disposal to determine who actually participated in philosophical conversations. This strategy corrects our tendency to dismiss those participants who are not canonical philosophers: "John Fiske? Never heard of him." "Edward Drinker Cope? Wasn't

---

15. For a more sympathetic view of Spencer, see Snait Gissis, "Spencer's Evolutionary Entanglement: From Liminal Individuals to Implicit Collectivities," in *Biological Individuality: Integrating Scientific, Philosophical, and Historical Perspectives*, ed. Scott Lidgard and Lynn K. Nyhart (Chicago: University of Chicago Press, 2017).

16. Sarah Hutton, "Intellectual History and the History of Philosophy," *History of European Ideas* 40 (2014): 935–37; see also Alan W. Richardson, "Occasions for an Empirical History of Philosophy of Science: American Philosophers of Science at Work in the 1950s and 1960s," *HOPOS* 2 (2012): 6–10.

17. Trevor Pearce, "'Protoplasm Feels': The Role of Physiology in Peirce's Evolutionary Metaphysics," *HOPOS* 8 (2018).

he a paleontologist?" "Jane Addams? Did she ever teach in a philosophy department?" It is now easier than ever to pursue Hutton's strategy, since almost all work published by the pragmatists and their associates prior to 1925 is freely available online, providing access to the conversation as it happened.[18]

I have been blithely using the words *pragmatism* and *pragmatist* without saying anything about what they mean. It is hard to find delimiting criteria that pick out even self-declared pragmatists, let alone all those who embraced something like the pragmatic approach. There were a few shared commitments: a focus on the practical effects of our philosophical theories; a championing of experimental inquiry; and as I will argue, a biological approach to philosophy.[19] But if you are inclined to think of *pragmatism* as indicating a particular theory of truth or meaning, or as privileging either experience or language, I would ask you to set the label aside; this then becomes a story of a specific group of interconnected philosophers who introduced biological ideas into their ongoing conversation, using them in relatively similar ways.[20] There is another respect in which the label itself is less important, at least for my purposes: this book covers the period from 1860 to 1910, and the idea of pragmatism was not even introduced to the broader philosophical community until 1898; thus although my focal actors are those philosophers who were later associated with pragmatism, I do not discuss the pragmatist movement of the early 1900s until the very last chapter.

The thinkers we group (very loosely) as pragmatists can be divided into cohorts based on their college graduation year. A standard concept in the social sciences, a *cohort* is a set of people who "experienced the same event within the same time interval"—for example, the Great Depression affected people differently depending on their age cohort.[21]

---

18. I have mainly used the HathiTrust Digital Library (https://www.hathitrust.org/), which has better search tools than the Internet Archive (https://archive.org/) or Google Books (https://books.google.com/). This also means that although I do use scholarly editions in my research, I normally cite them only when quoting from unpublished work.

19. For a similar list of pragmatism's "basic dimensions," see Erin McKenna, *Pets, People, and Pragmatism* (New York: Fordham University Press, 2013), 104.

20. On the experience/language debate, see Colin Koopman, "Conduct Pragmatism: Pressing beyond Experientialism and Lingualism," *European Journal of Pragmatism and American Philosophy* 6 (2014); Gregory Fernando Pappas, "The Narrative and Identity of Pragmatism in America: The History of a Dysfunctional Family?" *Pluralist* 9 (2014).

21. Norman B. Ryder, "The Cohort as a Concept in the Study of Social Change," *American Sociological Review* 30 (1965): 845. Some intellectual historians have made use of the cohort concept, although not by that name: for example, Claude Digeon, *La crise allemande de la pensée française (1870–1914)* (Paris: Presses Universitaires de France, 1959), 4–8; Lynn Nyhart, *Biology Takes Form: Animal Morphology and the German Universities, 1800–1900* (Chicago: University of Chicago Press, 1995), 20–28.

I use college graduation cohorts, given that the relevant events for this book are developments in biology, and individuals only began to engage seriously with these developments in college. Cohorts are a helpful tool in the history of philosophy, as they can help us generate and test hypotheses about extraphilosophical influences on particular groups of philosophers. For example, Louis Menand hypothesized that pragmatism was part of a struggle to replace those intellectual trends that had been "swept away" by the American Civil War.[22] Although it was not framed as such, this is a hypothesis about cohorts: Menand's claim was that those philosophers who were at a certain life stage during the Civil War were deeply affected by it. Because of the broad usefulness of the cohort concept, I have assembled in four tables a set of pragmatist cohorts, designed to reflect not only developments in biology but also a variety of overlapping external events: professionalization, World War I, the rise of Nazism, and so on.

Cohorts are determined not by shared doctrines but by shared experiences at a similar life stage; the tables thus include both self-declared pragmatists and some of their historical and conceptual associates. Tommy Curry has criticized historians of philosophy for ignoring black thinkers whose ideas do not mirror or extend those of canonical white philosophers. The cohort approach, by downgrading the importance of specific doctrines, helps direct our attention to these and other marginalized thinkers, since it is less concerned with who counts as a pragmatist and who counts as a philosopher.[23] Cohorts do not necessarily correspond with academic genealogies: Josiah Royce and George Herbert Mead, for example, belong to the same cohort even though Mead took a class with Royce at Harvard. However, there are more teacher-student links between cohorts than within them, and I note many of these below. For reasons of space, I have not included any of the new realists or critical realists apart from George Santayana, even though many of them were sympathetic critics of pragmatism.[24] I have also included only American pragmatists, although to the extent that the relevant events are global, cohorts can be global as well. Théodore Flournoy, Ferdinand Canning Scott Schiller, Bertrand Russell, and Carlos Vaz Ferreira would

---

22. Louis Menand, *The Metaphysical Club* (New York: Farrar, Straus & Giroux, 2001), x.
23. Tommy J. Curry, "The Derelictical Crisis of African-American Philosophy: How African American Philosophy Fails to Contribute to the Study of African-Descended People," *Journal of Black Studies* 42 (2011): 316–17.
24. Edwin B. Holt et al., *The New Realism: Coöperative Studies in Philosophy* (New York: Macmillan, 1912); Durant Drake et al., *Essays in Critical Realism: A Co-operative Study of the Problem of Knowledge* (New York: Macmillan, 1920).

be part of the second cohort; Heiji Oikawa, Stanislav Shatskii, Eugenio d'Ors, Giovanni Papini, Moisés Sáenz, Ludwig Wittgenstein, Pedro Zulen, Rudolf Carnap, and Hu Shih would be part of the third cohort; Anísio Teixeira, Frank Ramsey, Thomas Goudge, and Risieri Frondizi would be part of the fourth cohort. Finally, I have divided each cohort into rough subcohorts.[25]

The first cohort of pragmatists (see table 1) consists primarily of those connected to the famous Metaphysical Club of Cambridge, Massachusetts—that is, most of the Harvard pragmatists, along with some friends and colleagues.[26] Almost all the members of this cohort graduated from Harvard in the 1850s and '60s. James is included because although he did not earn a BA at Harvard, he studied chemistry and comparative anatomy there from 1861 to 1863 and is in the same age cohort as the others. The exceptions are Ella Flagg Young (also in the same age cohort), who did not go to college but earned her doctorate under Dewey at Chicago, and Christine Ladd-Franklin, who went to Vassar and later studied with Peirce at Johns Hopkins. This cohort is the one that Menand took to have been decisively influenced by the Civil War, although he did not discuss its younger members. More important for my purposes, most of the members of this cohort had only recently finished college when the evolutionary ideas of Spencer and Darwin came onto the scene, pulling many of them into debates about evolution at the very beginning of their careers (as will be discussed further in chapters 1–2).

By the time the second cohort of pragmatists (see table 2) got to college, however, evolution was in the textbooks (as will be discussed further in chapter 3). Many members of this cohort graduated from relatively low-prestige colleges, although both Mead and Du Bois completed second bachelor's degrees at Harvard. Charles Cooley, Robert Park, James Rowland Angell, and Amy Tanner were all undergraduate students at Michigan while Dewey was teaching there. Charlotte Perkins Gilman, who did not go to college, is included because she is a member of the same age cohort. Whereas only a few of those in the first cohort pursued doctoral research, almost everyone in this cohort did: Mary Whiton Calkins, Santayana, Alfred Henry Lloyd, Du Bois, John Elof Boodin, and William Henry Ferris did their postgraduate studies at Harvard (which

---

25. Except when indicated, biographies, college affiliations, and graduation dates for the philosophers listed are taken from John R. Shook, ed. *The Dictionary of Modern American Philosophers*, 4 vols. (Bristol: Thoemmes Continuum, 2005).

26. On "Harvard Pragmatism," see Bruce Kuklick, *The Rise of American Philosophy: Cambridge, Massachusetts, 1860–1930* (New Haven, CT: Yale University Press, 1977), 256–58. For a list of those connected with the Metaphysical Club, see Peirce, *Writings*, 3:xxx–xxxi.

Table 1. The first cohort of pragmatists, with college graduation dates between 1851 and 1869, divided into three subcohorts

Name	College	Graduation Year
Nicholas St. John Green	Harvard	1851
Chauncey Wright	Harvard	1852
Francis Ellingwood Abbot	Harvard	1859
Charles Sanders Peirce	Harvard	1859
Oliver Wendell Holmes Jr.	Harvard	1861
John Fiske	Harvard	1863
William James	*	—
Ella Flagg Young	†	—
Christine Ladd-Franklin	Vassar	1869
Francis Greenwood Peabody	Harvard	1869

* William James was not a BA student at Harvard, though he did study at the Lawrence Scientific School there from 1861 to 1863 before switching to medicine, receiving his MD in 1869.
† Ella Flagg Young, although part of this age cohort, did not earn a BA but eventually did doctoral work at the University of Chicago, receiving her PhD in 1900.

admitted its first doctoral students in 1872), where first-cohort members James and Francis Greenwood Peabody and second-cohort member Royce were teaching, and Edgar Singer Jr. did postdoctoral work there; Royce, Dewey, and Arthur Bentley studied at Johns Hopkins (founded in 1876); Mead, James Tufts, Du Bois, and Angell all worked toward German PhDs, although Tufts was the only one who actually received one; many of the younger members of the second cohort—Edward Ames, Addison Moore, Henry Waldgrave Stuart, Elizabeth Adams, Tanner, Henry Heath Bawden, Frederick Henke—studied at the University of Chicago (founded in 1892), where Dewey, Mead, Tufts, and Angell were teaching; William Kilpatrick and Savilla Elkus did doctoral work at Columbia, after Dewey had moved there; Boyd Henry Bode studied at Cornell, and was critical of pragmatism early in his career before allying himself with the movement. Cooley, Park, Bentley, and Mary Parker Follett were trained as social scientists but were influenced in various ways by pragmatism. Henke taught in China after receiving his PhD and translated the work of Wang Yangming, whose ideas anticipated those of the pragmatists.[27] Many members of this cohort—Dewey, Mead, Calkins, Moore,

---

27. Wang Yang-Ming, *The Philosophy of Wang Yang-Ming*, trans. Frederick Goodrich Henke (Chicago: Open Court, 1916); John Smith, "Some Pragmatic Tendencies in the Thought of Wang Yang-Ming," *Journal of Chinese Philosophy* 13 (1986). For Henke's graduation date, see *Bulletin of Northwestern University: Annual Catalogue, 1907–1908* (Evanston, IL: Northwestern University, 1908), 312.

Table 2. The second cohort of pragmatists, with college graduation dates between 1875 and 1898, divided into five subcohorts

Name	College	Graduation Year
Josiah Royce	California	1875
John Dewey	Vermont	1879
Jane Addams	Rockford	1881
Charlotte Perkins Gilman	*	—
George Herbert Mead	Oberlin	1883
James Hayden Tufts	Amherst	1884
Mary Whiton Calkins	Smith	1885
George Santayana	Harvard	1886
Alfred Henry Lloyd	Harvard	1886
Charles Horton Cooley	Michigan	1887
Robert Ezra Park	Michigan	1887
W. E. B. Du Bois	Fisk	1888
Edward Scribner Ames	Drake	1889
Addison Webster Moore	DePauw	1890
James Rowland Angell	Michigan	1890
William Heard Kilpatrick	Mercer	1891
Arthur Fisher Bentley	Johns Hopkins	1892
Edgar Arthur Singer Jr.	Pennsylvania	1892
Henry Waldgrave Stuart	California (Berkeley)	1893
Elizabeth Kemper Adams	Vassar	1893
Savilla Alice Elkus	Normal College (CUNY)	1893
Amy Eliza Tanner	Michigan	1893
Henry Heath Bawden	Denison	1893
John Elof Boodin	Brown	1895
William Henry Ferris	Yale	1895
Boyd Henry Bode	William Penn	1896
Frederick Goodrich Henke	Charles City	1897
Mary Parker Follett	Radcliffe	1898

* Charlotte Perkins Gilman did not earn a BA but is a member of the same age cohort.

Angell, Singer, Adams, Tanner, Bode, and Henke—also received advanced training in psychology, a subject taught in philosophy departments at the time. A large portion of them—Dewey, Mead, Tufts, Angell, Ames, Moore, Tanner—taught at Chicago and formed the Chicago school of philosophy and psychology, on which Addams had a pronounced influence.[28]

28. For the graduation dates of Tanner, Adams, Elkus, and Follett, see *Annual Register, July, 1894—July, 1895, with Announcements for 1895–6* (Chicago: University of Chicago Press, 1895), 23; *Annual Register, July, 1901–July, 1902, with Announcements for 1902–1903* (Chicago: University of Chicago, 1902), 38; *Normal College of the City of New York, Twenty-Fourth Annual Commencement, Thursday, June 22, 1893* (New York: Republic Press, 1893), 11; *Annual Reports of the President and Treasurer of Radcliffe College, 1897–1898* (n.p., 1898), 14.

In the hopes of spurring interest in the cohort methodology and in a broader array of American pragmatists—including more women and people of color—I also present two further cohorts whose members do not feature in this book. The third cohort of pragmatists (see table 3) finished college in the early years of the twentieth century, prior to the American entry into World War I. By this time, philosophy and psychology in the United States had gone through important institutional changes, with a series of professional associations and journals established in the 1890s and early 1900s: the American Psychological Association (1892), the *Philosophical Review* (1892), the *Psychological Review* (1894), the American Philosophical Association (1900), and the *Journal of Philosophy, Psychology and Scientific Methods* (1904).[29] Many members of this cohort—Morris Raphael Cohen, William Ernest Hocking, Horace Kallen, Clarence Irving Lewis, Alain Locke, Mary Coolidge—did graduate work at Harvard, where all but Coolidge (who was there too late) interacted with James and Royce.[30] Kallen and Locke both studied in 1907–8 at Oxford, where they became friends and saw James deliver some of the Hibbert Lectures that became *The Pluralistic Universe*.[31] Before Kallen moved to New York City in 1919 as a founding faculty member of the New School for Social Research, he taught with Max Otto at the University of Wisconsin. Kate Gordon, Julia Jessie Taft, Holly Estil Cunningham, Josef Roy Geiger, Clarence Edwin Ayres, Jacob Robert Kantor, and Charles Spurgeon Johnson studied with Mead and others at Chicago; Lucy Sprague Mitchell, Willystine Goodsell, Elsie Ripley Clapp, and Herbert Schneider worked with Dewey at Columbia, where John Childs also studied. Jerome Frank became a famous proponent of legal realism, a movement that had important historical connections with pragmatism. Although he was a professor of biochemistry, William Malisoff taught

---

29. Daniel J. Wilson, *Science, Community, and the Transformation of American Philosophy, 1860–1930* (Chicago: University of Chicago Press, 1990); James Campbell, *A Thoughtful Profession: The Early Years of the American Philosophical Association* (Chicago: Open Court, 2006); Christopher D. Green et al., "Bridge over Troubled Waters? The Most 'Central' Members of Psychology and Philosophy Associations ca. 1900," *Journal of the History of the Behavioral Sciences* 52 (2016); Christopher D. Green and Ingo Feinerer, "How the Launch of a New Journal in 1904 May Have Changed the Relationship between Psychology and Philosophy," *History of Psychology* 20 (2017).

30. Cohen is not often counted as a pragmatist, as he was very critical of Dewey, but he is included here because of his praise for Peirce and his mentoring of fourth-cohort pragmatists at City College: see Morris R. Cohen, "Charles S. Peirce and a Tentative Bibliography of His Published Writings," *Journal of Philosophy, Psychology and Scientific Methods* 13 (1916); Morris R. Cohen, introduction to *Chance, Love and Logic: Philosophical Essays*, by Charles Sanders Peirce (New York: Harcourt, Brace, 1923).

31. Jeffrey C. Stewart, *The New Negro: The Life of Alain Locke* (New York: Oxford University Press, 2018), 153–55; William James, *A Pluralistic Universe: Hibbert Lectures at Manchester College on the Present Situation in Philosophy* (New York: Longmans, Green, 1909).

Table 3. The third cohort of pragmatists, with college graduation dates between 1900 and 1916, divided into three subcohorts

Name	College	Graduation Year
Lucy Sprague Mitchell	Radcliffe	1900
Kate Gordon	Chicago	1900
Morris Raphael Cohen	City College of New York	1900
Julia Jessie Taft	Drake	1900
William Ernest Hocking	Harvard	1901
Grace Andrus de Laguna	Cornell	1903
Horace Meyer Kallen	Harvard	1903
Willystine Goodsell	Columbia	1905
Clarence Irving Lewis	Harvard	1905
Max Carl Otto	Wisconsin	1906
Alain LeRoy Locke	Harvard	1907
Elsie Ripley Clapp	Barnard	1908
Holly Estil Cunningham	Lebanon (Ohio)	1909
Josef Roy Geiger	Furman	1909
Jerome New Frank	Chicago	1909
John Lawrence Childs	Wisconsin	1911
Clarence Edwin Ayres	Brown	1912
Jacob Robert Kantor	Chicago	1914
Mary Lowell Coolidge	Bryn Mawr	1914
Herbert Wallace Schneider	Columbia	1915
William Marias Malisoff	Columbia	1915
Charles Spurgeon Johnson	Virginia Union	1916

philosophy alongside Singer at the University of Pennsylvania and was the founding editor of *Philosophy of Science*.[32]

The fourth cohort of pragmatists (see table 4) finished college in the 1920s and '30s, between the two world wars. Several members of this cohort attempted to bring pragmatism into conversation with the views of European philosophers who fled the continent during the rise of Nazism in the 1930s. Nelson Goodman and Susanne Langer were influenced by the neo-Kantian philosopher Ernst Cassirer, and many of the others

---

32. For the college graduation dates of Gordon, Goodsell, Clapp, Cunningham, Geiger, Malisoff, and Johnson, see *Annual Register, July, 1900–July, 1901, with Announcements for 1901–1902* (Chicago: University of Chicago, 1901), 37; *Catalogue and General Announcement, 1906–1907* (New York: Columbia University, 1906), 401; Pam Hackbart-Dean, "Elsie Ripley Clapp Papers, 1910–1959," Special Collections Research Center, Southern Illinois University; *Annual Register, Covering the Academic Year Ending June 30, 1913, with Announcements for the Year 1913–1914* (Chicago: University of Chicago Press, 1913), 499; *Annual Register, Covering the Academic Year Ending June 30, 1914, with Announcements for the Year 1914–1915* (Chicago: University of Chicago Press, 1914), 535; *Catalogue, 1915–1916* (New York: Columbia University), 255; Patrick J. Gilpin and Marybeth Gasman, *Charles S. Johnson: Leadership beyond the Veil in the Age of Jim Crow* (Albany: State University of New York Press, 2003), 3.

Table 4. The fourth cohort of pragmatists, with college graduation dates between 1919 and 1939, divided into four subcohorts

Name	College	Graduation Year
Donald Ayres Piatt	Chicago	1919
Susanne Katherina Langer	Radcliffe	1920
Richard Peter McKeon	Columbia	1920
Herbert George Blumer	Missouri	1921
Harold Newton Lee	Oregon	1922
Charles William Morris	Northwestern	1922
Joseph Ratner	City College of New York	1922
Ernest Nagel	City College of New York	1923
Sidney Hook	City College of New York	1923
George Edward Axtelle	Washington	1923
Carolyn Eisele	Hunter	1923
Max Harold Fisch	Butler	1924
Theodore Thomas Lafferty	Chicago	1925
Philip Paul Wiener	City College of New York	1925
Paul Weiss	City College of New York	1927
David Louis Miller	College of Emporia	1927
Abraham Edel	McGill	1927
Nelson Goodman	Harvard	1928
Willard Van Orman Quine	Oberlin	1930
William Thomas Fontaine	Lincoln	1930
Wilfrid Stalker Sellars	Michigan	1933
Charles West Churchman	Pennsylvania	1934
Elizabeth Farquhar Flower	Wilson	1935
Thelma Zeno Lavine	Radcliffe	1936
Morton Gabriel White	City College of New York	1936
Charles Wright Mills	Texas	1938
Samuel Morris Eames	Culver-Stockton	1939
Donald Davidson	Harvard	1939

were involved with the logical empiricist movement, which also had neo-Kantian roots. Some in the cohort—Ernest Nagel, Goodman, Willard Van Orman Quine, Wilfrid Sellars—were key players in the history of analytic philosophy. Others—Joseph Ratner, George Axtelle, Carolyn Eisele, Max Fisch, Philip Wiener, Paul Weiss, David Louis Miller, Elizabeth Flower, Thelma Lavine, Morton White, Charles Wright Mills, Samuel Morris Eames—were important scholars of first- and second-cohort pragmatists. Ratner, Nagel, Sidney Hook, and White all studied with Cohen at City College before moving on to doctoral work at Columbia, where Dewey was teaching; conversely, Wiener and Abraham Edel both taught with Cohen at City College after graduate work at Columbia.[33]

---

33. Like Cohen, Nagel praised Peirce and was critical of Dewey, though he did teach Dewey's logic at Columbia: see Ernest Nagel, "Charles S. Peirce, Pioneer of Modern Empiricism," *Philosophy*

Donald Ayres Piatt, Herbert Blumer, Charles Morris, Theodore Lafferty, and Miller worked with Mead at Chicago, and Eames studied with Morris there. Mills did his MA in philosophy with Miller at Texas and his PhD in sociology at Wisconsin, where Otto was chair of the philosophy department. William Fontaine, Charles West Churchman, and Flower all worked with Singer at the University of Pennsylvania, where they (along with Goodman) went on to teach. The remainder of the cohort—Langer, Harold Newton Lee, Weiss, Goodman, Quine, Sellars, Lavine, Donald Davidson—did graduate work at Harvard, coming into contact with Lewis and the British mathematician-turned-metaphysician Alfred North Whitehead.[34]

Although I will not go any further in this book, it is perhaps worth noting that those associated with the 1980s revival of pragmatism mentioned previously—Bernstein, Rorty, and Putnam—along with others such as Sidney Morgenbesser, John Edwin Smith, Elizabeth Ramsden Eames, Joseph Margolis, Jo Ann Boydston, Richard Rudner, Isaac Levi, Nicholas Rescher, Murray Murphey, and John McDermott, would make up a fifth cohort, consisting of those who finished college during or shortly after World War II.

I have presented these later cohorts of pragmatists primarily as an aid to future research. But viewing the whole array also reveals a difference between the early cohorts, with graduation dates in the nineteenth century, and the later cohorts, with graduation dates in the twentieth century: as Randall Auxier has noted, biology featured much less prominently in the writings of the later pragmatists, even though some of them were self-declared naturalists. For example, although there are vague echoes of the organism-environment perspective in the work of

---

of Science 7 (1940); Ernest Nagel, "Dewey's Theory of Natural Science," in *John Dewey: Philosopher of Science and Freedom*, ed. Sidney Hook (New York: Dial, 1950); Patrick Suppes, "Nagel's Lectures on Dewey's Logic," in *Philosophy, Science, and Method: Essays in Honor of Ernest Nagel*, ed. Sidney Morgenbesser, Patrick Suppes, and Morton White (New York: St. Martin's, 1969).

34. On the importance of Lewis for members of this cohort, see Cheryl Misak, *The American Pragmatists* (Oxford: Oxford University Press, 2013), chaps. 11–12; Jacquelyn Ann K. Kegley, "C. I. Lewis? A Significant Figure in American Pragmatism: Tracing Lines of Influence and Affinities of Themes and Ideas," in *Pragmatism in Transition: Contemporary Perspectives on C. I. Lewis*, ed. Peter Olen and Carl Sachs (Cham, Switz.: Palgrave Macmillan, 2017). For the graduation dates of Lafferty (inferred), Flower, and Eames, see Rosamond Kent Sprague, "Theodore Thomas Lafferty, 1901–1970," *Proceedings and Addresses of the American Philosophical Association* 43 (1969–70), 204; John R. Shook, "Lafferty, Theodore Thomas (1901–70)," in *The Dictionary of Modern American Philosophers*, ed. John R. Shook, vol. 3 (Bristol, UK: Thoemmes Continuum, 2005), 1401; Kaiyi Chen, "Elizabeth F. Flower Papers, 1929–2001," University Archives, University of Pennsylvania; "S. Morris Eames, 1916–1986," *Proceedings and Addresses of the American Philosophical Association* 60 (1986): 260.

Alain Locke, he associated this approach with the social sciences rather than with philosophy.[35] (There are some exceptions to this pattern: Taft and Gordon, two of the youngest members of the third cohort, adopted the organism-environment framework of their Chicago professors; the language research of De Laguna, Langer, and Morris was grounded in biology; and Ayres took an evolutionary approach to economics.) An explanation of this difference is beyond the scope of this book, but I can at least propose a hypothesis. Dewey argued that philosophers receive their problems "from the world of action," and thus he connected historical developments in both epistemology and ethics to changes in the broader social environment.[36] Along these lines, I would suggest that in the last four decades of the nineteenth century, the ideas of evolution and organism-environment interaction were so popular and prominent that scientifically inclined philosophers could not help but absorb and transform them. All the members of the first two cohorts completed their undergraduate and graduate education during this decisive period, and many of them were also trained in psychology, which was closely associated with biology and the organism-environment approach.[37] The twentieth-century pragmatists, on the other hand, tended not to have scientific training—or if they did, it was in mathematics and logic. As Suzanne Cunningham has shown, the founders of analytic philosophy and phenomenology deliberately excluded evolution from philosophy, and it is notable that Gottlob Frege, Bertrand Russell, and Edmund Husserl were all trained in mathematics. This educational difference, along with the new developments in mathematics, logic, and physics that became important topics of discussion in the early twentieth century,

---

35. Randall E. Auxier, "The Decline of Evolutionary Naturalism in Later Pragmatism," in *Pragmatism: From Progressivism to Postmodernism*, ed. Robert Hollinger and David Depew (Westport, CT: Praeger, 1995); Alain Locke and Bernhard J. Stern, eds., *When Peoples Meet: A Study in Race and Culture Contacts* (New York: Progressive Education Association, 1942), 232–34; Jacoby Adeshei Carter, *African American Contributions to the Americas' Cultures: A Critical Edition of Lectures by Alain Locke* (New York: Palgrave Macmillan, 2016), 37.

36. John Dewey, *The Significance of the Problem of Knowledge* (Chicago: University of Chicago Press, 1897), 4; John Dewey, "Moral Philosophy," in *Johnson's Universal Cyclopaedia: A New Edition*, ed. Charles Kendall Adams, vol. 5 (New York: A. J. Johnson, 1894).

37. On the rise of the idea of organism-environment interaction, see Trevor Pearce, "From 'Circumstances' to 'Environment': Herbert Spencer and the Origins of the Idea of Organism-Environment Interaction," *Studies in History and Philosophy of Biological and Biomedical Sciences* 41 (2010); Trevor Pearce, "The Origins and Development of the Idea of Organism-Environment Interaction," in *Entangled Life: Organism and Environment in the Biological and Social Sciences*, ed. Gillian Barker, Eric Desjardins, and Trevor Pearce (Dordrecht, Neth.: Springer, 2014). On psychology's association with the organism-environment perspective, see Josiah Royce, *The World and the Individual: Gifford Lectures, Delivered before the University of Aberdeen*, 2 vols. (New York: Macmillan, 1900–1901), 1:21.

may be partly responsible for the decline of a more biological pragmatism. Thus, my tentative hypothesis is that although twentieth-century pragmatism was still for the most part a scientific philosophy, the model science had changed.[38]

Some members of the second cohort—Dewey and Du Bois, for example—were still publishing books in the 1930s and even later. Does this not undermine any split between nineteenth- and twentieth-century pragmatists? After all, the pragmatist movement was not even a going concern until the early 1900s. This is where the cohort approach helps clarify the situation: members of the earlier cohorts were still active in the twentieth century, but their critical period of engagement with biology was over. It is hard to imagine that Dewey was entirely unaware of later developments in the life sciences, some of which were closely linked to his home institution of Columbia University: for example, the rise of mutationism in the early 1900s or the "modern synthesis" of evolution and genetics in the 1930s and '40s.[39] Nevertheless, his later work was still based on the organism-environment approach he had developed in the 1890s. The views of pragmatists such as Dewey and Du Bois changed over the course of their long careers, but they kept one foot in the nineteenth century—the century of history, evolution, and progress.

For reasons of space, I will focus on a subset of the early pragmatists—namely, Chauncey Wright, Peirce, Francis Ellingwood Abbot, Fiske, and James from the first cohort; and Royce, Dewey, Addams, Mead, and Du Bois from the second cohort. But my hypothesis about early cohorts and biological ideas suggests that we should also expect other members of these cohorts to have engaged with evolution and organism-environment interaction. Oliver Wendell Holmes Jr., in the first cohort, confirms this expectation: he was impressed by Spencer's work and drew from Edward Burnett Tylor's evolutionary anthropology in the early chapters of *The Common Law*. Likewise, Christine Ladd-Franklin's theory of color vision, developed in the early 1890s, linked adaptive chemical changes with an evolutionary history of vision, and she later called it "the evolution theory of color-sensation." According to Peabody, another member of this cohort, even Jesus's teaching recognized "that the problem of adjusting to the social environment must be a new problem

---

38. Suzanne Cunningham, *Philosophy and the Darwinian Legacy* (Rochester, NY: University of Rochester Press, 1996). On scientific philosophy, see Alan W. Richardson, "Toward a History of Scientific Philosophy," *Perspectives on Science* 5 (1997).

39. On mutationism and pragmatism, see Beth L. Eddy, *Evolutionary Pragmatism and Ethics* (Lanham, MD: Lexington, 2016), 96–100; Trevor Pearce, review of *Evolutionary Pragmatism and Ethics*, by Beth L. Eddy, *Transactions of the Charles S. Peirce Society* 53 (2017): 497–98.

with each new age."⁴⁰ The biological interests of some of those in the second cohort—Tufts and Angell at Chicago, Lloyd at Michigan, Santayana at Harvard—are well known, but the rest appear to fit the pattern as well: for example, Gilman adopted the organism-environment perspective in her book *Women and Economics*, published in 1898, and Ferris's *The African Abroad*, published in 1913, contained a detailed analysis of evolution.⁴¹ Biological concepts were less prominent in the work of Singer, but they were there in the background: his discussions of pragmatism did not mention its links to biology, but he claimed elsewhere that life essentially involved "adjustment and adaptation."⁴² I suspect further investigation would reveal that most of the pragmatists who began their careers in the nineteenth century were in conversation with biology and evolution.

This book is designed to speak not only to historians of philosophy but also to historians and philosophers of biology. Since the 1980s, historians of biology have investigated a series of topics that were of particular interest to the pragmatists: the relation between development and evolution, the debate over the causes of evolution, the interplay of habit and instinct, and the evolutionary views of Herbert Spencer.⁴³ I have

---

40. Leslie Stephen to Oliver Wendell Holmes Jr., 7 December 1866, in Frederic William Maitland, *The Life and Letters of Leslie Stephen* (London: Duckworth, 1906), 188; Oliver Wendell Holmes Jr. to Georgina Harriet Deffell Pollock, 2 July 1895, in Oliver Wendell Holmes and Frederick Pollock, *Holmes–Pollock Letters: The Correspondence of Mr. Justice Holmes and Sir Frederick Pollock, 1874–1932*, ed. Mark DeWolfe Howe, 2 vols. (Cambridge, MA: Harvard University Press, 1941), 1:57–58; Oliver Wendell Holmes, *The Common Law* (Boston: Little, Brown, 1881), 11, 19; Christine Ladd-Franklin, "On Theories of Light-Sensation," *Mind*, n.s., 2 (1893); Christine Ladd-Franklin, "Evolution Theory of Colour Vision," in *The American Encyclopedia and Dictionary of Ophthalmology*, ed. Casey A. Wood, vol. 6 (Chicago: Cleveland Press, 1915); Francis Greenwood Peabody, *Jesus Christ and the Social Question: An Examination of the Teachings of Jesus in Its Relation to Some of the Problems of Modern Social Life* (New York: Macmillan, 1900), 113.

41. Charlotte Perkins Stetson, *Women and Economics: A Study of the Economic Relation between Men and Women as a Factor in Social Evolution* (Boston: Small, Maynard, 1898) [the name Charlotte Perkins Stetson reflects her first marriage to Charles Walter Stetson; she changed it to Charlotte Perkins Gilman after wedding Houghton Gilman in 1900]; William H. Ferris, *The African Abroad, or His Evolution in Western Civilization, Tracing His Development under Caucasian Milieu*, 2 vols. (New Haven, CT: Tuttle, Morehouse & Taylor, 1913), chap. 3.

42. Edgar A. Singer Jr., "The Pulse of Life," *Journal of Philosophy, Psychology and Scientific Methods* 11 (1914): 650. On pragmatism, see Edgar A. Singer Jr., "Mind as an Observable Object," *Journal of Philosophy, Psychology and Scientific Methods* 8 (1911); Edgar A. Singer Jr., "The Empiricism of William James," in *University Lectures Delivered by Members of the Faculty in the Free Public Lecture Course, 1917–1918*, vol. 5 (Philadelphia: University of Pennsylvania, 1918).

43. For example, Peter J. Bowler, *The Eclipse of Darwinism: Anti-Darwinian Theories in the Decades around 1900* (Baltimore: Johns Hopkins University Press, 1983); Philip J. Pauly, *Controlling Life: Jacques Loeb and the Engineering Ideal in Biology* (Oxford: Oxford University Press, 1987); Jane Maienschein, *Transforming Traditions in American Biology, 1880–1915* (Baltimore: Johns Hopkins University Press, 1991); Ronald Rainger, *An Agenda for Antiquity: Henry Fairfield Osborn and Vertebrate Paleontology at the American Museum of Natural History, 1890–1935* (Tuscaloosa: University of Alabama Press,

attempted to contribute to this growing literature, focusing especially on generational differences in the uptake of evolutionary ideas by non-biologists, Spencer's importance for biology and philosophy, and the 1890s debates over the "factors" of evolution. Philosophers of biology have also shown a renewed interest in pragmatism, with Peter Godfrey-Smith's revival of the organism-environment perspective, Philip Kitcher's elaboration of pragmatist evolutionary ethics, and Lucas McGranahan's portrayal of pragmatism as a corrective to neo-Darwinism and sociobiology.[44] My own account of the pragmatists as philosophers of biology (before there was such a thing) is indebted to these recent discussions and seeks to place them in a richer historical context.

Although the seven chapters that follow are thematic, they also proceed in rough chronological order, with the themes corresponding to a series of historical periods: the reaction to Darwin and Spencer in the 1860s and '70s (chapters 1–2); the education of the second cohort of pragmatists in the 1880s (chapter 3); the idealist appropriation of evolutionary thought in the 1880s and '90s (chapter 4); the debate over the factors of evolution in the 1890s (chapter 5); and the pragmatist approach to ethics and logic in the 1890s and early 1900s (chapters 6–7).

In chapter 1, I describe how the philosophers connected to the Metaphysical Club—a Cambridge, Massachusetts, discussion group that met in the early 1870s—reacted to the publication of the *Origin of Species*. Surprisingly, given their later evolutionary outlook, there is no clear evidence that James or Peirce—who both worked for Darwin's opponent Louis Agassiz—had adopted evolutionary views prior to the late 1860s or even the 1870s. I suggest that it was Wright and Fiske who ultimately convinced the other club members to embrace evolution. Both of them publicly advocated for evolution in the 1860s and Fiske gave lectures popularizing the views of Darwin and Spencer starting in the early 1870s. Wright and Fiske also defended Darwin against the attacks of the zoologist St. George Mivart, attacks which coincided with the first meetings of the Metaphysical Club.

1991); Robert J. Richards, *The Meaning of Evolution: The Morphological Construction and Ideological Reconstruction of Darwin's Theory* (Chicago: University of Chicago Press, 1992); Bruce H. Weber and David J. Depew, *Evolution and Learning: The Baldwin Effect Reconsidered* (Cambridge, MA: MIT Press, 2003); Mark Francis, *Herbert Spencer and the Invention of Modern Life* (Ithaca, NY: Cornell University Press, 2007); Mark Francis and Michael Taylor, eds., *Herbert Spencer: Legacies* (New York: Routledge, 2015).

44. Peter Godfrey-Smith, *Complexity and the Function of Mind in Nature* (Cambridge: Cambridge University Press, 1996); Philip Kitcher, *The Ethical Project* (Cambridge, MA: Harvard University Press, 2011); Lucas McGranahan, *Darwinism and Pragmatism: William James on Evolution and Self-Transformation* (New York: Routledge, 2017).

Spencer introduced the idea of organism-environment interaction to the English-speaking world and popularized the term *environment*. In chapter 2, I argue that his evolutionary philosophy was of central importance for the philosophers of the Metaphysical Club, describing how Fiske, Wright, Peirce, and Abbot all engaged directly with Spencer's ideas. Fiske was enthusiastic but the others were skeptical. I then argue that James's early work was a response to Spencer's evolutionism from the standpoint of a broader naturalism: according to James, Spencer not only ignored important mental phenomena but also misconstrued the role of the environment in evolution. I suggest, however, that James's attack on Spencer's approach to social evolution was ultimately unsuccessful.

Turning to the second cohort of pragmatists, I demonstrate in chapter 3 that Royce, Dewey, Addams, Mead, and Du Bois were exposed to evolutionary ideas in college by their teachers and textbooks. They each developed a serious interest in biology and the sciences, reading and writing about evolution during both college and graduate school. All of them went on to teach about evolution in their own classes, with many of them using Spencer's works as textbooks.

The pragmatists had a much more interactive picture of the organism-environment relationship than that usually attributed to Spencer. In chapter 4, I explore the origins of this picture, arguing that it was produced against the background of an established connection between evolution and idealism. Dewey, in particular, inherited his dialectical account of organism and environment from a subset of British idealist philosophers who were trying to reconcile evolutionary ideas with a critique of Spencer's environmentalist theories of human thought and action. These idealists insisted that adaptation or adjustment results from the reciprocal action of organism and environment: just as the environment affects the organism, the organism affects the environment. They also claimed that organism and environment are best seen as two aspects of one thing—life. Royce and Du Bois also saw evolution and idealism as inextricably linked, and Du Bois's early work frequently invoked the broader evolutionary spirit of the nineteenth century.

In chapter 5, I examine how both first- and second-cohort pragmatists participated in the debates over the causal factors of evolution that accompanied the reception of August Weismann's work in the 1890s. Weismann, a biologist, argued in the mid-1880s that the hereditary substance was confined to what he called the "germ-plasm," which was isolated from the rest of the body. One implication was that acquired characters could not be inherited, undermining the neo-Lamarckian theories of American scientists such as Edward Drinker Cope. These discussions were of interest to James and Peirce because of their opposition to Spencer (also a neo-Lamarckian), and Peirce's 1893 essay "Evolutionary Love" should be interpreted as contributing to the "factors" debates. Dewey also followed the debates, applying key concepts from an 1893–95 dispute between Spencer and Weismann to his early work in ethics and social psychology.

The pragmatists often highlighted the parallels between moral and scientific inquiry. In the final two chapters, I argue that in the years around 1900, they associated both of these with experimental-evolutionary progress. I demonstrate in chapter 6 that Dewey, Mead, Addams, and Du Bois, building on the experimental approach defended by philosophers such as William James and economists such as William Stanley Jevons, applied the ideas of evolution and experiment to ethics and social reform. Each of them constructed experimental field sites for social inquiry and relied on Spencer's organism-environment framework. They also shared a vision of moral progress: evolution guided by experimental science.

The texts most famously associated with the pragmatist movement of the early 1900s—Dewey's *Studies in Logical Theory*, Peirce's "What Pragmatism Is," and James's *Pragmatism*—finally make their appearance in chapter 7, in which I argue that Peirce, James, and Dewey, despite their differences, all embraced a view of epistemic progress that was both experimental and evolutionary. Despite Peirce's claims to the contrary, they each developed a "natural history" approach to logical inquiry and framed logic as fundamentally experimental. Although Dewey and Peirce, unlike James, were sympathetic to the idea of directed variation in evolution, Dewey and his students linked it to individual and social goals whereas Peirce connected it to a broader cosmic destiny.

Gérard Deledalle was right to claim that "experimental science and evolutionary theory made pragmatism possible."[45] This book shows how that possibility was realized.

---

45. Gérard Deledalle, *La philosophie américaine*, 3rd ed. (Paris: De Boeck & Larcier, 1998), 51. Unless otherwise noted, all translations in this book are my own.

ONE

The Metaphysical Club and the *Origin of Species*

Late in life, Charles Sanders Peirce recalled that in the early 1870s "a knot of us young men in Old Cambridge, calling ourselves, half-ironically, half-defiantly, 'The Metaphysical Club,'—for agnosticism was then riding its high horse, and was frowning superbly upon all metaphysics,—used to meet, sometimes in my study, sometimes in that of William James."[1] It was at a meeting of this club in 1872 that Peirce first presented what James later called the principle of pragmatism: "Our idea of anything *is* our idea of its sensible effects." The Metaphysical Club has thus been dubbed "the birthplace of pragmatism."[2] If we believe Peirce's testimony, the club was a success: "It proved quite the most successfully organized body of students for genuine educative efficiency, in contradistinction to saw-dust-stuffing, that ever I had the good fortune to be placed in."[3]

The various people who participated in the Metaphysical Club were part of the intellectual community of Cambridge, Massachusetts. All of them apart from James had attended Harvard as undergraduates in the 1850s and '60s, and many went on to teach there. Most of the members of the Metaphysical Club were also members of the first cohort of pragmatists (shown in table 1 of the introduction).

---

1. MS 318 (1907), in Peirce, *Essential Peirce*, 2:399.
2. Charles Sanders Peirce, "How to Make Our Ideas Clear," *Popular Science Monthly* 12 (1878): 293; Philip P. Wiener, *Evolution and the Founders of Pragmatism* (Cambridge, MA: Harvard University Press, 1949), 18–30.
3. MS 620 (1909), in Peirce, *Illustrations*, 187.

Table 5. Individuals connected to the Metaphysical Club of Cambridge, Massachusetts, divided into three subcohorts by Harvard class year

Club Member	Harvard Class	Field
Nicholas St. John Green	1851	Law
Chauncey Wright	1852	Mathematics/Philosophy
Charles Sanders Peirce	1859	Physics/Philosophy
Francis Ellingwood Abbot	1859	Theology/Philosophy
Oliver Wendell Holmes Jr.	1861	Law
William James	*	Psychology/Philosophy
John Fiske	1863	History/Philosophy
Joseph Bangs Warner	1869	Law
Francis Greenwood Peabody	1869	Theology
Henry Ware Putnam	1869	Law
William Pepperell Montague	1869	Law

* William James was not a BA student at Harvard, though he did study at the Lawrence Scientific School there from 1861 to 1863 before switching to medicine, receiving his MD in 1869.

The full list of those affiliated with the club, which also included many future lawyers, is presented here in table 5 and can be divided into three subcohorts: Nicholas St. John Green and Chauncey Wright graduated in the early 1850s; Peirce, Francis Ellingwood Abbot, Oliver Wendell Holmes Jr., and John Fiske in the years around 1860; and Joseph Bangs Warner, Francis Greenwood Peabody, Henry Ware Putnam, and William Pepperell Montague in 1869.[4] As table 5 indicates, many of those involved in the club finished college either shortly before or shortly after Charles Darwin's *On the Origin of Species* appeared, late in 1859. Their early intellectual development thus coincided with the reception of Darwinian ideas in the United States.

Many historians have surveyed the reactions of Americans to Darwin's book. James Moore and others have shown that religious views were no great impediment to the acceptance of evolution.[5] Fiske and Ab-

---

4. Peirce, *Writings*, 3:xxx–xxxi. For Harvard graduation dates, see the *Catalogue of the Officers and Students of Harvard University* for the relevant years. William James was not a BA student at Harvard, though he did study chemistry and then physiology and comparative anatomy at the Lawrence Scientific School there from 1861 to 1863 before switching to medicine, finally receiving his degree in 1869: see Robert D. Richardson, *William James: In the Maelstrom of American Modernism* (Boston: Houghton Mifflin, 2006), 101–3.

5. James R. Moore, *The Post-Darwinian Controversies: A Study of the Protestant Struggle to Come to Terms with Darwin in Great Britain and America, 1870–1900* (Cambridge: Cambridge University Press, 1979); David Livingstone, *Darwin's Forgotten Defenders* (Vancouver, BC: Regent College, 1984); Jon H. Roberts, *Darwinism and the Divine in America: Protestant Intellectuals and Organic Evolution, 1859–1900* (Madison: University of Wisconsin Press, 1988).

bot, confirming this claim, eventually treated evolution and religion as mutually reinforcing rather than antagonistic. By the mid-1870s, most *naturalists*—the standard term at the time for what we now call *natural scientists*—had adopted evolution as the correct account of the history of life on earth. At this point, as Ronald Numbers summarized, "scientific opposition in North America had diminished to a whisper," although debates over the various factors involved in evolution continued for decades, as we will see in chapter 5.[6] The story of evolution and the Metaphysical Club is thus in most ways typical: by the time the club began meeting in the early 1870s, and by the end of that decade at the very latest, its members had all embraced evolutionary ideas. Nevertheless, philosophical and personal differences among the club members led to differences in their initial attitudes toward Darwin's theories.

In this chapter, I will demonstrate that discussions of Darwinian ideas in the 1860s and early 1870s, following the publication of the *Origin of Species*, had a major impact on the philosophers associated with the Metaphysical Club. Despite their personal connections to Louis Agassiz, a prominent defender of orthodoxy in biology, Wright and the others all went over to the evolutionists' camp. This relatively painless acceptance of evolution should perhaps not surprise us, for even Agassiz's own students—many of whom went on to become important naturalists—eventually abandoned their teacher's views.[7] But unlike Agassiz's students, the philosophers of the Metaphysical Club participated in the Darwin debates as philosophers rather than as naturalists: that is, they were primarily concerned with the theological, epistemological, and metaphysical arguments surrounding evolution rather than with its empirical vindication. Some of the club members—Wright, Abbot, and Fiske—endorsed evolution relatively quickly, probably because of their shared commitment to a broad form of positivism.[8] But Peirce and James, with their close personal connections to Agassiz, were more reticent. It was probably the positivist trio of Wright, Abbot, and Fiske, with their public defenses of Darwin and their attacks on his orthodox religious opponents, who pushed James and Peirce in the direction of evolution in the late 1860s and early 1870s. Darwin famously called himself a

---

6. Ronald L. Numbers, *Darwinism Comes to America* (Cambridge, MA: Harvard University Press, 1998), 24.
7. Edward Lurie, *Louis Agassiz: A Life in Science* (Chicago: University of Chicago Press, 1960), chap. 8; Mary P. Winsor, *Reading the Shape of Nature: Comparative Zoology at the Agassiz Museum* (Chicago: University of Chicago Press, 1991), 37–42.
8. Trevor Pearce, "'Science Organized': Positivism and the Metaphysical Club, 1865–1875," *Journal of the History of Ideas* 76 (2015).

CHAPTER ONE

philosophical naturalist; Wright, Abbot, and Fiske were philosophers of natural history.[9]

In 1857, a decade after he became professor of zoology and geology in the newly founded Lawrence Scientific School at Harvard University, the Swiss naturalist Louis Agassiz published a work "written in America, and more especially for America." He expected his *Contributions to the Natural History of the United States*, which opened with an "Essay on Classification," to be read "by operatives, by fisherman, by farmers, quite as extensively as by the students in our colleges, or by the learned professors." In this work and others, Agassiz showed little patience for the "development hypothesis" (what we would now call *evolution*), which had been more frequently discussed after the 1844 publication of the anonymous *Vestiges of the Natural History of Creation*. Agassiz argued that the systematic relationships of plants and animals were evidence of "premeditation prior to the act of creation"; thus we could "have done, once and for ever, with the desolate theory which refers us to the laws of matter as accounting for all the wonders of the universe." Agassiz, perhaps the most prominent scientist in the United States in the 1850s, was an implacable foe of evolutionary ideas.[10]

Several future members of the Metaphysical Club had close ties to Agassiz. Some of their parents were his Harvard colleagues: Benjamin Peirce, father of Charles, was professor of astronomy and mathematics, and Oliver Wendell Holmes Sr. was professor of anatomy and physiology. Both were early members along with Agassiz of a group known as the Saturday Club, which met each month beginning in the mid-1850s for dinner at the Parker House in Boston (other prominent members included Ralph Waldo Emerson and Henry Wadsworth Longfellow). The elder Holmes later recalled that "the most jovial man at table was Agas-

---

9. Charles Darwin, *Journal of Researches into the Geology and Natural History of the Various Countries Visited by H.M.S. Beagle, under the Command of Captain Fitzroy, R.N. from 1832 to 1836* (London: Henry Colburn, 1839), 210.

10. Louis Agassiz, *Contributions to the Natural History of the United States of America*, vol. 1 (Boston: Little, Brown, 1857), x, 9. The first part of this volume was later published separately as Louis Agassiz, *An Essay on Classification* (London: Longman, Brown, Green, Longmans, & Roberts, 1859). For the development hypothesis, see Robert Chambers, *Vestiges of the Natural History of Creation* (London: John Churchill, 1844), 191–235; Herbert Spencer, "The Development Hypothesis," *Leader* (London), March 20, 1852. On the importance of *Vestiges*, see James A. Secord, *Victorian Sensation: The Extraordinary Publication, Reception, and Secret Authorship of "Vestiges of the Natural History of Creation"* (Chicago: University of Chicago Press, 2000).

siz; his laugh was that of a big giant." In fact, the club was known in some quarters simply as "Agassiz's club." Holmes greatly admired Agassiz and reportedly called him "a 'Liebig's Extract' of the wisdom of ages," referring to the concentrated meat stock invented by chemist Justus Liebig.[11] Like Holmes, Benjamin Peirce was a personal friend of Agassiz: Charles later recalled that "Agassiz came in every day without ringing, and standing in the large hall, would call 'Ben!'"[12] Benjamin Peirce and Agassiz were also connected professionally, serving as successive presidents of the American Association for the Advancement of Science (AAAS) in the early 1850s; Agassiz referred to Peirce in lecture as "our great mathematician." Edward Lurie has shown that Agassiz and Peirce—along with Alexander Dallas Bache, who preceded them as AAAS president and led the United States Coast Survey—were part of a small group attempting "to exert a dominant influence over the entire structure of American science." Such influence was even noted across the Atlantic: as Darwin wrote in an 1854 letter to the botanist Joseph Hooker, "I seldom see a Zoological paper from N. America, without observing the impress of Agassiz's doctrines."[13]

The older members of the Metaphysical Club would have encountered Agassiz's ideas at Harvard if not before. The required natural history class that both Green and Wright took as sophomores used Agassiz and Augustus Addison Gould's recently published *Principles of Zoology* as a textbook. Agassiz and Gould divided the geological record into four great ages: the Reign of Fishes, the Reign of Reptiles, the Reign of Mammals, and the Reign of Man (figure 1). Despite this progressive picture, they insisted that the findings of science "unequivocally indicate the direct interventions of creative power":

---

11. Edward Waldo Emerson, *The Early Years of the Saturday Club, 1855–1870* (Boston: Houghton Mifflin, 1918), 19, 24, 30, 32–33. Holmes's "most jovial man" letter was first quoted in Charles Francis Adams, *Richard Henry Dana: A Biography*, 2 vols. (Boston: Houghton Mifflin, 1890), 2:168. For the recipe of Liebig's extract, including a suggestion that it be "injected by the rectum" to avoid its "raw, disagreeable taste," see "Reports of Societies: Medical Society of London," *Medical Times and Gazette* (London), December 16, 1854, 625.

12. MS 619 (1909), Peirce Papers (see full citation of this and other such archival sources in the Abbreviations of Manuscript Sources list in the front of this book).

13. *Proceedings of the American Association for the Advancement of Science: Sixth Meeting, Held at Albany (N. Y.), August 1851* (Washington City [Washington, DC]: S. F. Baird, 1852), iii; *Proceedings of the American Association for the Advancement of Science: Seventh Meeting, Held at Cleveland, Ohio, July, 1853* (Cambridge, MA: Joseph Lovering, 1856), vii; Louis Agassiz, *Methods of Study in Natural History* (Boston: Ticknor & Fields, 1863), 117; Edward Lurie, *Louis Agassiz: A Life in Science* (Chicago: University of Chicago Press, 1960), 183; Charles Darwin to Joseph Hooker, 26 March 1854, in Darwin, *Correspondence*, 5:187.

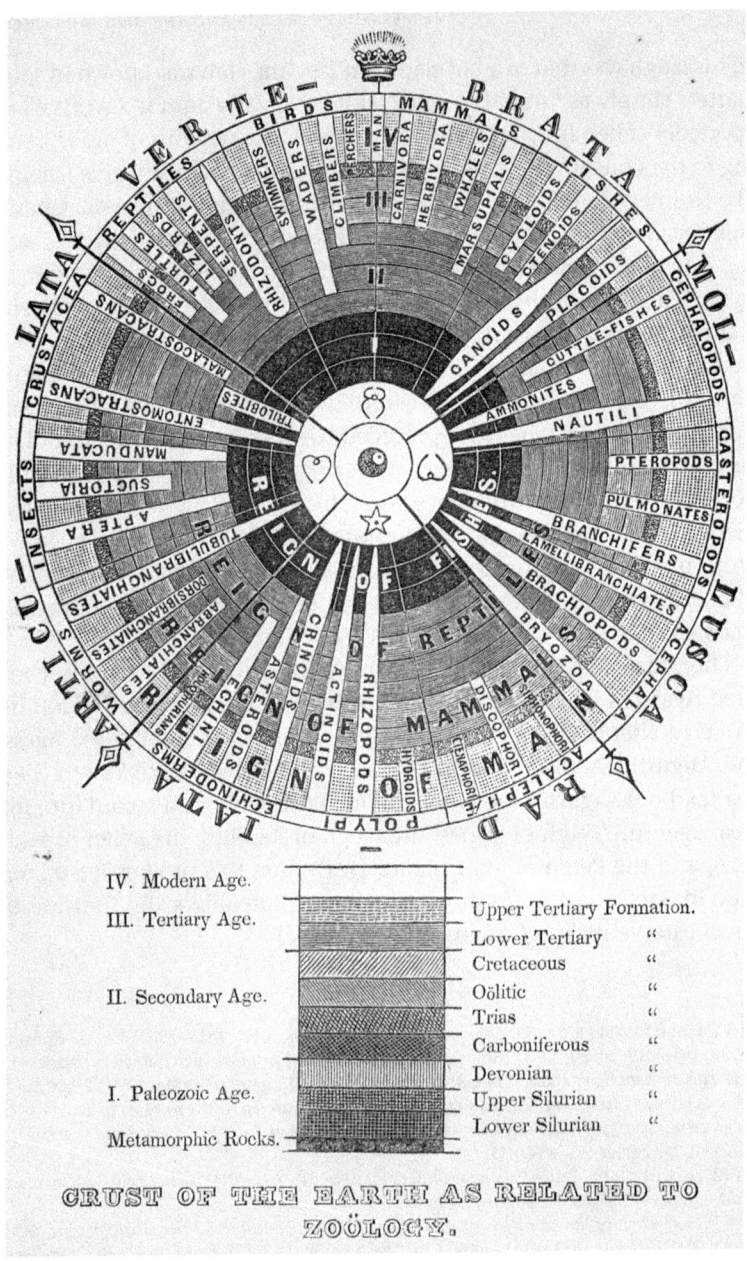

Figure 1 "Crust of the Earth as Related to Zoology," depicting the fossil record of the four main types of animals (radiates, mollusks, articulates, vertebrates). The successive appearances of vertebrate groups define four ages, dominated by fish, reptiles, mammals, and finally humans. Frontispiece of Louis Agassiz and Augustus A. Gould, *Principles of Zoölogy* (Boston: Gould, Kendall & Lincoln, 1848). Reproduced courtesy of the University of Chicago Library.

It is evident that there is a manifest progress in the succession of beings on the surface of the earth.... But this connection is not the consequence of a direct link between the faunas of different ages. There is nothing like parental descent connecting them.... The link by which they are connected is of a higher and immaterial nature; and their connection is to be sought in the view of the Creator himself.

Peirce and Abbot, as seniors at Harvard in 1858–59, may have elected to take Agassiz's lectures on geology or zoology—lectures which would have made similar points.[14]

Abbot and Peirce graduated in 1859, having learned natural history at least in part from Agassiz. That summer, Fiske (not yet in college) praised Agassiz's *Contributions to the Natural History of the United States* in a letter to his mother. In mid-November, Wright began teaching natural philosophy at a school for college-age women run by Agassiz and his wife Elizabeth Cabot Agassiz. Peirce's future wife, Harriet Melusina Fay, began studying at the Agassiz school that same month. But everything was about to change: late in 1859, Darwin's *On the Origin of Species* appeared.[15]

Chauncey Wright wasted no time embracing Darwin's theory. He declared himself a Darwinian in a February 1860 letter:

The idea of ["On the Insensible Gradation of Words"] is a very attractive one, and closely resembles the argument in that new book on "The Origin of Species,"—Darwin's—which I have just finished reading, and to which I have become a convert, so far as I can judge in the matter.

Agassiz comes out against its conclusions, of course, since they are directly opposed to his favorite doctrines on the subject; and, if true, they render his essay on Classification a useless and mistaken speculation.[16]

---

14. *A Catalogue of the Officers and Students of the University at Cambridge, for the Academical Year 1848–49* (Cambridge, MA: Metcalf, 1848), 33, 41; *A Catalogue of the Officers and Students of Harvard College, for the Academical Year 1849–50* (Cambridge, MA: Metcalf, 1849), 34, 43; Louis Agassiz and Augustus A. Gould, *Principles of Zoölogy: Touching the Structure, Development, Distribution, and Natural Arrangement of the Races of Animals, Living and Extinct* (Boston: Gould, Kendall, & Lincoln, 1848), 182, 190, 205–6; *A Catalogue of the Officers and Students of Harvard University, for the Academical Year 1858–59, First Term* (Cambridge, MA: John Bartlett, 1858), 10–13, 30–32.

15. John Fiske to Mary Fisk Green Stoughton, 16 July 1859, Box 1, Fiske Papers; Elizabeth Cabot Agassiz to Chauncey Wright, 11 November 1859, Wright Papers; Chauncey Wright to Susan Inches Lesley, 12 February 1860, in Wright, *Letters*, 42; Norma P. Atkinson, "An Examination of the Life and Thought of Zina Fay Peirce: An American Reformer and Feminist" (PhD diss., Ball State University, 1983), 17. On the vagaries of John Fiske's name, see John Spencer Clark, *The Life and Letters of John Fiske*, 2 vols. (Boston: Houghton Mifflin, 1917), 1:55.

16. Wright to Lesley, 12 February 1860, in Wright, *Letters*, 43. Lesley's husband, in the article mentioned by Wright, claimed that "in philology, as in palaeontology, . . . organic forms pass

Wright had thus privately accepted species evolution only a few months after Darwin's book appeared.

He soon joined the public controversy over Darwin's views. In March, Benjamin Peirce moved that a special meeting of the AAAS (which met in Cambridge) be held to discuss the *Origin*. At this meeting, on March 27, 1860, "the hypothesis of the origin of species through variation and natural selection" was criticized by both Agassiz and Francis Bowen, the Harvard philosophy professor of Peirce, Abbot, and Fiske.[17] At the next monthly meeting on April 10, the botanist Asa Gray—another Harvard professor and a confidant of Darwin even before the *Origin* was published—replied to the general tendency of these criticisms, arguing that "the theory of derivation of one species or sort of animal from another" did not necessarily conflict with "the doctrines of final cause, utility, special design, or whatever other teleological view." Interested students were following the debate: "Gray advocates the views of Darwin in regard to the 'Origin of Species,'" wrote Fiske in a letter, and "he and Agassiz have some warm controversies on the subject."[18]

Wright's contribution to the AAAS discussion, a response to Bowen, came during the monthly meeting of May 8. In the April issue of the *North American Review*, Bowen had published an account of the *Origin* that Darwin described as "clever & dead against me." Several pages of the review focused on instincts, in particular the cell-making instinct of bees. Darwin had sought to show that "the most wonderful of all known instincts, that of the hive-bee, can be explained by natural selection having taken advantage of numerous, successive, slight modifications of simpler instincts." The structure of honeycomb consists of two offset layers of hexagonal prisms, with the top of each prism open and three rhombuses converging at its base (figure 2). According to Darwin, this apparently complex cell structure is the result of simple instincts to sweep out closely packed spherical hollows and then to "build up and excavate the wax" where these hollows meet. The bees have not actually "solved a recondite problem": it is possible to see the cell-making

---

into each other by almost insensible gradations." See Peter Lesley, "On the Insensible Gradation of Words," *Proceedings of the American Philosophical Society* 7 (1859): 129.

17. *Proceedings of the American Academy of Arts and Sciences* 4 (1860): 410; *A Catalogue of the Officers and Students of Harvard University, for the Academical Year 1858–59* (Cambridge, MA: John Bartlett, 1858), 30–32; *A Catalogue of the Officers and Students of Harvard University, for the Academical Year 1862–63* (Cambridge, MA: Sever and Francis, 1862), 32–34.

18. *Proceedings*, 414; John Fiske to Lizzie Wilcox, 16 September 1860, Box 2, Fiske Papers.

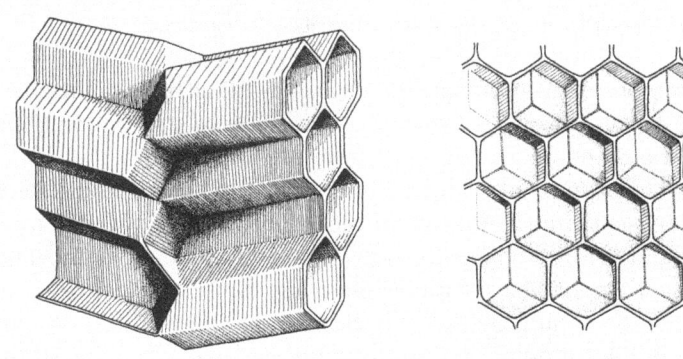

Figure 2 Structure of honeycomb, showing two layers of hexagonal cells (left), each of which has a base formed by three rhombuses (right).
Drawing from Karl von Frisch and Otto von Frisch, *Animal Architecture*, trans. Lisbeth Gombrich (New York: Harcourt Brace Jovanovich, 1974), 86. Copyright © 1974 by Karl von Frisch and Otto von Frisch. Illustrations copyright © 1974 by Turid Hölldobler. Reproduced by permission of the Houghton Mifflin Harcourt Publishing Company and the estate executors of Turid Hölldobler-Forsyth. All rights reserved.

instinct as a composite result of selection for more economical use of wax via the modification of simple instincts.[19]

Bowen, in his review, sarcastically summarized Darwin's argument:

The construction, then, according to Mr. Darwin, is very simple. We have only to suppose that several hundred or thousand bees, beginning work on the opposite faces of a thin plate of wax, excavate at once many hemispheres, with the centre of each at the distance of radius × 1.41421 from the centres of all the adjacent spheres both in the same layer and in the other and parallel layer. It is only necessary to add, that the bees then economize their precious wax by biting away every particle of it which is not absolutely needed, and the work is practically done. The problem of constructing the marvellous cells is solved.

As it seems to us, Mr. Darwin's explanation only makes the work of the bees appear more wonderful than ever. Not only do they build cells having the marvellous properties

---

19. Francis Bowen, "Darwin on the Origin of Species," *North American Review* 90 (1860); Charles Darwin to Joseph Hooker, 18 April 1860, in Darwin, *Correspondence*, 8:162; Charles Darwin, *On the Origin of Species by Means of Natural Selection, or the Preservation of Favoured Races in the Struggle for Life* (London: John Murray, 1859), 224, 235.

first described, but the *modus operandi*—the process of building them . . . —rivals in beauty and simplicity any solution that mathematicians ever effected.

Bowen's most telling criticism, as indicated by his citation of the precise figure 1.41421, was to ask how the bees are able to start building their cells an exact and uniform distance from one another.[20]

Wright, responding to Bowen without naming him, did not directly address this criticism; instead, he focused on the comparison with mathematics and a related ambiguity in the notion of economy. "Mathematicians," said Wright, "have regarded the economical characteristics of the honey-cell too exclusively, to the neglect of . . . symmetries."[21] Wright's account depended on a distinction between rational economy and sensible economy: the former involves rational foresight that "forestalls waste," whereas the latter "remedies waste or simply saves." The central part of Wright's argument, as set out in a June article deriving from his AAAS remarks, thus opened with a pair of questions relating to economy and symmetry: "Of what advantage is elegant symmetry to the bee, unless it also economizes labor and material? And what therefore could have fashioned the instinct of the bee except a supersensible principle of rational foresight, superior to mere sensible perception?" In other words, how does the bee's approach to the problem differ from the mathematician's?[22]

Wright's answer was that the symmetry of the cells arises not through rational foresight but through simple modification of the kinds of spherical and cylindrical structures that characterize nests and cocoons more generally. The bees' cells must be open at one end; thus the "natural type" (i.e., primitive form) of such a cell is a cylinder with a hemispherical base—a common structure in nature. Wright pointed out that if one starts by excavating many such cylinders, placing them as closely together as possible, a slight change to the boundaries transforms the cylinders into regular hexagonal prisms "by simple saving, or by the economy of afterthought." Wright illustrated this transformation from circles to hexagons, in two offset layers, with an image in his paper (figure 3). Thus, he concluded:

An unreflective and unforeseeing economy, which, without reference to an end, simply saves, through sensuous preference, what the conditions of life render useful and costly

---

20. Bowen, "Darwin on the Origin," 495. This review probably repeated some of what was said at the March 27 meeting: see *Proceedings of the American Academy of Arts and Sciences* 4 (1860): 411.

21. *Proceedings*, 432.

22. Chauncey Wright, "The Economy and Symmetry of the Honey-Bees' Cells," *Mathematical Monthly* 2 (1860): 304–5, 308; see also *Proceedings*, 432–33.

THE METAPHYSICAL CLUB AND THE *ORIGIN OF SPECIES*

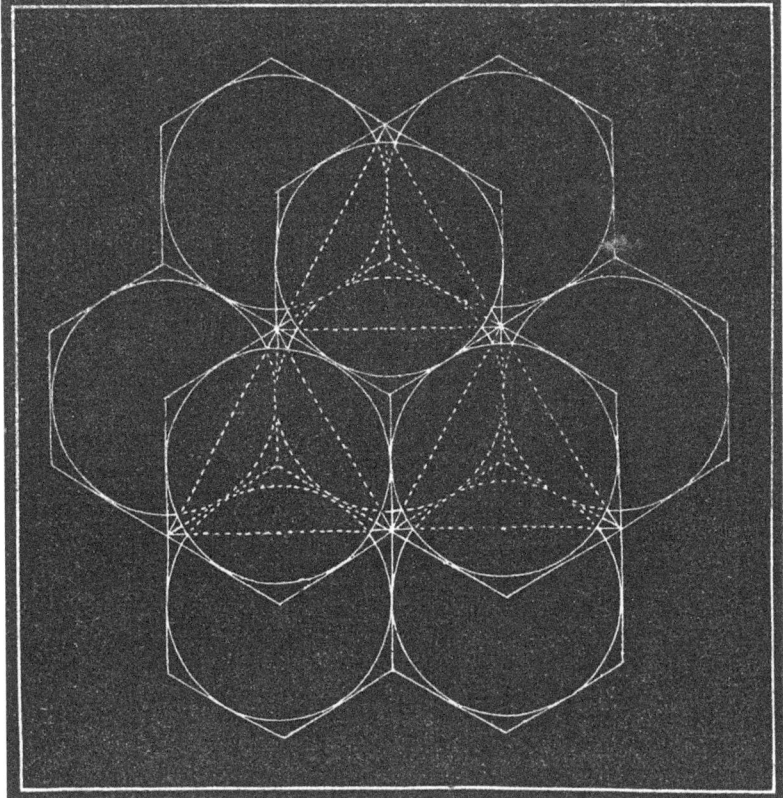

Figure 3 Comparison of circular and hexagonal cells, showing material saved when the former are transformed into the latter.
From Chauncey Wright, "The Economy and Symmetry of the Honeybees' Cells," *Mathematical Monthly* 2 (1860): 309. Reproduced courtesy of the University of Chicago Library.

to the race, characterizes the whole animal kingdom. . . . The bee's instinct ought not therefore to be regarded as an exception to animal instincts in general.

Wright agreed with Darwin that the bees do not have to solve a "recondite problem" using some kind of rational foresight. But he used his greater mathematical training to show more precisely how two layers of hexagonal prisms with pyramidal bases can result from minor modifications to "a pile of equal spheres."[23]

23. Wright, "Economy and Symmetry," 309, 319.

Wright sent a copy of the paper to Bowen, who wrote back in a letter that he agreed with Wright's mathematical conclusions. However, Bowen accused Wright of attacking a straw man:

> Your general remarks about instinct, also, seem to me well founded, excepting perhaps that you appear occasionally to argue against a doctrine which no one thinks of maintaining. Certainly, no competent psychologist would attribute to the bee any perception of the supersensible properties of form, or even any "rational" economy acting by foresight.

Bowen also reiterated his point about "exquisite precision," which Wright had not directly addressed. Although we do not know whether or how Wright replied to Bowen, he would probably have appealed again to the way in which the mathematical symmetries are easily attainable when starting with two layers of hemispheres—greater and greater precision can then be attributed to the gradual action of natural selection. (Darwin did eventually read Wright's paper, sent to him by Asa Gray, though he professed not to understand it.)[24]

The other individuals involved with the Metaphysical Club did not move as swiftly as Wright to Darwin's side, but Fiske and Abbot were almost as quick. In a letter from the spring of 1860, Fiske embraced "the law of 'natural selection,' so richly illustrated by Darwin, which furnishes us with a new stand-point from which to contemplate the history of the human race." He later told Darwin, remembering his 1860 encounter with the *Origin of Species*, "I hailed your book with exultation, reading and re-reading it till I almost knew it by heart."[25] It is more difficult to determine exactly when Abbot became an evolutionist, but his biographers suggest that it happened while he was studying at the Meadville Theological School in Pennsylvania from 1860 to 1863.[26] Abbot's mentor at Meadville was Oliver Stearns, who endorsed some evolutionary ideas at the time but perhaps not what we now think of as

---

24. Francis Bowen to Chauncey Wright, 25 June 1860, Wright Papers; Charles Darwin to Asa Gray, 3 July 1860, and Charles Darwin to William Miller, 1 December 1860, in Darwin, *Correspondence*, 8:273–75, 506.

25. John Fiske to Jonathan Ebenezer Barnes, [April/May/June 1860], Box 2, Fiske Papers; John Fiske to Charles Darwin, 23 October 1871, in Darwin, *Correspondence*, 19:648–50. Fiske's letter to Barnes can be roughly dated, as it is a reply to a letter from Barnes (9 April 1860) and precedes Barnes's subsequent reply (9 June 1860).

26. Sydney E. Ahlstrom and Robert Bruce Mullin, *The Scientific Theist: A Life of Francis Ellingwood Abbot* (Macon, GA: Mercer University Press, 1987), 38–39; W. Creighton Pedan, *The Philosopher of Free Religion: Francis Ellingwood Abbot, 1836–1903* (New York: Peter Lang, 1992), 12.

biological evolution. Stearns cited one of Agassiz's students in an 1856 address:

> A living lecturer upon the natural history of the earth and its inhabitants has indicated that in the evolution of nature [i.e., ontogeny], the point of departure is a *homogeneous unit*, that the progress is *diversification*, that the end is an *organic* or *harmonic unit*, that all life is mutual exchange, and that all condition of more active life is a greater variety of forms of nature, of relative situations, of contrast.

Arnold Guyot, the lecturer in question, had applied this view to social evolution. Stearns applied it in turn to the development of religion, arguing that "the history of all living Christianity is the history of controversy" and concluding that "diversification . . . is the law of the Christian evolution."[27] Thus, although Stearns may not have embraced the theory presented in the *Origin of Species*, which appeared just before Abbot's arrival at Meadville, his interest in evolution more broadly may have nudged Abbot in that direction. By 1860, the school's library also contained copies of the *Vestiges*, Herbert Spencer's *Principles of Psychology*, and Darwin's *Origin*; thus Meadville students would have had access to evolutionary ideas. As one such student (a tutee of Abbot) recalled, "I accepted Darwin [in 1860] without mental reservation and went forward to study the Bible and the Christian religion."[28]

Wright, Fiske, and Abbot engaged, often critically, with Spencer's evolutionary philosophy throughout the 1860s, as we will see in the next chapter. They were also prominent public defenders of Darwin's views in the early 1870s, just prior to the first meetings of the Metaphysical Club. Wright claimed, in an 1870 review, that there were fundamental flaws in Alfred Russel Wallace's argument that higher human capacities could not be explained by evolution. "The metaphysical isolation of human nature," Wright concluded, is based on "barbaric conceptions of dignity, which are restricted in their application by every step forward in the progress of science." Whereas Wright's discussions of evolution were usually somewhat technical, Fiske and Abbot were popularizers, defending evolution from the attacks of more orthodox Christians. Fiske compared the doctrines of special creation and evolution in an 1871

---

27. Oliver Stearns, "The Written Word and Christian Consciousness," *Christian Examiner and Religious Miscellany* 60 (1856): 174, 176; citing Arnold Guyot, *The Earth and Man: Lectures on Comparative Physical Geography, in Its Relation to the History of Mankind*, trans. C. C. Felton (Boston: Gould, Kendall, & Lincoln, 1849), 75–78.

28. *Catalogue of the Library of the Meadville Theological School* (Meadville, PA: Republican Printing House, 1870), 39, 108; George Batchelor, *Personal Reminiscences* (Boston: Geo. H. Ellis, 1916), 17–18.

lecture at Harvard, later published in the *New York World*. He dismissed the former for invoking "a hypothetical assumption as to divine interposition which is incapable of scientific verification" and praised the latter for being "a purely scientific theory, since it appeals to no agencies which are not known to be in operation, and involves no hypothetical assumptions which cannot, sooner or later, be subjected to a crucial test." Abbot made a similar point the next year, in a lecture published in his weekly free religion magazine the *Index*:

> Can we account for the appearance of new species of animals and plants, and the disappearance of fossil species, without miracle? The Biblical Theory says, "No." The Development Theory says, "Yes." And the battle is hot and fierce between the two. But day by day faith in miracle loses ground, while faith in law gains ground; and it requires little prophetic insight to foretell on which banner Victory will perch at last. Miracle is passing away into the same limbo which has received witchcraft and kindred delusions; and law is seen more and more clearly to be the true explanation of all seeming anomalies.

Abbot argued that, rather than undermining people's faith, this victory should strengthen it: "Law means cosmos, order, reason, miracle means chaos, disorder, unreason"; thus a person who believes in God "should believe in law, and reject miracle." This attitude explains his otherwise surprising declaration: "I make the Development Theory an essential part of the gospel."[29]

Wright, Abbot, and Fiske were immediately enthusiastic about Darwin's ideas. Charles Peirce and James were more reticent, possibly because of their close connection to Agassiz. Peirce, many years later, recalled hearing about the *Origin* in 1860 while working for the United States Coast Survey, charting the complex coastline of Louisiana near Breton Sound (figure 4): "In the course of the winter, a letter from my mother told me what a sensation the book had made; and thereupon I wrote to my friend Mr. Chauncey Wright that I felt confident that Darwin had received a hint of his idea from Malthus *On Population*." Despite this apparent attunement to Darwin's ideas, Peirce also had strong family connections to Agassiz. Another letter from his mother reveals that he was actually collecting specimens for the Harvard naturalist in 1859–

---

29. Chauncey Wright, "Limits of Natural Selection," *North American Review* 111 (1870): 310; John Fiske, "The Evolution of Life," *New York World*, June 12, 1871; Francis Ellingwood Abbot, "The Development Theory," *Index*, April 13, 1872, 114. On free religion, see Francis Ellingwood Abbot, "Fifty Affirmations," *Index*, January 1, 1870, 1.

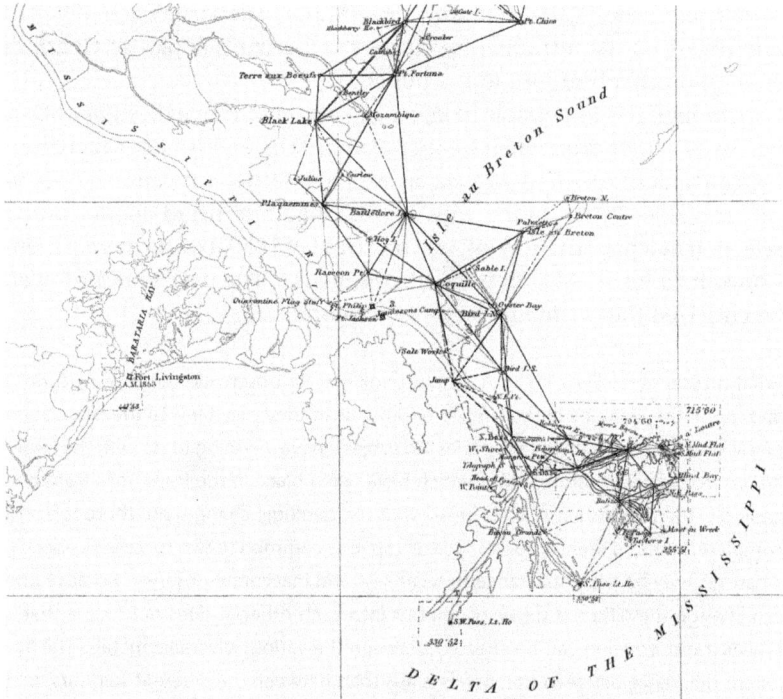

Figure 4 Detail of United States Coast Survey map showing the progress of Section 8 up to 1860, including Isle au Breton Sound, which Charles Sanders Peirce helped survey that year. Peirce received letters at both Raccoon Point and Oyster Bay in 1860 (see Cuddeback Letter Book).
From Sketch H, in *Report of the Superintendent of the Coast Survey, showing the Progress of the Survey during the Year 1860* (Washington, DC: Government Printing Office, 1861), 84, 198; https://historicalcharts.noaa.gov/image=AR18-00-1861. Reproduced courtesy of Historical Maps and Charts, Office of Coast Survey, National Oceanic and Atmospheric Administration.

60 while working in Louisiana: "Mr Agassiz says if you cannot get fresh water specimens he shall be thankful for any thing you can get from the salt water there—star fish—sea anemone—anything that swims as he has nothing at all from that region." Peirce's younger brother Benjamin Mills Peirce, who had just turned sixteen, was also very interested in biology, and Charles had promised him some specimens for his local student group, the Agassiz Natural History Society. Benjamin Mills worried that he would lose his animals to Agassiz unless Charles took care to separate them: "Please put *our* specimens in a different bottle from Mr Agassiz' for, if they are together, the society will get none at all, as I know

from experience." Later that same month, the little society actually read and discussed the introductory chapter of Darwin's book—so Charles's teenage brother beat him to the punch.[30]

Although it is impossible to know exactly what Peirce thought of Darwin early on, or even when he first read the *Origin*, his father Benjamin Peirce—a close friend of Agassiz, as we have seen—was cautiously critical. The elder Peirce, along with Agassiz, had first heard about natural selection at a presentation by Asa Gray in May 1859, but he merely mentioned it to his wife Sarah without comment.[31] The next year, however, he criticized the *Origin* in a letter to a friend:

What do you say to Darwin? . . . I cannot think that his observations, however curious and useful they may be in themselves and however they may tend to the elucidation of the laws of change to which species are subject, have anything to do with the larger and more radical transformations which have taken place in the transitions from one geologic age to another. Agassiz insists that the geologic changes are thorough and complete, and that there is no instance of a species common to two successive epochs, whatever may be the appearance to a careless and inaccurate observer. Because one can change the different forms of sulphur into each other, it does not follow that a similar transformation can be effected between the various silver-like metals. The apparent difference between the metals is less than between the different sulphurs, and yet to him, who understands the true fact, the difference is radical and impassable. Is not this a true analogy? The transitions of the successive ages of geology have their own laws, which are to be studied by themselves. They came from God, and so did gravitation and the one not less directly than the other. Both are divine messages, intended for man.[32]

Benjamin Peirce was clearly siding with Agassiz against Darwin, at least in denying that species evolution was an accurate description of the overall progress of life in the fossil record. He argued, drawing an analogy with chemistry, that the existence of local transmutation did not imply global transmutation. Moreover, appearances could be misleading

---

30. MS 706 (1909), in Peirce, *New Elements*, 3:155; Sarah Mills Peirce to Charles Sanders Peirce, 16 February 1860, and Benjamin Mills Peirce to Charles Sanders Peirce, 4 April 1860, Cuddeback Letter Book; *Records of the Agassiz Natural History Society* (MCZ 073), 2:118, Special Collections, Museum of Comparative Zoology, Harvard University. For C. S. Peirce's survey work, see Alexander Dallas Bache, *Report of the Superintendent of the Coast Survey, Showing the Progress of the Survey during the Year 1860* (Washington, DC: Government Printing Office, 1861), 84–91.

31. Benjamin Peirce to Sarah Mills Peirce, 13 May 1859, "Peirce, Sarah Hunt Mills," Box 11, Benjamin Peirce Correspondence (MS Am 2368), Houghton Library, Harvard University.

32. Benjamin Peirce to John LeConte, 11 March 1860, Folder 2, Box 1, LeConte Family Papers: Additions, 1856–1916 (BANC MSS C-B 1014), Bancroft Library, University of California–Berkeley.

when it came to the possibility of transmutation, as shown by the contrast between sulfur and "the various silver-like metals": tin and silver may look similar, but they cannot be transformed into each other the way the various forms of sulfur can.[33]

Having returned in May 1860 from Louisiana, presumably with Agassiz's specimens in tow, Charles Peirce ended up studying with the Swiss professor as a private student for the remainder of the year. On June 24, Peirce's brother Benjamin told their father that Charles was "studying with Mr Agassiz now for a while." A few months later, in August, a former classmate supposed him to be "busily engaged at the Museum"— meaning the Museum of Comparative Zoology, directed by Agassiz. Fifty years later, Charles recalled difficulties due to lack of guidance: apparently Agassiz set him "to sorting fossil brachiopods without knowing anything about them."[34] He ended up studying chemistry instead, a subject in which he was already interested. Earlier that year, his older brother James Mills Peirce had assumed he was heading in that direction: "I have often thought what a fine thing it would be if you and Benjy [Benjamin Mills Peirce] and I should go into different departments of science: Chemistry, Natural History, and Mathematics."[35]

Joseph Brent points out that given its timing, Peirce's brief period of study with Agassiz would have "put him at the center of the arguments about evolution."[36] However, despite his later statements—for example, that Darwin's *Origin* produced a "tremendous upheaval" and ushered in "the greatest mental awakening since Newton and Leibniz"—Peirce's writings of the 1860s and early 1870s do not reveal which side he had chosen (if any) at that stage.[37] Nevertheless, several passing remarks do

---

33. There are many allotropes of sulfur, several of which were known in Peirce's day: see Jöns Jacob Berzelius, *Rapport annuel sur les progrès des sciences physiques et chimiques*, trans. Philippe Plantamour (Paris: Fortin, Masson, 1841), 5–7.

34. Benjamin Mills Peirce to Benjamin Peirce, 24 June 1860, L667, Peirce Papers; John Howland Ricketson to Charles Sanders Peirce, 1 August 1860, Cuddeback Letter Book; MS 902 (1910), Peirce Papers; see also MS 865 (1897), Peirce Papers. In both of these manuscripts, Peirce misdated his study with Agassiz to 1863 (following his MA in Chemistry at the Lawrence Scientific School), probably confusing his two Harvard graduations.

35. James Mills Peirce to Charles Sanders Peirce, 10 January 1860, Cuddeback Letter Book. Although he was then practicing as a minister, James Peirce did end up becoming a mathematician.

36. Joseph Brent, *Charles Sanders Peirce: A Life*, 2nd ed. (Bloomington: Indiana University Press, 1998), 60; see also Paul Forster, "The Logical Foundations of Peirce's Indeterminism," in *The Rule of Reason: The Philosophy of Charles Sanders Peirce*, ed. Jacqueline Brunning and Paul Forster (Toronto: University of Toronto Press, 1997), 71.

37. MS 706 (1909), in Peirce, *New Elements*, 3:150; MS 620 (1909), in Peirce, *Illustrations*, 187. On Peirce's early hesitancy about evolution, see also Philip P. Wiener, *Evolution and the Founders of Pragmatism* (Cambridge, MA: Harvard University Press, 1949), 77–78; Murray G. Murphey, *The Development of Peirce's Philosophy* (Cambridge, MA: Harvard University Press, 1961), 323.

CHAPTER ONE

suggest some sympathy with Darwin. In one of his Lowell lectures of 1866 on "The Logic of Science," Peirce claimed that "the great disputes of science have usually been between those who ask for causes and those who ask for classification; and the Darwinian controversy is a case in point." Although he listed natural history (including zoology and botany) among the classificatory sciences, which harmonizes well with Agassiz's view, his comment about the controversy suggests a belief that Darwin may have partly transformed biology into a hypothetic (or causal) science such as history or geology. Peirce also explicitly praised Darwin in his Harvard lectures of 1869 on "British Logicians," remarking that what made him so admired was "his minute, systematic, extensive, and strict scientific researches which have given his theories a more favorable reception—theories which in themselves would barely command scientific respect." But despite this backhanded compliment, there is no explicit evidence that Peirce embraced evolutionary ideas until 1877, when he claimed that natural selection had made human beings logical "in regard to practical matters."[38] (I will discuss Peirce's later evolutionary metaphysics and his 1870s discussion of natural selection in chapters 5 and 7, respectively.)

James was also professionally connected to Agassiz. Like Peirce, James had enrolled at the Lawrence Scientific School at Harvard in the fall of 1861. Although both of them were studying chemistry, James still traveled to Boston to hear Agassiz's lectures on Methods of Study in Natural History. James was impressed by the Swiss naturalist: "He is an admirable, earnest lecturer, clear as day and his accent is most fascinating. I should like to study under him."[39] Agassiz described these lectures, when they were published as a book in 1863, as entering an "earnest protest against the transmutation theory [i.e., evolution], revived of late with so much ability." In this later preface, Agassiz pulled no punches: Darwin and naturalists like him were "chasing a phantom"; moreover, there was "a repulsive poverty" in their explanation of life. Apart from a short argument against Darwin's move from artificial to natural selection, however, the lectures themselves contained few criticisms of evolutionary ideas. James was likely more fascinated by the content of Agassiz's lectures than their context: they covered everything from the general classification of organisms to the complex life cycles of marine

38. "Lecture IX," MS 130 (November 1866), in Peirce, *Writings*, 1:487–88; "Lecture I. Early Nominalism and Realism," MS 158 (November–December 1869), in Peirce, *Writings*, 2:314; Charles Sanders Peirce, "The Fixation of Belief," *Popular Science Monthly* 12 (1877): 3.

39. William James to Mary James, 10 September 1861, and William James to James Family, 16 September 1861, in James, *Correspondence*, 4:41–43.

invertebrates. James soon embraced the persona of budding naturalist, submitting a "future history" to his family in November: "1 year Study Chemistry, then spend one term at home, then 1 year with [the anatomist Jeffries] Wyman, then a medical education, then 5 or 6 years with Agassiz, then probably death, death, death with inflation and plethora of knowledge."[40]

James started another year of chemistry at Harvard in the fall of 1862 but switched to comparative anatomy the following fall to begin studying—a bit later than predicted—with Wyman.[41] James was at this stage still unsure about his future profession and told his cousin that he had four alternatives: "Natural History, Medicine, Printing, Beggary." He was drawn to natural history and bragged to his sister of working "in a vast museum, at a table all alone, surrounded by skeletons of mastodons, crocodiles, and the like." By December, however, he had chosen medicine—which seemed to combine his scientific interests with the necessity of making money—and he began attending medical lectures in 1864.[42]

James read and took notes on Darwin's *Origin of Species* in September 1863, while deciding whether or not to pursue a career in natural history. Unfortunately, these notes are lost.[43] Thus, the earliest hint of his opinion of evolutionary ideas comes in his very first publication: a review, written in the fall of 1864, of Thomas Henry Huxley's *Elements of Comparative Anatomy*. James did not explicitly endorse evolution in this review. However, referring to Huxley's earlier book *Evidence as to Man's Place in Nature*, he suggested that much of the opposition to evolution was emotional rather than scientific:

[Huxley] jovially says that, if we admit the transmutation hypothesis at all, we must apply it even unto majestic man, and see in him the offspring of some great ape, pregnant with Futurity. Probably our feeling on this point, more than anything else, will make many of us refuse to accept any theory of transmutation. This is indeed not the place to discuss the question, but we think it could be easily proved that such a feeling has

---

40. Louis Agassiz, *Methods of Study in Natural History* (Boston: Ticknor & Fields, 1863), iii–iv, 141–47; William James to James Family, 10 November 1861, in James, *Correspondence*, 4:52.

41. William James to Katherine James Prince, 12 September 1863, in James, *Correspondence*, 4:81. For James's enrollments at Harvard, see the *Catalogue of the Officers and Students of Harvard University* for the relevant years.

42. James to Katherine James Prince, 12 September 1863, William James to Alice James, 13 September 1863, and William James to Katherine James Prince, 13 December 1863, in James, *Correspondence*, 4:81–87.

43. Richardson, *William James: In the Maelstrom of American Modernism* (Boston: Houghton Mifflin, 2006), 57.

CHAPTER ONE

even less foundation than any other aristocratic prejudice. . . . Perhaps, by accustoming our imagination to contemplate the possibility of our ape descent now and then, as a precautionary measure, the dire prospect, should it ever really burst upon us, will appear shorn of some of its novel horrors, and our humanity appear no less worthy than it was before.

James gave a list of recent converts to the evolutionary hypothesis, including Asa Gray and Charles Lyell, and offered an amusingly reticent prediction as to its future success: "We may well doubt whether it may not be destined eventually to prevail." Thus, late in 1864, James was at least somewhat attracted to Darwin's views—perhaps not surprising given that his teacher Wyman supported evolution and had publicly endorsed it the year before.[44]

Despite these sympathies, James would soon become—like Wright and Peirce before him—an employee of Agassiz. By April 1865, a few months after the Huxley review was published, James was heading to Brazil as one of Agassiz's assistants, and one of the objects of the expedition was to find evidence against species evolution.[45] During the voyage south, Agassiz gave a series of scientific lectures to James and the other assistants to prepare them for their work in Brazil. The last of these, on April 20, concerned "the development theory" (i.e., evolution). Although Agassiz was clearly critical, he ended with an appeal to empiricism:

I bring this subject before you now, not to urge upon you this or that theory, strong as my own convictions are. I wish only to warn you, not against the development theory itself, but against the looseness in the methods of study upon which it is based. Whatever be your ultimate opinions on the subject, let them rest on facts and not on arguments, however plausible. This is not a question to be argued, it is one to be investigated.

In a letter written the day after this lecture, James expressed skepticism that Agassiz was really employing this open-minded approach, alluding to the religious aspect of the Swiss naturalist's scientific views:

---

44. Thomas Henry Huxley, *Lectures on the Elements of Comparative Anatomy: On the Classification of Animals and On the Vertebrate Skull* (London: John Churchill & Sons, 1864); Thomas Henry Huxley, *Evidence as to Man's Place in Nature* (New York: D. Appleton, 1863), 125–32; William James, "Huxley's Comparative Anatomy," *North American Review* 100 (1865): 290–91. For the date of composition of this review, see William James to Charles Eliot Norton, 3 September 1864, and William James to Charles Eliot Norton, 14 November 1864, in James, *Correspondence*, 4:92–94. On Wyman, see Toby A. Appel, "Jeffries Wyman, Philosophical Anatomy, and the Scientific Reception of Darwin in America," *Journal of the History of Biology* 21 (1988): 84–85.

45. Edward Lurie, *Louis Agassiz: A Life in Science* (Chicago: University of Chicago Press, 1960), 345.

Last Sunday, [Bishop Alonzo Potter] preached a sermon particularly to us "savans" as the outsiders call us, and told us we must try to imitate the simple child like devotion to truth of our great leader [i.e., Agassiz]. We must give up our pet theories of transmutation . . . and seek in nature what God has put there rather than try to put there some system wh. our imagination has devised &c &c. (Vide Agassiz passim.) The good old Prof. was melted to tears, and wept profusely.[46]

As in the Huxley review, it was feeling and not fact that turned people away from transmutation. The theological basis of Agassiz's position on evolution was obvious, according to James: "Vide Agassiz passim" means "see Agassiz's works, throughout."

The main duty of James and the other assistants on the expedition was the collection of specimens, and the party often split up to collect fish and other animals from a broader range of sites. For almost a month in the early autumn of 1865, while the group was exploring the Amazonas region of northwest Brazil, James and a Brazilian guide traveled east down the Amazon by canoe from São Paulo de Olivença to Tefé. Their main task along the way was to collect fish from the Içá and Jutai Rivers, tributaries of the Amazon (figure 5). James reported shortly after this side trip that his collections were "very satisfactory to the Prof, as they contained almost 100 new species." Elizabeth Agassiz agreed:

The commission could not have been better executed, and the result raises the number of species from the Amazonian waters to more than six hundred, every day showing more clearly how distinctly the species are localized, and that this immense basin is divided into numerous zoölogical areas, each one of which has its own combination of fishes.[47]

As Louis Menand has noted, James's collecting provided support for the trip's anti-Darwinian narrative, for Agassiz believed that the distinct fish populations in the different parts of the Amazon proved that species had

---

46. Louis Agassiz and Elizabeth Cabot Agassiz, *A Journey in Brazil* (Boston: Ticknor & Fields, 1868), 3, 43–44; William James to Henry James Sr. and Mary Robertson Walsh James, 21 April 1865, in James, *Correspondence*, 4:101.

47. William James to Henry James Sr. and Mary Robertson Walsh James, 21 October 1865, in James, *Correspondence*, 4:126–27; William James, "A Month on the Solimoens," in *Brazil through the Eyes of William James: Letters, Diaries, and Drawings, 1865–1866*, ed. Maria Helena P. T. Machado (Cambridge, MA: Harvard University Press, 2006); "Special Report of the Director," in *Annual Report of the Trustees of the Museum of Comparative Zoölogy, at Harvard College, in Cambridge, Together with the Report of the Director, 1866* (Boston: Wright & Potter, 1867), 14; Agassiz and Agassiz, *Journey in Brazil*, 208–9, 241–42. The section of the Amazon River between Tabatinga (at the Peru-Brazil border) and Manaus (where the Rio Negro enters the Amazon) is called the Solimões.

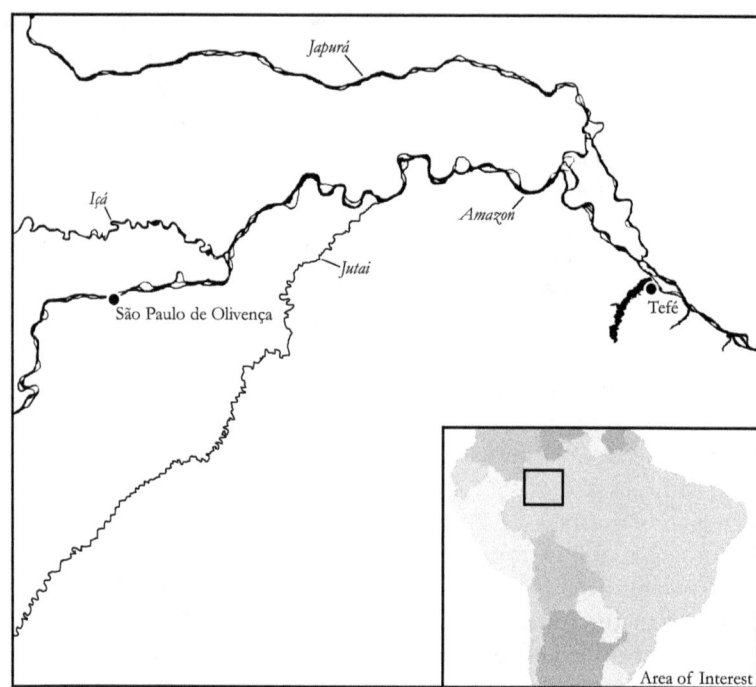

Figure 5 Amazon River in northwest Brazil. William James traveled down the Amazon from São Paulo de Olivença to Tefé in 1865 as part of Louis Agassiz's expedition, collecting fish from the Içá and Jutai Rivers.
Map by Patrick Jones.

been independently created to occupy those regions.[48] The expedition also, as Agassiz proudly announced in a letter to James, overturned the conventional picture of aquatic diversity:

I look forward to land in Pará with over 2000 species. This will be double the number of the Mediterranean, or any other circumscribed marine basin; and yet thus far the sea has been looked upon as the real home of the Fishes and the freshwaters as containing comparatively few. All the ideas now prevailing upon the intensity of life in the waters will have to be modified.[49]

---

48. Agassiz and Agassiz, *Journey in Brazil*, 7–12; Louis Menand, *The Metaphysical Club* (New York: Farrar, Straus & Giroux, 2001), 131. For a reconstruction of why Agassiz took these facts to be evidence against Darwin's theory, see Mary P. Winsor, *Reading the Shape of Nature: Comparative Zoology at the Agassiz Museum* (Chicago: University of Chicago Press, 1991), 72–76.

49. Louis Agassiz to William James, 8 December 1865, Item 11, Letters to William James from various correspondents (MS Am 1092), Houghton Library, Harvard University. See also the drawing

Thus, at least from Agassiz's point of view, the trip had been a scientific success.

James had great respect for Agassiz as a naturalist but thought him closed-minded and biased when it came to evolution. In a letter to his brother Henry during the trip, James attributed this antievolutionary bias to a general character flaw: "[Agassiz] is doubtless a man of some wonderful mental faculties, but such a politician & so self-seeking & illiberal to others that it sadly diminishes one's respect for him." After returning home, James complained to a friend that the whole endeavor had been

> more profitable in the way of general experience than of Science.—For the manual labor of collecting and packing took so much time and energy that little was left for dissecting and studying specimens and "the principal light of modern science" [i.e., Agassiz] is not exceedingly communicative of his learning except in the way of damning the Darwinians, wh. though instructive is open to the charge of being monotonous.[50]

Although James enjoyed poking fun at Agassiz's animosity toward Darwin's views, it is not obvious whether James counted himself among "the Darwinians" in 1866. But as his letters during and after the Brazilian expedition show, he found Agassiz's criticisms unconvincing.

After being back in Cambridge for just over a year, James left to study in Europe, not returning until November 1868. While in Berlin, he attended physiology lectures at the university by Emil du Bois-Reymond and others. In a letter to his father written about that time, James presciently outlined his academic interests: "As a central point of study I imagine that the border ground of physiology & psychology, overlapping both, wd. be as fruitful as any, and I am now working on it."[51] Despite health problems including depression and severe lower back pain, James read widely while in Europe. In the early autumn of 1868, for example, he asked his aunt to send him Spencer's *Principles of Biology* and Charles Brace's *Races of the Old World*. Both of these books defended evolution: Brace opposed Agassiz by tracing human races to one common

---

reproduced in Paul Jerome Croce, *Young William James Thinking* (Baltimore: Johns Hopkins University Press, 2018), 64. Pará is the state where the Amazon River reaches the Atlantic Ocean, directly east of Amazonas.

50. William James to Henry James, 3 May 1865, and William James to Frederick George Bromberg, 30 September 1866, in James, *Correspondence*, 1:8, 4:142.

51. William James to Henry James Sr., 26 December 1867, in James, *Correspondence*, 4:243.

CHAPTER ONE

origin using "the Darwinian theory," and Spencer rejected the special creation hypothesis as "worthless."[52]

James's continuing interest in evolution was also evident in two reviews of Darwin's *Variation of Animals and Plants under Domestication*. As soon as it was published in 1868, James asked his brother Henry to see whether Charles Eliot Norton, editor of the *North American Review*, would be interested in a notice of the work.[53] The answer must have been positive, for James's review appeared in the July issue. A different review by James of the same book also appeared in the *Atlantic Monthly*. He was complimentary, but he still did not explicitly endorse Darwin's theory. James wrote of "the great value of [Darwin's] hypotheses in setting naturalists to work, and sharpening their eyes for new facts and relations," but he also emphasized the hypothetical nature of Darwin's theory: "it may never be any more possible to give a strict proof of it, complete in every link, than it now is to give a logically binding disproof of it." Nevertheless, James argued that this might end up benefiting the theory:

It removes the matter from the jurisdiction of critics who are not zoologists, but mere reasoners (and who have already written nonsense enough about it), and leaves it to the learned tact of experts, which alone is able to weigh delicate facts against each other, and to decide how many possibilities make a probability, and how many small probabilities make an almost certainty.[54]

Just as Agassiz reminded his charges in 1865 to let their views of the history of life "rest on facts and not on arguments," James insisted that Darwin's theories must be tested by experts weighing facts and probabilities, and not by logic or abstract reasoning.

James, although keenly interested in Darwin and evolution, did not consider himself one of these experts. He wrote to his brother Henry shortly after finishing the first of his two reviews:

---

52. William James to Catharine Walsh, 13 September 1868, in James, *Correspondence*, 4:336; Charles Brace, *The Races of the Old World: A Manual of Ethnology* (London: John Murray, 1863), 390; Herbert Spencer, *The Principles of Biology*, vol. 1 (London: Williams & Norgate, 1864), 378.

53. William James to Alice James, 9 January 1868, in James, *Correspondence*, 4:254. Although this letter was to his sister Alice, James added a P.P.S. "To Harry."

54. William James, "Darwin's Variation of Animals and Plants," *North American Review* 107 (1868): 367; William James, review of *Variation of Animals and Plants under Domestication*, by Charles Darwin, *Atlantic Monthly* 22 (1868): 124. For James's authorship of these reviews, see William James to Thomas Wren Ward, 24 May 1868, in James, *Correspondence*, 4:310.

The more I think of Darwin's ideas the more weighty do they appear to me—tho' of course my opinion is worth very little—still I *believe* that that scoundrel Agassiz is unworthy either intellectually or morally for him to wipe his shoes on, & I find a certain pleasure in yielding to the feeling.[55]

Other letters from 1868, which I will discuss in chapter 2, indicate that James had almost certainly embraced evolution by then. But it had taken him a while, especially when compared with friends such as Wright, Abbot, and Fiske. Moreover, despite his private support of the theory, James was reluctant to make the kind of explicit public endorsement of evolution that his three friends all made in the late 1860s. Thus, although I agree with Paul Croce that James was seeking to distinguish himself "from both antagonists and enthusiastic supporters of evolution," I interpret James's position during the 1860s as even more tentative than Croce suggests.[56]

Wright, Abbot, and Fiske, in contrast, were enthusiastic about evolution and wore the expert's mantle comfortably. Wright confidently declared in 1870 that natural selection "had conquered the opposition of the great majority of students of natural history, as well as of the students of general philosophy"—and all in less than a decade. The next year, he admitted that natural selection as a particular causal factor in evolution was still regarded with suspicion by religious "students of science." But these same individuals, he said, "have found means of reconciling the general doctrine of evolution with the dogmas they regard as essential to religion."[57] General statements of this kind, absent in James's reviews, indicate that Wright felt able to speak more authoritatively about the state of play in natural history than his younger friend.

Wright, like James, attacked Agassiz in private during the 1860s, complaining in 1866 that his old teacher had "repeated yesterday what he has said at every scientific meeting at which I have heard him speak; and he said it with as much animation, as if the world were not weary of it."[58] Fiske was willing to make such feelings public. Reacting to a

---

55. William James to Henry James, 9 March 1868, in James, *Correspondence*, 1:38–39; see also William James to Henry Pickering Bowditch, 8 April 1871, in James, *Correspondence*, 4:416.

56. Paul Jerome Croce, *Young William James Thinking* (Baltimore: Johns Hopkins University Press, 2018), 74.

57. Chauncey Wright, "Limits of Natural Selection," *North American Review* 111 (1870): 283; Chauncey Wright, "The Genesis of Species," *North American Review* 113 (1871): 64–65.

58. Chauncey Wright to Charles Eliot Norton, 10 August 1866, Item 8280, Charles Eliot Norton Papers, Houghton Library, Harvard University. Agassiz's name was redacted in the published version: Wright, *Letters*, 88.

CHAPTER ONE

newspaper's announcement that the Darwinian theory had been "utterly demolished" by "Agassiz himself," Fiske asked,

> Can it be that we have, after all, a sort of scientific pope among us? Has it come to this, that the dicta of some one "servant and interpreter of nature" are to be accepted as final, even against the better judgment of the majority of his compeers? In short, who is Agassiz himself, that he should thus single-handed have demolished the stoutest edifice which observation and deduction have reared since the day when Newton built to such good purpose?

Agassiz had commented in 1866, shortly after the Brazil expedition, that "he preferred the theory which makes man out a fallen angel to the theory which makes him out an improved monkey." Fiske replied that "a scientific inquirer has no business to have 'preferences'":

> What matters it in the end whether we are pleased with the notion of a monkey-ancestry or not? The end of scientific research is the discovery of truth, and not the satisfaction of our whims or fancies, or even of what we are pleased to call our finer feelings. The proper reason for refusing to accept any doctrine is, that it is inconsistent with observed facts, or with some other doctrine which has been firmly established on a basis of fact. The refusal to entertain a theory because it seems disagreeable or degrading, is a mark of intellectual cowardice and insincerity.

Although Fiske denied that he was calling Agassiz a coward, his public attack was condemned by many Americans as polemical and unfair—an opinion exacerbated by Agassiz's death a few months later.[59]

Both Fiske and Wright assumed the role of philosophers (rather than naturalists) in their defense of Darwin's theories. It was as a philosopher that Agassiz was deficient, according to Fiske: "[He] philosophizes on unsound principles . . . because his philosophizing is not a natural outgrowth from the facts of Nature which lie at his disposal, but is made up out of sundry traditions of his youth."[60] During a visit to England in 1873–74, Fiske discovered that others agreed with this assessment of Agassiz. As he reported to his wife Abby, "[John] Tyndall and [Thomas

---

59. John Fiske, "Agassiz and Darwinism," *Popular Science Monthly* 3 (1873): 693, 697–98. The phrase "servant and interpreter of nature" is from the first aphorism of Francis Bacon's *New Organon*: see *The Works of Francis Bacon*, ed. James Spedding, Robert Leslie Ellis, and Douglas Denon Heath, 15 vols. (Boston: Brown and Taggard, 1860–64), 8:67. On the American condemnation of Fiske's essay, see John Fiske to Abby Brooks Fiske, 31 December 1873, Box 6, Fiske Papers; John Fiske to Mary Fisk Green Stoughton, 13 June 1878, Box 7, Fiske Papers.

60. Fiske, "Agassiz and Darwinism," 696.

Henry] Huxley were very much pleased with the Agassiz-article. But Huxley said that in Europe Agassiz would hardly be thought worth so much [gun]powder. Not but what he was a good naturalist,—but as *a thinker,*—and here his shoulders gave a shrug which said more than my article." Ten days later, Fiske told her that Herbert Spencer was of a similar opinion: "While Agassiz deserves great credit as an indefatigable collector and observer, he is of no weight at all as a philosophical naturalist."[61] Similarly, Wright's review of Wallace had claimed "that his metaphysical views, carefully excluded from his scientific work, are the results of an earlier and less severe training than that which has secured to us his valuable positive contributions to the theory of Natural Selection."[62] Thus Fiske and Wright, as expert philosophers, attacked the inexpert philosophical views of naturalists such as Agassiz and Wallace.

Wright's role as a philosophical authority on evolutionary ideas was probably one he played often at Metaphysical Club meetings in the early 1870s. James and the other men connected with the club developed their opinions about Darwin's theories in conversation with Wright, who was several years older. Fiske, for example, discussed evolution and phyllotaxis (i.e., the arrangement of leaves around the stems of plants) with Wright in 1873 and then brought up the subject when he first met Darwin a few months later.[63] Peirce famously described Wright as "our boxing-master, whom we,—I, particularly—used to face to be severely pummelled." Wright reflected on his parental attitude toward James in a letter to Grace Norton (sister of Charles Eliot Norton):

"Boyish" is a well-chosen word to express both our common judgment of his present and mine in particular of his future; for I imagine that by laboring with him I shall get him into better shape by and by. One remains a boy longer in philosophy than in any other direction. . . . You see that my interest in him is like that of the preacher in the sinner.[64]

---

61. John Fiske to Abby Brooks Fiske, 13 November 1873 and 23 November 1873, Box 6, Fiske Papers.
62. Chauncey Wright, "Limits of Natural Selection," *North American Review* 111 (1870): 310.
63. For the conversation with Wright, see John Fiske to Abby Brooks Fiske, 11 August 1873, Box 6, Fiske Papers; John Fiske to Charles Darwin, 31 October 1873, in Darwin, *Correspondence*, 21:472–73. For Fiske's meeting with Darwin, see John Fiske to Mary Fisk Green Stoughton, 13 November 1873, Box 6, Fiske Papers. Fiske had reviewed Wright's paper on phyllotaxis the year before: Chauncey Wright, "The Uses and Origin of the Arrangement of Leaves in Plants," *Memoirs of the American Academy of Arts and Sciences* 9 (1873); John Fiske, "Science," *Atlantic Monthly* 30 (1872): 125–26.
64. MS 318 (1907), in Peirce, *Essential Peirce*, 2:399; Chauncey Wright to Grace Norton, 18 July 1875, Norton Letters. Although the letter was included in editor Thayer's volume, the section I have quoted was omitted and James's name was redacted: Wright, *Letters*, 341–43.

CHAPTER ONE

When Wright died a few months later, James's obituary alluded to such interactions: "His best work has been done in conversation; and in the acts and writings of the many friends he influenced his spirit will, in one way or another, as the years roll on, be more operative than it ever was in direct production." Fiske agreed: "An evening's talk with Mr. Wright always seemed to me one of the richest of intellectual entertainments." Peirce recalled Wright's early conversion to Darwinism: "I was away surveying in the wilds of Louisiana when Darwin's great work appeared, and though I learned by letters of the immense sensation it had created, I did not return until early in the following summer when I found Wright all enthusiasm for Darwin." This enthusiasm continued throughout the 1860s and early 1870s, as Wright made a habit of reviewing books related to evolution by Spencer, Wallace, Darwin, and others. As Fiske wrote in 1878, Wright's "most important literary work was done in elucidation and defense of [the Darwinian] theory." A few years earlier, Fiske had summed up Wright's importance for anyone thinking about philosophy and evolution in Cambridge: "To have known such a man is an experience one cannot forget or outlive. To have had him pass away, leaving so scanty a record of what he had it in him to utter, is nothing less than a great public calamity."[65]

The most prominent example of Wright's work "in elucidation and defense" of Darwin was a critical review of St. George Mivart's *On the Genesis of Species*. This book appeared in 1871, coinciding with or shortly preceding the first meetings of the Metaphysical Club, and was reviewed by Fiske and Wright that same year.[66] Both of them criticized Mivart on positivist grounds, arguing that he did not provide verifiable empirical evidence for his claims. Mivart was one of those religious "students of science" mentioned earlier: he accepted evolution but opposed natural selection. The thesis of his book was that "in the genesis of species an *internal* force or tendency intervenes, co-operating with and controlling the action of external conditions." He allowed that natural selection might play a subordinate role but argued that the "internal power is a great, perhaps the main, determining agent."[67] To illustrate

65. William James, "Chauncey Wright," *Nation*, September 23, 1875, 194; John Fiske, "Chauncey Wright," *Radical Review* 1 (1878): 703–4; "On Phenomenology," MSS 305–306 (1903), in Peirce, *Essential Peirce*, 2:158; John Fiske, "Chauncey Wright," *Harvard Advocate* 20 (1875): 9.

66. Henry James reported in a letter of January 1872 that William had "just helped to found a metaphysical club, in Cambridge, (consisting of Chauncey Wright, C. Pierce [sic] etc.)," but Peirce sometimes assigned an earlier date to the first meetings: see Henry James to Elizabeth Boott, 24 January 1872, in Henry James, *Henry James Letters*, 4 vols., ed. Leon Edel (Cambridge, MA: Harvard University Press, 1974–84), 1:269; Peirce, *Writings*, 3:xxx.

67. St. George Mivart, *On the Genesis of Species* (London: Macmillan, 1871), 259.

the role of this power, Mivart adapted a metaphor from Francis Galton (the other famous grandson of Erasmus Darwin), who "compares the development of species with a many facetted spheroid tumbling over from one facet, or stable equilibrium, to another." As Mivart later explained, "the internal tendency of an organism to certain considerable and definite changes would correspond to the facets on the surface of the spheroid."[68] He gave at least three kinds of evidence for this notion of an internal tendency in evolution: the stability of species combined with the sudden development of specific differences; the appearance of similar structures in groups not closely connected by descent; and the absence of transitional fossils.

Wright began his review by criticizing Mivart's general point. He noted that natural selection is not meant to explain the "facts of variation" on which Mivart focused and that the naturalist does not require knowledge of their causes. Wright thus suggested that individual variations could "be taken as ultimate facts," though this should not be interpreted "as denying the existence of any real determining causes and more ultimate laws." He admitted that internal tendencies may be responsible for "reversional and correlated variations" but said that these variations "are far from accounting for, or bearing any relations to, the adaptive characters of the organism."[69]

Turning to Mivart's emphasis on the stability of species, Wright argued that there was no reason to attribute this stability to internal rather than external causes. Describing what biologists today call *stabilizing selection*, Wright stated:

Utility..., in conjunction with the laws of inheritance, determines not only the middle line or safest way of a race, but also the bounding limits of its path of life; and so long as the conditions and principles of utility embodied in a form of life remain unchanged, they will, together with the laws of inheritance, maintain a race unchanged in its average characters. "Specific stability," therefore, for which theological and descriptive naturalists have speculated a transcendental cause, is even more readily and directly accounted for by the causes which the theory of Natural Selection regards than is specific change.

That is, natural selection provides a clearer explanation of species stability than it does of species change, at least given stable conditions.

---

68. Mivart, *On the Genesis of Species*, 24, 109–10, 261; citing Francis Galton, *Hereditary Genius: An Inquiry into Its Laws and Consequences* (London: Macmillan, 1869), 369.
69. Chauncey Wright, "The Genesis of Species," *North American Review* 113 (1871): 66–67, 69–70.

As Wright wrote in another exchange with Mivart, "what fixes species (when they are fixed) is the continuance of the same advantages in their structures and habits." Darwin appreciated this point, telling Wright, "nothing can be clearer than the way in which you discuss the permanence or fixity of species."[70]

Although Wright was apparently arguing that the stability of characters is always the result of natural selection or present utility—"the continuance of the same advantages"—he had sketched a more nuanced position in his 1871 essay on phyllotaxis. Wright distinguished between two types of characters, "genetic" and "adaptive": the former serve no present purpose and are merely inherited, whereas the latter are presently useful to the organism. Although he confessed that genetic characters could be "the result of a physiological necessity among the laws of growth," Wright also suggested that they were often the product not of present but of *former* advantage"—that is, even though they appear useless, they may at some earlier time "have stood in more immediate and important relations to the conditions of the plant's existence." Thus, genetic characters "are related principally to past and generally unknown adaptations" and adaptive characters "to present and more obvious ones." In a letter to Darwin, Wright elaborated on the relationship between these types of characters: "Adaptive characters are generally superposed on genetic ones, . . . thus giving them an indirect utility and preserving them." In other words, genetic characters will stick around to the extent that adaptive characters depend on them.[71] Thus, although Wright disputed Mivart's "internal tendency" account of the stability of species, he did not attribute this stability merely to presently acting natural selection.

Just as a "transcendental cause" (as Wright termed Mivart's "internal tendency") is not required for species stability, neither is it needed for "independent similarities of structure," what we would today call *evolutionary convergence*. Such independent similarities could be explained, according to Wright, by natural selection acting via "similar means and

---

70. Wright, "Genesis of Species," 79–80; Chauncey Wright, "Evolution by Natural Selection," *North American Review* 115 (1872): 14; Charles Darwin to Chauncey Wright, 3 June 1872, in Darwin, *Correspondence*, 20:241.

71. Chauncey Wright, "The Uses and Origin of the Arrangement of Leaves in Plants," *Memoirs of the American Academy of Arts and Sciences* 9 (1873): 408–9, 412; Chauncey Wright to Charles Darwin, 24 May 1872, in Darwin, *Correspondence*, 20:226. Although Wright's "Arrangement of Leaves" paper is in the 1873 volume of the *Memoirs*, it was submitted in 1871, and he sent Darwin an offprint in the spring of 1872: see Darwin to Wright, 6 April 1872, in Darwin, *Correspondence*, 20:149. On parallels between Wright's views and those of modern biologists, see Andrea Parravicini, "A New Use for an Old Theory: Chauncey Wright between Darwinism and Pragmatism," *Cognitio-Estudos* 6 (2009).

conditions . . . of such a general sort that they belong to wide ranges of life."[72] Fiske also insisted that no appeal to transcendental causes was needed to explain convergence, treating it as the result of directed variation, which we will examine in greater detail in chapter 5. He began by noting that "an inherent capacity for adaptive changes is possessed by all organisms." Unlike Mivart, however, Fiske was not using the phrase *inherent capacity* "to insinuate the existence of any *occulta vis* [hidden force], or metaphysical 'innate power,' of which no scientific account is to be given in terms of matter and motion," but rather to capture "the expression of tendencies due to the co-operation of countless ancestral forces," akin to the tendencies of minerals to assume certain crystal structures. Fiske argued that the "parallel variations" required for evolutionary convergence are the result of the "direct adaptation" of different organisms—often with similar inherent capacities—to the same environmental factors. (Like many others at this time, including Darwin and Spencer, Fiske assumed that acquired characters could be passed to offspring.) He explained cases in which convergence was unlikely to involve shared capacities, such as the similar eyes of mollusks and vertebrates, as those in which "such variations as occur must be in a preeminent degree directly adaptive" and in which natural selection has "but very few directions in which to act."[73] Thus both Wright and Fiske claimed that naturalists could explain similarities of structure in unrelated groups of organisms as results of ordinary evolutionary processes, without any need for transcendental causes.

Those familiar with modern evolutionary biology may be tempted to defend Mivart retrospectively against the attacks of Wright and Fiske. After all, the idea of developmental constraints (developed in the 1980s) now plays a role comparable to that of Mivart's notion of internal tendencies, and similar traits that evolve independently are often the result of both internal and external factors, as Fiske himself granted.[74] However, Wright and Fiske were not opposed to the idea of internal causes in evolution but only to Mivart's "transcendental" or "metaphysical" approach. Wright admitted that "even Mr. Galton's hypothetical 'facets,' or internal conditions of abrupt changes and successions of stable equilibriums," could be among the causes of evolution "if there were

---

72. Chauncey Wright, "The Genesis of Species," *North American Review* 113 (1871): 96.
73. John Fiske, "Law of Organic Evolution," *New York World*, June 30, 1871.
74. Trevor Pearce, "Evolution and Constraints on Variation: Variant Specification and Range of Assessment," *Philosophy of Science* 78 (2011); Trevor Pearce, "Convergence and Parallelism in Evolution: A Neo-Gouldian Account," *British Journal for the Philosophy of Science* 63 (2012).

any good inductive grounds for supposing their existence."[75] Wright did not reject the idea of internal causes but only Mivart's nonempirical account of such causes. Mivart described the innate within organisms as "an internal law or 'substantial form,'" whereas for Wright the innate was concrete rather than abstract: "The general resemblances of animals or plants of any species, their agreements in specific characters, are doubtless due, in the main, to the properties of what is innate in them, yet not to any abstraction." According to Wright, we cannot just postulate some internal form; the job of a "general physiology" is to discover the concrete internal processes that, along with external conditions, produce a particular organism. He thus accused Mivart of not following the "principles of method which the examples of inductive and experimental science have established"—that is, "the rigorous rules of experimental philosophy." According to Wright:

[These rules] require us to assume no causes that are not true or phenomenally known, and known in some other way than in the effect to be explained; and to prove the sufficiency of those we do assume in some other way than by putting an abstract name or description of an effect for its cause. . . . It is enough for the present that Darwinians do not rest, like their opponents, contented with framing what Newton would have called, if he had lived after Kant, "*transcendental hypotheses*," which have no place in experimental philosophy.[76]

It is simply inappropriate, said Wright, to look for something that transcends the empirical when engaged in "inductive and experimental science." Fiske made a similar point in his discussion of the supposed lack of transitional forms in the fossil record, criticizing Mivart for postulating "sudden jumps [in evolution], occurring at rhythmical intervals," without alleging any "physical agencies competent to cause the sudden jumps from one specific form to another."[77] The problems with Mivart's arguments, according to Wright and Fiske, were not scientific but philosophical: it is not that Mivart had the facts wrong; he was confused as to the enterprise.

Fiske and Wright entered the Darwin debates not as naturalists but as philosophers—expert reasoners. Wright was explicitly cast in this role

---

75. Chauncey Wright, "The Genesis of Species," *North American Review* 113 (1871): 69.
76. Wright, "Genesis of Species," 72–73; cf. St. George Mivart, *On the Genesis of Species* (London: Macmillan, 1871), 208–10.
77. John Fiske, "Law of Organic Evolution," *New York World*, June 30, 1871.

at the time. He sent Darwin the page proofs of his Mivart review in June 1871, shortly before its publication. Darwin thought it "admirable" and eventually decided to reprint the piece separately at his own expense. It appeared in September under a new title chosen by Wright: *Darwinism: Being an Examination of Mr. St. George Mivart's "Genesis of Species."*[78] Wright was seen as a sort of metaphysical soldier for Darwin, more able than a naturalist to refute philosophical criticisms. William Winwood Reade, for example, alluding to Mivart's claim that Darwin subscribed to a "radically false metaphysical system," commented that "it is amusing to read such remarks [as Mivart's] when one knows that you have satisfied the logical requirements of J. S. Mill, & of a man like Wright who is a mathematician of high order, and perhaps the best *reasoner* in the U.S." Darwin echoed Reade's assessment a week later in a terse exchange with his critic: Mivart asked him for a copy of Wright's pamphlet, commenting that the original review had contained "misapprehensions and misunderstandings"; Darwin replied that the review seemed fair to him and pointed out that Wright was "highly esteemed in his own country, (as a mathematician & a sound reasoner)."[79] Whereas James had dismissed "critics [of evolution] who are not zoologists, but mere reasoners," Wright and Fiske demonstrated that reasoning had a role to play in evolutionary discussions.

The debates over evolution that followed the appearance of Darwin's *Origin of Species* had a substantial impact on the members of the Metaphysical Club, as they did on countless other American intellectuals. Despite their various personal and professional connections to Agassiz, those connected with the club eventually turned against him and embraced evolution. Wright, Fiske, and Abbot, however, had a special relationship with evolutionary ideas. They were among the first to defend Darwin publicly against detractors such as Bowen, Agassiz, Mivart, and other voices of religious orthodoxy. They also all corresponded with Darwin

---

78. Chauncey Wright to Charles Darwin, 21 June 1871, Darwin to Alfred Russel Wallace, 9 July 1871, Darwin to Wright, 13–14 July 1871, Darwin to Wright, 17 July 1871, Wright to Darwin, 1 August 1871, Darwin to John Murray, 17 August 1871, R. F. Cooke to Darwin, 18 August 1871, and Darwin to Murray, 13 September 1871, in Darwin, *Correspondence*, 19:452–53, 478, 487–88, 495, 513–16, 536–38, 572; Chauncey Wright, *Darwinism: Being an Examination of Mr. St. George Mivart's "Genesis of Species"* (London: John Murray, 1871).

79. St. George Mivart, "Darwin's Descent of Man," *Quarterly Review* 131 (1871): 48; W. W. Reade to Charles Darwin, 18 September 1871, St. George Mivart to Darwin, 26 September 1871, and Darwin to Mivart, 27 September 1871, in Darwin, *Correspondence*, 19:580–81, 600–601.

in the early 1870s. As we have just seen, Darwin deployed Wright as a "sound reasoner" in his 1871 battle with Mivart. That same year, in their initial letters to Darwin, both Abbot and Fiske also presented themselves as philosophers rather than naturalists. Abbot admitted that he was not "entitled to the praise of being a scientific man" before mentioning his editorial work for the *Index* and his several contributions to philosophy. Fiske took a similar approach: "Though I am no naturalist, and cannot claim any ability to support your discovery by original observations of my own, yet I have striven, to the best of my ability, to point out the strong points of your theory of natural selection, and to help win for it acceptance on philosophic grounds." Darwin responded positively to both of these letters, subscribing to the *Index* and praising Fiske's criticisms of Mivart.[80] Wright and Fiske even managed to visit Darwin at his home in England in the next few years. Wright wrote excitedly to a friend about his meeting in 1872 with the sixty-three-year-old Darwin: "If you can imagine me enthusiastic,—absolutely and unqualifiedly so, without a *but* or criticism,—then think of my last evening's and this morning's talks with Mr. Darwin as realizing that beatific condition."[81] Peirce and James, although they were very interested in evolution, were more reluctant to enter the fray. Neither wrote to Darwin, and both of them were tentative in their acceptance of evolutionary ideas. For Peirce, this tentativeness may have derived in part from one of his basic maxims of logic, set out in 1869–70: "It is folly for me not to doubt what men as capable as myself of forming a correct conclusion doubt. For Agassiz to attach no weight to the opinion of Darwin or for Darwin to attach no weight to that of Agassiz, would show a narrow-mindedness, most fatal to the sober investigation of truth."[82]

Although James disagreed with Fiske and Wright about the role of "reasoners" in evolutionary debates, Peirce ultimately sided with them on the point, declaring in the late 1870s that "the Darwinian theory is, in large part, a question of logic." In debates over evolution, said Peirce, "questions of fact and questions of logic are curiously interlaced."[83] It was thus to be expected that philosophers as well as naturalists would contribute to the conversation about evolution in the 1860s and '70s, and there was one philosopher who towered over all others in this

80. Francis Ellingwood Abbot to Darwin, 11 May 1871; Darwin to Abbot, 27 May 1871 and 6 June 1871; John Fiske to Darwin, 23 October 1871; Darwin to Fiske, 9 November 1871, in Darwin, *Correspondence*, 19:368, 391, 427, 649, 678.
81. Chauncey Wright to Sara Sedgwick, 5 September 1872, in Wright, *Letters*, 248.
82. MS 166 (1869–70), in Peirce, *Writings*, 2:357.
83. Peirce, "The Fixation of Belief," *Popular Science Monthly* 12 (1877): 2–3.

arena: Herbert Spencer. For the thinkers connected with the Metaphysical Club and for most everyone else, Spencer's system was synonymous with the philosophy of evolution. But as we will see in the next chapter, the club members were split in their assessment of Spencer: Fiske was his foremost American popularizer, and Wright and James were his two most trenchant critics.

TWO

Products of the Environment: Spencer's Challenge

Herbert Spencer was "the greatest Englishman since Shakespeare"—at least according to a letter published in the *Washington Post* after his death in 1903. A decade later, the biologist D'Arcy Thompson declared that "no philosopher of modern times, not Kant himself, has exercised in his lifetime so wide a dominion." Spencer was arguably the most famous philosopher of the late nineteenth century, and his influence on American "social Darwinism" is standard fare in high school history classes. Nevertheless, we still often neglect his central role in scientific and philosophical debates. In the last decades of the nineteenth century, it was almost impossible to discuss evolution without addressing Spencer's system of philosophy. Grant Allen, a popularizer of evolution we will meet later this chapter, summed up his significance in 1890: "Mr. Spencer is the inventor and patentee of Evolution. And as our age is essentially the age of evolution, Mr. Spencer may fairly claim to rank as its truest prophet."[1]

---

1. Herman E. Kittredge, "Appreciation of Spencer," *Washington Post*, December 14, 1903; D'Arcy Wentworth Thompson, *On Aristotle as a Biologist, with a Prooemion on Herbert Spencer* (Oxford: Clarendon Press, 1913), 3; Grant Allen, "The Gospel According to Herbert Spencer: I.—His Cardinal Ideas," *Pall Mall Gazette*, April 26, 1890, 1. For some exceptions to Spencer's neglect, see Robert M. Young, *Mind, Brain and Adaptation in the Nineteenth Century* (Oxford: Oxford University Press, 1970); Robert J. Richards, *Darwin and the Emergence of Evolutionary Theories*

Spencer was particularly important for the first two cohorts of pragmatists. Among those in the first cohort, Chauncey Wright, William James, and Charles Sanders Peirce were among his fiercest critics; Wright, Francis Ellingwood Abbot, and John Fiske all reviewed his books in the 1860s, with Fiske becoming a card-carrying Spencerian. Among those in the second cohort, Josiah Royce, John Dewey, and George Herbert Mead, like James before them, taught classes based around Spencer's books; and traces of his ideas are scattered through the writings of Jane Addams and W. E. B. Du Bois. Although they often highlighted his inferiority as a writer and a philosopher, the first-cohort pragmatists granted that Spencer's theories were socially significant and historically momentous. Here is Oliver Wendell Holmes Jr., writing in 1895: "He is dull. He writes an ugly uncharming style, his ideals are those of a lower middle class British Philistine. And yet after all abatements I doubt if any writer of English except Darwin has done so much to affect our whole way of thinking about the universe." Peirce called him a "master of abductive reasoning," referring to his "extraordinary skill and ingenuity in forming theories that deserved examination." James pronounced Spencer's "whole system wooden, as if knocked together out of cracked hemlock boards," but admitted that "we feel his heart to be *in the right place* philosophically. His principles may be all skin and bone, but at any rate his books try to mould themselves upon the particular shape of this particular world's carcase."[2]

In this chapter, which focuses on the initial response to Spencer by the first cohort of pragmatists, I will argue that Spencer's evolutionary philosophy was "a programme full of suggestiveness" for Wright, Abbot, Fiske, and James.[3] Apart from Fiske, each of these thinkers presented detailed criticisms of Spencer's philosophy. But although these criticisms were often convincing, some were manifestly unfair, as both Wright and Fiske noted at the time. The chapter has three parts. First, after a brief overview of Spencer's theory of evolution and his organism-environment framework, I will discuss Fiske's enthusiastic endorsement and Wright's

---

of Mind and Behavior (Chicago: University of Chicago Press, 1987); Mark Francis, *Herbert Spencer and the Invention of Modern Life* (Ithaca, NY: Cornell University Press, 2007); Mark Francis and Michael Taylor, eds., *Herbert Spencer: Legacies* (New York: Routledge, 2015); Bernard Lightman, ed. *Global Spencerism: The Communication and Appropriation of a British Evolutionist* (Leiden, Neth.: Brill, 2015).

2. Oliver Wendell Holmes Jr. to Georgina Harriet Deffell Pollock, 2 July 1895, in Oliver Wendell Holmes and Frederick Pollock, *Holmes–Pollock Letters: The Correspondence of Mr. Justice Holmes and Sir Frederick Pollock, 1874–1932*, ed. Mark DeWolfe Howe, 2 vols. (Cambridge, MA: Harvard University Press, 1941), 1:58; MS 470 (10 December 1903), p. 156, Peirce Papers; William James, *Pragmatism: A New Name for Some Old Ways of Thinking* (New York: Longmans, Green, 1907), 39–40.

3. William James, "The Sentiment of Rationality," *Mind* 4 (1879): 321.

fierce criticism of Spencer's progressive evolutionism in the 1860s. Second, I will analyze James's polemic against Spencer's psychology in the 1870s, which emphasized the limitations of Spencer's externalist perspective, drawing on points already made by Abbot and others.[4] Third, I will examine James's 1878–81 critique of Fiske and Spencer's evolutionary sociology, arguing that he unfairly portrayed them as advocates of a simplistic environmental determinism. I will conclude by emphasizing that the pragmatists' criticisms of Spencer were naturalistic—that is, they presented a scientific response to his challenge, putting aside metaphysical disagreements.

Spencerian Evolution

The pragmatists acknowledged Herbert Spencer's role as evolution's prophet: "To Spencer," said William James, "is certainly due the immense credit of having been the first to see in evolution an absolutely universal principle." Dewey likewise spoke of the "thorough-going identification in the popular mind of Spencer's system with the very idea and name of evolution." As figure 6 indicates, Spencer's system was often simply termed "The Philosophy of Evolution," which was also the title of Mead's 1892 class on Spencer at the University of Michigan.[5]

What were the main tenets of this evolutionary philosophy? Spencer himself traced it all the way back to a strange blend of agnosticism and the conservation of energy—namely, "the persistence of that Unknown Cause, Power, or Force, which is manifested to us through all phenomena."[6] But the philosophy of evolution was most famously associated with specific conceptions of evolution and life, both of which Spencer originally developed in the 1850s. These two conceptions were featured in James's obituary of Spencer: "Who, since he wrote, is not vividly able to conceive of the world as a thing evolved from a primitive fire mist, by progressive integrations and differentiations, and increases in heterogeneity and coherence of texture and organization? Who can fail

---

4. On Spencer's externalism, see Peter Godfrey-Smith, *Complexity and the Function of Mind in Nature* (Cambridge: Cambridge University Press, 1996), chap. 3.

5. William James, "Herbert Spencer," *Atlantic Monthly* 94 (1904): 103; John Dewey, "The Philosophical Work of Herbert Spencer," *Philosophical Review* 13 (1904): 172; *Calendar of the University of Michigan for 1891–92* (Ann Arbor: University of Michigan, 1892), 61.

6. Herbert Spencer, *First Principles* (London: Williams & Norgate, 1862), 258. On Spencer's agnosticism and his related doctrine of "the Unknowable," see Bernard Lightman, *The Origins of Agnosticism: Victorian Unbelief and the Limits of Knowledge* (Baltimore: Johns Hopkins University Press, 1987), chap. 3.

SPENCER'S SYSTEM OF PHILOSOPHY.

THE PHILOSOPHY OF EVOLUTION.

By HERBERT SPENCER.

This great system of scientific thought, the most original and important mental undertaking of the age, to which Mr. Spencer has devoted his life, is now well advanced, the published volumes being: *First Principles*, *The Principles of Biology*, two volumes, and *The Principles of Psychology*, vol. i., which will be shortly printed.

This philosophical system differs from all its predecessors in being solidly based on the sciences of observation and induction; in representing the order and course of Nature; in bringing Nature and man, life, mind, and society, under one great law of action; and in developing a method of thought which may serve for practical guidance in dealing with the affairs of life. That Mr. Spencer is the man for this great work will be evident from the following statements:

"The only complete and systematic statement of the doctrine of Evolution with which I am acquainted is that contained in Mr. Herbert Spencer's 'System of Philosophy;' a work which should be carefully studied by all who desire to know whither scientific thought is tending."—T. H. HUXLEY.

"Of all our thinkers, he is the one who has formed to himself the largest new scheme of a systematic philosophy."—Prof. MASSON.

"If any individual influence is visibly encroaching on Mills in this country, it is his."—*Ibid*.

"Mr. Spencer is one of the most vigorous as well as boldest thinkers that English speculation has yet produced."—JOHN STUART MILL.

"One of the acutest metaphysicians of modern times."—*Ibid*.

"One of our deepest thinkers."—Dr. JOSEPH D. HOOKER.

It is questionable if any thinker of finer calibre has appeared in our country."—GEORGE HENRY LEWES.

"He alone, of all British thinkers, has organized a philosophy."—*Ibid*.

"He is as keen an analyst as is known in the history of philosophy; I do not except either Aristotle or Kant."—GEORGE RIPLEY.

"If we were to give our own judgment, we should say that, since Newton, there has not in England been a philosopher of more remarkable speculative and systematizing talent than (in spite of some errors and some narrowness) Mr. Herbert Spencer."—*London Saturday Review*.

"We cannot refrain from offering our tribute of respect to one who, whether for the extent of his positive knowledge, or for the profundity of his speculative insight, has already achieved a name second to none in the whole range of English philosophy, and whose works will worthily sustain the credit of English thought in the present generation."—*Westminster Review*.

Figure 6 Advertisement for *Spencer's System of Philosophy*.
From St. George Mivart, *On the Genesis of Species* (New York: D. Appleton, 1871), back matter. Reproduced courtesy of the University of Chicago Library.

to think of life, both bodily and mental, as a set of ever-changing ways of meeting the 'environment'?"[7] The first conception was officially presented in *First Principles*, published in 1862, where *evolution* was defined as "a change from an indefinite, incoherent homogeneity, to a definite, coherent heterogeneity; through continuous differentiations and integrations." The second conception was officially presented in the first volume of *The Principles of Biology*, published in 1864, where *life* was defined as "the continuous adjustment of internal relations to external relations." Since Spencer treated mind as merely a higher form of life, the latter conception was also the basis of his theory of intelligence: "Alike in the simplest inferences of the child, and the most complex ones of the man of science, we find a correspondence between simultaneous and successive changes in the organism, and coexistences and sequences in its environment."[8]

According to Dewey, it was these conceptions of evolution and life that Spencer "furnished [to] the common consciousness of his day, so that it could appropriate to its ordinary use in matters of 'life, mind, and society,' the most fundamental generalizations which had been worked out in the abstract regions of both philosophy and science." According to Peirce, it was these ideas that "did in the beginning influence a collection of men as remarkable for their intellect as for their great numbers, and influenced them to such a degree that all their subsequent opinions were built upon that basis."[9] The first cohort of pragmatists were part of this collection, notwithstanding their frequent criticisms of Spencer. In this section I will focus on their response to Spencer's notion of evolution. In subsequent sections I will examine their treatment of the idea of life as a correspondence between organism and environment.

Fiske and Wright both discovered Spencer's work in the early 1860s, coincident with their reception of Darwin (described in chapter 1). Although they were both enthusiastic about Darwin's ideas, they diverged in their assessment of Spencer: Fiske was unreservedly positive and Wright extremely negative, despite their shared commitment to positiv-

---

7. William James, "Herbert Spencer Dead," *Evening Post* (New York), December 8, 1903.

8. Herbert Spencer, *First Principles* (London: Williams & Norgate, 1862), 216; Herbert Spencer, *The Principles of Biology*, vol. 1 (London: Williams & Norgate, 1864), 77, 80; see also Herbert Spencer, *The Principles of Psychology* (London: Longman, Brown, Green, & Longmans, 1855), 334, 371, 472. For more details, see Trevor Pearce, "From 'Circumstances' to 'Environment': Herbert Spencer and the Origins of the Idea of Organism-Environment Interaction," *Studies in History and Philosophy of Biological and Biomedical Sciences* 41 (2010).

9. John Dewey, "The Philosophical Work of Herbert Spencer," *Philosophical Review* 13 (1904): 172; MS 470 (10 December 1903), p. 158, Peirce Papers.

ism. In June 1860, the summer before he started college at Harvard, Fiske came across the recently published "prospectus" of Spencer's *System of Philosophy*, a massive series projected to include *First Principles, The Principles of Biology, The Principles of Psychology, The Principles of Sociology,* and *The Principles of Morality.* Fiske wrote excitedly to his mother, "If I had $2,000,000, I would lay $1,000,000 at Mr. Spencer's feet to help him execute this great work." Although the first volumes would not appear until a few years later, Fiske read the 1855 edition of *Principles of Psychology* (which contained the system's main ideas) in 1861, calling it "the profoundest work I ever read" and declaring "I have had an 'intellectual drunk' over it."[10]

Although Spencer's conception of evolution came to be associated primarily with *First Principles*, it was also presented—two years before the *Origin of Species* appeared—in an 1857 essay titled "Progress: Its Law and Cause." Citing several embryologists, Spencer claimed that "organic progress consists in a change from the homogeneous to the heterogeneous." His own contribution was to generalize this idea:

This law of organic progress is the law of all progress. Whether it be in the development of the Earth, in the development of Life upon its surface, in the development of Society, of Government, of Manufactures, of Commerce, of Language, Literature, Science, Art, this same evolution of the simple into the complex, through a process of continuous differentiation, holds throughout.[11]

In a preview of his later arguments in *First Principles*, Spencer argued that evolutionary progress characterized the development of the cosmos, life, society, and science.

Fiske, inspired by this essay, deployed Spencer's views in his first published article, an 1861 review of Henry Thomas Buckle's *History of Civilization in England.* Buckle, whose *History* was remembered (along with Darwin's *Origin*) as one of the "two great intellectual shocks" of late-1850s England, had claimed in his "General Introduction" that "in the present state of our knowledge, we cannot safely assume that there has been any permanent improvement in the moral or intellectual faculties of man, nor have we any decisive ground for saying that those faculties are likely to be greater in an infant born in the most civilized part of

---

10. John Fiske to Mary Fisk Green Stoughton, 24 June 1860 and 21 July 1861, Box 2, Fiske Papers.
11. Herbert Spencer, "Progress: Its Law and Cause," *Westminster Review* 67 (1857): 446; also in Herbert Spencer, *First Principles* (London: Williams & Norgate, 1862), 148.

Europe, than in one born in the wildest region of a barbarous country."[12] Citing this passage, Fiske accused Buckle of claiming that humans were somehow excepted from the physiological truth that "every organism is constantly advancing in the vigor and complexity of its functions, in relation to the conditions which surround it." Like all living beings, said Fiske, humans were products of evolution: "If we are to accept the development theory at all, we must accept it without limitations. We might as well say that the human race forms an exception to the operation of the laws of gravitation or chemical affinity, as to say that it forms an exception in the case of the law of evolution." Fiske explicitly endorsed Spencer's "fundamental law of human evolution," which he saw as explaining "all the phenomena of man's history, and all those of external nature." According to Fiske, following Spencer, this was a law not only of change but of progressive change in that "retrogression nowhere meets us; progress meets us everywhere." Human evolution, for Fiske—again following Spencer and opposing Buckle's equal faculties picture—supposedly involved a progression from "the lower races" to "the Europeans," more evidence (he thought) that "the human species is in a course of evolution from the less perfect to the more perfect." Hence, by the time the first installment of the *System of Philosophy* appeared in 1862, Spencer's work was already an integral part of Fiske's thinking. It even played a role in his romance: Fiske and his fiancée read *First Principles* together in the Massachusetts countryside, leading "a life of intimate thought-communion."[13]

Fiske loved Spencer's progressive account of evolution, but Wright hated it. Although he grudgingly admitted in that "in psychology, and in the physiology of familiar facts, we regard his contributions to philosophy as of real and lasting value," Wright was highly critical of Spencer's treatment of evolution and progress.[14] But Wright had rejected progressivism well before his mid-1860s critique of Spencer. For example, in a short 1857 essay not yet noticed by scholars, he argued against a progressive account of growth: "If simple progress were the law of growth, the tree would be a failure"; the "forms and orders of nature may be said

---

12. Henry Thomas Buckle, *History of Civilization in England*, vol. 1 (London: John W. Parker & Son, 1857), 161. For the "intellectual shocks," see Leslie Stephen, "An Attempted Philosophy of History," *Fortnightly Review* 33 (1880): 672.

13. John Fiske, "Fallacies of Buckle's Theory of Civilization," *National Quarterly Review* 4 (1861): 32, 35–37, 41; cf. Herbert Spencer, "Progress: Its Law and Cause," *Westminster Review* 67 (1857): 451–52; John Fiske to James Willson Brooks, 26 May 1862, and John Fiske to Mary Fisk Green Stoughton, 3 June 1862, Box 2, Fiske Papers.

14. Chauncey Wright, "A Physical Theory of the Universe," *North American Review* 99 (1864): 12.

to depend as much upon destructive agencies as upon the progressive energies of life." Focusing on the example of a tree and its falling leaves, Wright claimed that growth involved both death and life: "Growth may be defined as this perpetual interchange of offices, through the union of the good and evil principles; the co-operation of formation and destruction, development and decay." Natural growth involved both progress and decline, according to Wright.[15]

Wright's antagonism toward orderly progress was also evident in his early writings on meteorology and astronomy, which personified nature as fickle. He wrote an 1857 column for the *Evening Post* on "wandering comets," noting "the gregarious habits of these long-tailed monsters" and also the irregularity of their orbits. Wright highlighted the capriciousness of Lexell's Comet, the return of which was "predicted from observations" made in 1770 but which "entirely changed its course" after passing near Jupiter in the late 1770s "and was never recognized again."[16] He returned to the theme of unpredictability the following year in an article on "The Winds and the Weather." Predicting "the direction of the weather" is difficult, said Wright, because the operation of "an innumerable host of minor causes" is "so complicated, that the repetition of similar phenomena or similar combinations of causes, to any great extent, is the most improbable of events." The weather thus exhibits "a most inconsequent and incalculable fickleness." At the end of the article, Wright recalled the theme of his "Growth" essay: "Progression in new directions is effected by retrogression in previous modes of growth." He even presented a speculative account of the history of organic life that reflected decline rather than progress, claiming that the present system of organisms is "not the structure of a regular though incomplete development, but the broken and fragmentary form of a ruin." Present conditions, said Wright, since they are "no longer able to develope, much less to create new forms, can only sustain those that are left to its care."[17]

---

15. Chauncey Wright, "Growth," *Monthly Religious Magazine and Independent Journal* 18 (1857): 181–83.

16. Chauncey Wright, "The Comet and its 'Ten Billion Leagues of Tail,'" *Evening Post* (New York), May 5, 1857. For evidence of Wright's authorship, see William Sydney Thayer to Chauncey Wright, 5 May 1857, and Joseph W. Sprague to Chauncey Wright, 8 May 1857, Wright Papers. Wright's title refers to "The Comet," a poem in Oliver Wendell Holmes, *Poems* (Boston: Otis, Broaders, 1836), 141–44. Long after Wright's day, Lexell's Comet remained "lost"; see Quan-Zhi Ye, Paul A. Weigert, and Man-To Hui, "Finding Long Lost Lexell's Comet: The Fate of the First Discovered Near-Earth Object," *Astronomical Journal* 155 (2018).

17. Chauncey Wright, "The Winds and the Weather," *Atlantic Monthly* 1 (1858): 272–73, 279; see also Chauncey Wright, "Climatology," *Christian Examiner* 63 (1857); Louis Menand, *The Metaphysical Club* (New York: Farrar, Straus & Giroux, 2001), 207–9.

When he began to review Spencer's books, Wright used these earlier ideas to criticize the English philosopher. He agreed with Spencer that people should seek "natural explanations" in cosmology and biology but insisted that these need not be progressive: "Alternations of progress and regress relatively to any standard of ends or excellence which we might apply, is to us the most probable hypothesis that the general analogy of natural operations warrants." Evolution, said Wright in another review, "implies more in Mr. Spencer's philosophy than the transmutation hypothesis postulates. It implies and necessitates progress, a progress which is inherent in the order of things, and is more than the continuity and community of causation which the physical sciences postulate. It borrows an idea from the moral sciences, the idea of an end." Spencer's account of evolution was thus, as Wright argued in his longest essay on Spencer, "tainted by teleological biases." The idea of inevitable evolutionary progress was simply the scientific expression of a "faith that moral perfectibility is possible, not in remote times and places, not in the millennium, not in heaven, but in the furtherance of a present progress":

Progress is a grand idea,—Universal Progress is a still grander idea. It strikes the keynote of modern civilization. . . . What the ideas God, the One and the All, the Infinite First Cause, were to an earlier civilization, such are Progress and Universal Progress to the modern world,—a reflex of its moral ideas and feelings.

Despite this modern enthusiasm for progress, Wright thought the picture sketched in his own earlier essays came closer to the truth: "Nothing is exempt from change. Worlds are formed and dissipated. Races of organic beings grow up like their constituent individual members, and disappear like these. Nothing shows a trace of an original, immutable nature, except the unchangeable laws of change." For Wright, change—and not progress—governed nature and evolution.[18]

Like Fiske and Wright, James first encountered Spencer in the early 1860s:

I read [*First Principles*] as a youth when it was still appearing in numbers, and was carried away with enthusiasm by the intellectual perspectives which it seemed to open. When

---

18. Chauncey Wright, "A Physical Theory of the Universe," *North American Review* 99 (1864): 8, 16; Chauncey Wright, "Spencer's Biology," *Nation*, June 8, 1866, 725; Chauncey Wright, "The Philosophy of Herbert Spencer," *North American Review* 100 (1865): 450, 452, 454–55.

a mature companion, Mr. Charles S. Peirce, attacked it in my presence, I felt spiritually wounded, as by the defacement of a sacred image or picture.

James probably also read Spencer's later works "in numbers"—that is, in serially published sections. He seems to have had a subscription to the *System*, and if so, would have received the individual parts of *Principles of Biology* and *Principles of Psychology* as they appeared.[19] The attack recalled by James may have been an earlier version of a comment Peirce made in an 1869 lecture, contrasting Spencer's "grander and more comprehensive theories" with Darwin's "minute, systematic, extensive, and strict scientific researches." Peirce suggested that, given this difference in approach, "the followers of Herbert Spencer" should not be surprised when "scientific men" hold Darwin in higher esteem.[20] Peirce, however, did not begin to criticize Spencer in earnest until the late 1880s, and thus I defer discussion of those criticisms until chapter 5. Unlike Wright and Peirce, who were primarily critical of Spencer's conception of evolution, James was focused on his conception of life as a correspondence between organism and environment. This conception was the foundation of Spencerian psychology and sociology—the topics of the next two sections.

Spencerian Psychology

Even before he engaged directly with Herbert Spencer's notion of life, William James was interested in the ability of organisms to adapt to changes in their external environments. After reading John William Draper's *Human Physiology* in 1863, when he was still studying chemistry, James noted the author's "argument from similarity of development in the individual & in the geological history of the animal series[,] attended with similarity of physical conditions in each[,] to prove that the career of development is guided by external physical causes."[21] This

---

19. William James, "Herbert Spencer," *Atlantic Monthly* 94 (1904): 104; William James to Catharine Walsh, 13 September 1868, and William James to Henry Pickering Bowditch, 30 November 1868, in James, *Correspondence*, 4:336, 350. Sections of *First Principles* were distributed to subscribers from 1860 to 1862; sections of *Principles of Biology* from 1863 to 1867; and sections of *Principles of Psychology* from 1868 to 1872 (dates for specific sections are listed in the prefaces of the final published volumes).
20. MS 158 (November–December 1869), in Peirce, *Writings*, 2:314.
21. "[Notebook 3]," 12 February 1863, Item 4497, James Papers; citing John William Draper, *Human Physiology, Statical and Dynamical; Or, The Conditions and Course of the Life of Man* (New York: Harper & Brothers, 1856), 502–4. As the publication date of *Human Physiology* indicates, Draper was

CHAPTER TWO

interest persisted throughout the 1860s. James's 1868 review of Armand de Quatrefages's *Report on the Progress of Anthropology in France*, for example, highlighted a terminological dispute over how to talk about environmental influence: should we, James asked, include the period prior to birth?

> Through the whole work we find the modifying effect of external circumstances largely insisted on. There has always been a great deal of controversy as to the extent of this modifying influence, and much of it has arisen from an ambiguity in the use of terms. When Darwin, for instance, says external circumstances have little effect, he means that they produce little change in the visible characters of a given animal taken from birth upwards. . . . Quatrefages [in contrast] uses the term *actions de milieu* as including everything that happens to a creature from the ovum upwards. The peculiar circumstances in which his parents may be plunged, and their physiological reactions, become thus a part of his external medium. This seems the best and most consistent way to use the term in question.[22]

James thus endorsed Quatrefages's broad sense of *milieu* (literally "medium" but often translated as "environment") rather than Darwin's narrow one, although as I discuss in the next section, James's later critique of Spencerian sociology implies that he eventually changed his mind. That same year, James also praised Claude Bernard's notion of an "interior medium": according to Bernard, as described by James, the life of "elementary cells . . . has no essential connection with that of the organism, and could continue anywhere where their appropriate medium was supplied."[23]

James thus highlighted the biological importance of the environment throughout the 1860s. He also emphasized the related notion of *plasticity*—the ability to respond flexibly to environmental changes. According to James, the upshot of Darwin's *Variation of Animals and Plants*

---

an evolutionist before the *Origin* was published: see Gregory A. Wickliff, "Draper, Darwin, and the Oxford Evolution Debate of 1860," *Earth Sciences History* 34 (2015).

22. William James, "The Progress of Anthropology," *Nation*, February 6, 1868, 114; see Armand de Quatrefages, *Rapport sur les progrès de l'anthropologie* (Paris: Imprimerie Impériale, 1867), 143. For more on the term *milieu* in nineteenth-century French biology, see Georges Canguilhem, "Le vivant et son milieu," in *La connaissance de la vie* (Paris: Hachette, 1952); Trevor Pearce, "The Origins and Development of the Idea of Organism–Environment Interaction," in *Entangled Life: Organism and Environment in the Biological and Social Sciences*, ed. Gillian Barker, Eric Desjardins, and Trevor Pearce (Dordrecht, Neth.: Springer, 2014), 15–16.

23. William James, "Bernard's Rapport," *North American Review* 107 (1868): 324; see Claude Bernard, *Rapport sur les progrès et la marche de la physiologie générale en France* (Paris: Imprimerie Impériale, 1867), 40–41.

under Domestication was "a conviction of the endlessly fluctuating character, or, to use Mr. Darwin's words, of the 'plasticity of the whole organization.'" Darwin devoted a whole chapter to "Direct and Definite Action of the External Conditions of Life," although he still maintained as he had in the *Origin* that that the main driver of evolution was "natural selection of serviceable variations which have arisen independently of the nature of the conditions." Before his attack on Spencer's psychology, James had already joined the ongoing conversation about the dynamic relationship between organisms and their environments.[24]

As mentioned in chapter 1, James seems by this time to have embraced evolution despite his earlier caution, having been influenced by the naturalistic outlook of Darwin, Spencer, and other empirically oriented writers. For example, James told Oliver Wendell Holmes Jr. in 1868 that he was "tending strongly to an empiristic view of life":

I shall continue to apply empirical principles to my experience as I go on and see how much they fit. One thing makes me uneasy. *If* the end of all is to be that we must take our sensations as simply given or as preserved by natural selection for us, and interpret this rich and delicate overgrowth of ideas, moral, artistic, religious & social, as a mere mask, a tissue spun in happy hours by creative individuals and adopted by other men in the interests of their sensations—how long is it going to be well for us not to "let on" all we know to the public?[25]

This general outlook may have been inspired by Spencer's *First Principles*, which placed evolution in the background of humankind's entire mental and social existence. Holmes had read this book, as well as *Principles of Biology*, in the mid-1860s and was sympathetic to Spencer's perspective despite having also read Wright's critical reviews (an English friend—only half jokingly—had in 1866 called Holmes "an admirer of H. Spencer"). James also corresponded frequently in the late 1860s with Thomas Wren Ward, a friend from the Louis Agassiz expedition

---

24. James, "Darwin's Variation of Animals and Plants," *North American Review* 107 (1868): 362; quoting Charles Darwin, *The Variation of Animals and Plants under Domestication*, 2 vols. (London: John Murray, 1868), 2:406. For the point about "serviceable variations," see Darwin, *Variation of Animals and Plants under Domestication*, 2:290; see also Darwin, *On the Origin of Species by Means of Natural Selection, or the Preservation of Favoured Races in the Struggle for Life* (London: John Murray, 1859), 134.

25. William James to Oliver Wendell Holmes Jr., 15 May 1868, in James, *Correspondence*, 4:302. For more on the personal impact of James's encounter with Spencer, see Robert J. Richards, *Darwin and the Emergence of Evolutionary Theories of Mind and Behavior* (Chicago: University of Chicago Press, 1987), 409–50.

whose "casual reading of Herbert Spencer" had confirmed him as an "unbeliever."[26]

Francis Ellingwood Abbot reviewed the initial volumes of Spencer's system during this same period. He focused on Spencer's internal-external dichotomy rather than his notion of evolution, foreshadowing James's later criticisms of Spencer's psychology. In an 1866 review of *First Principles*, for example, Abbot classified various philosophical positions according to their view of the organism-environment relation:

> Is the Organism purely the product of the Environment? Then we have Empiricism, Sensationalism, Materialism. . . . Is the Environment the product of the Organism? Then we have Transcendentalism, Egoism, Idealism. . . . Are the Organism and Environment both products of some underlying and active Unity? Then we have Identity or Pantheism. . . . Are the Organism and Environment given simply in the co-ordination and correlation of actual knowledge? Then we have Dualism, Natural Realism, Positivism.

Abbot sided with dualism and positivism, criticizing Spencer's empiricism for ignoring the fact that there are "two radically distinct orders of phenomena presented to [science's] observation and study—the one material, the other mental." But he also admitted that "the world is waiting for a creative and organizing intellect which shall integrate Empiricism and Transcendentalism in a deeper and wider synthesis than any yet attempted, and thus inaugurate the reign of a truly stable philosophy." Although Abbot was thinking of himself, this image also anticipated James's approach to psychology and philosophy.[27]

Abbot continued his criticism of Spencer's one-sided empiricism in an 1868 review of *Principles of Biology*. According to Abbot, one of the great questions of biology was "What are the causes of evolution in general?" Anticipating the views of Mivart, discussed in chapter 1, he sketched two basic classes of answers: "The one class finds the causes of organic evolution solely in the direct or indirect action of cosmical forces external to the organism; the other class, fully recognizing the action of these external forces, finds a concurrent cause in forces which manifest them-

---

26. Eleanor N. Little, "The Early Reading of Justice Oliver Wendell Holmes," *Harvard Library Bulletin* 8 (1954): 170; Leslie Stephen to Oliver Wendell Holmes, 7 December 1866, in Frederic William Maitland, *The Life and Letters of Leslie Stephen* (London: Duckworth, 1906), 188; Thomas Wren Ward to William James, 18 June 1868, in James, *Correspondence*, 4:322. For Spencer's view of natural selection, see Herbert Spencer, *First Principles* (London: Williams & Norgate, 1862), 297–98; Herbert Spencer, *The Principles of Biology*, vol. 1 (London: Williams & Norgate, 1864), 443–63.

27. Francis Ellingwood Abbot, "Positivism in Theology," *Christian Examiner* 80 (1866): 249–50, 254–55.

selves in the organism alone, and are therefore irreducible to known cosmical forces." Abbot called the first set of answers "mechanism" and the second "vitalism," siding with the latter while warning against "the fanciful speculations . . . in which some vitalists indulge." Spencer was a mechanist on this classification and thus committed to the claim that "any reaction of the organism on the environment, however seemingly spontaneous, is merely part of the multiplication of effects produced in the first instance by the incident forces."[28]

One place where mechanism ran into trouble, said Abbot, was in accounting "for the morphological development of each organism according to its own specific type." Why do horses give birth to horses and not fish? To answer this question, Spencer appealed to "formative tendencies," apparently undermining his mechanist rejection of such forces. Abbot granted that Spencer had a reasonable response to this problem—namely, to attribute such tendencies "to natural inheritance from ancestral organisms." But Abbot thought Spencer would then face the difficult question of where the first tendencies came from, especially since he seemed to deny the possibility of spontaneous generation.[29] Spencer actually wrote a public reply to this particular criticism, claiming that the very notion of spontaneity was opposed to that of evolution, rejecting the idea of a "first organism" and arguing that even in the chemical realm, "progress toward higher types of organic molecules is effected by modifications upon modifications," each of which "is a change of the molecule into equilibrium with its environment." Spencer thus continued to maintain his emphasis on external forces, or inherited tendencies caused by ancestral versions of these forces, even in the face of Abbot's criticisms.[30]

Like Abbot, the French philosopher Charles Renouvier—who had a marked influence on James beginning in the late 1860s—emphasized internal factors in his criticisms of the philosophy of evolution. In an essay that James read in the early autumn of 1868, Renouvier seemed to echo both Wright and Abbot, although he was probably not familiar with their work. After reviewing the work of Henri de Saint-Simon and various forms of positivism and socialism, he criticized what all of

---

28. Francis Ellingwood Abbot, "Philosophical Biology," *North American Review* 107 (1868): 382–83, 400–401.

29. Abbot, "Philosophical Biology," 403–4; see also Herbert Spencer, *The Principles of Biology*, vol. 2 (London: Williams & Norgate, 1867), 8.

30. Herbert Spencer, *Spontaneous Generation, and the Hypothesis of Physiological Units: A Reply to the North American Review* (New York: D. Appleton, 1870), 4, 6–7.

these approaches had in common—not only a "belief in the natural and necessary progress of humanity" but also an "avowed determinism, the negation of freedom, and the substitution of the idea of evolution for that of fixed laws." This nineteenth-century "evolution school," said Renouvier,

> is forced to explain the constitution of apparent individualities by the action of environments [*milieux*] exclusively. . . . It must explain their faculties and their acts via suggestions coming from outside. Empiricism and sensualism work for this school to establish the laws of these suggestions, reducing almost to nothing the internally given.

Although Renouvier only briefly mentioned Spencer, he did connect him to "ideas of development and progress," and James would almost certainly have associated the English philosopher with the "evolution school" criticized by Renouvier.[31]

In the early 1870s, as Spencer's work became more widely known in France, Renouvier and his collaborator François Pillon began to attack it in their new journal *Critique Philosophique*, which James read avidly.[32] For example, Pillon argued in a May 1872 article that Spencer's psychology "systematically reduces innateness to the inheritance of acquired modifications" and thus represents "the negation of all mental nature, of all intellectual constitution, of all mental law, of all psychological specificity."[33] James was happy to have found an alternative to Spencer's "empiristic view of life," as he observed in a letter to Renouvier a few months later:

> Over here, it is the philosophy of [John Stuart] Mill, [Alexander] Bain, & Spencer that presently carries all before it. This philosophy produces excellent works in psychology, but from the practical point of view it is determinist and materialist. . . . Your phenomenist philosophy seems well suited to make an impression on the elevated minds of the English empirical school.

---

31. Charles Renouvier, "De la philosophie de XIXe siècle en France," *Année Philosophique* 1 (1868): 7, 88, 90, 92–93. For James's reading and praise of this essay, see William James to Henry James Sr., 5 October 1868, in James, *Correspondence*, 4:342.

32. William James, "[Note on *La Critique Philosophique*]," *Nation*, February 6, 1873, 94; see also William James to Charles Renouvier, 29 July 1876, in James, *Correspondence*, 4:542; William James, *The Principles of Psychology*, 2 vols. (New York: Henry Holt, 1890), 1:iv. On the French reception of Spencer, see Daniel Becquemont and Laurent Mucchielli, *Le cas Spencer: Religion, science et politique* (Paris: Presses Universitaires de France, 1998), 257–74; Naomi Beck, *La gauche évolutionniste: Spencer et ses lecteurs en France et en Italie* (Besançon, Fr.: Presses universitaires de Franche-Comté, 2014), 49–64.

33. François Pillon, "L'innéité selon M. Herbert Spencer," *Critique Philosophique* 1 (1872): 214.

By the early 1870s, James had moved away from Spencer's empirical outlook and adopted a version of Renouvier's neo-Kantian "phenomenism," according to which "the knowable universe is . . . a system of phenomena" but still contains within it the possibility of freedom.[34]

Abbot, Renouvier, and Pillon were all attempting to undermine what Peter Godfrey-Smith has called Spencer's "externalism"—the view that organisms are primarily shaped by their external environments and those of their ancestors, rather than by internal tendencies unrelated to those environments. As implied by these criticisms, Spencer had argued that both physical and mental life should be viewed primarily from this perspective:

That particular kind of Life which we distinguish as intelligence, including as it does the various developments of the correspondence in Space, in Time, in Speciality, in Complexity, &c.; it necessarily follows that the changes of which this intelligence consists, must, in their general mode of coordination, harmonize with the co-ordination of phenomena in the environment.[35]

Although I will complicate this story in the next section, Fiske seemed to offer a straightforward defense of Spencer's externalism in a course of lectures on the Positive Philosophy given at Harvard in 1869. Two future members of the Metaphysical Club, Francis Greenwood Peabody and Joseph Bangs Warner, attended the course, and much of its content—along with material from a related 1871 course on the Doctrine of Evolution—was later included in Fiske's *Outlines of Cosmic Philosophy*, a book discussed at a Metaphysical Club meeting in 1874. As James recalled, "when Chauncey Wright, Cha[rle]s P[eirce], [Nicholas] St. John Green, Warner, & I appointed an evening to discuss the Cosmic Phil., just out, J.F. went to sleep under our noses."[36]

---

34. William James to Charles Renouvier, 2 November 1872, in James, *Correspondence*, 4:430–31; William James, "[Note on *La Critique Philosophique*]," *Nation*, February 6, 1873, 94.

35. Peter Godfrey-Smith, *Complexity and the Function of Mind in Nature* (Cambridge: Cambridge University Press, 1996), chap. 3; Herbert Spencer, *The Principles of Psychology* (London: Longman, Brown, Green, & Longmans, 1855), 506; see also Herbert Spencer, *The Principles of Psychology*, 2nd ed., vol. 1 (London: Williams & Norgate, 1870), 385.

36. *Catalogue of the Officers and Students of Harvard University, for the Academical Year 1869–70* (Cambridge, MA: Sever, Francis, 1869), 102–3; William James to Thomas Sergeant Perry, 24 August 1905, in James, *Correspondence*, 11:94. Fiske's Positive Philosophy lectures were published in the *New York World* from November 13, 1869, to January 10, 1870; his Doctrine of Evolution lectures were published in the same newspaper from May 1 to September 1, 1871. For handy scrapbooks containing the relevant articles, see Box 2, John Fiske Papers, Special Collections, Charles E. Young Research Library, University of California–Los Angeles.

Defending psychological externalism, Fiske declared in the Positive Philosophy lectures that "the intelligence of any man must consist, partly of internal relations continually adjusted in conformity with the relations present in his own environment, and partly of organized and integrated relations bequeathed him by countless generations of ancestors, and adjusted to the relations present in innumerable ancestral environments." According to Fiske, although the design argument made famous by natural theologians was an apparently reasonable inference from the correspondence between mind and environment, it got things exactly backward: "It is not the intelligence which has made the environment, but it is the environment which has shaped the intelligence. In the mint of nature, the coin mind has been stamped; and man, perceiving the likeness of the die to its impression, has unwittingly inverted the causal relation of the two." For Fiske, as for Spencer, the environment was not designed by a mind; rather, human minds were designed by the environment in the course of evolution.[37]

James's criticism of Spencer's psychology was an attack on precisely this view, according to which the mind is formed by the environment like a coin struck by a die. Although it paralleled the earlier attacks of Abbot and Renouvier, its special weapons were provided by Shadworth Hodgson and Wilhelm Wundt. In the work of these philosopher-psychologists, James discovered the two concepts that would frame his critique of Spencer: interest and attention. Hodgson had introduced the notion of interest in his 1865 book *Time and Space* as part of a chapter on "Spontaneous Redintegration"—the involuntary restoration of a past state of consciousness. When we have a new experience, it sometimes calls to mind a past experience. But why some particular past experience? According to Hodgson, redintegration has two stages: first, those parts of an experienced object that are uninteresting fade from consciousness; then the remaining interesting parts of the object combine with those past objects with which they have been habitually associated, yielding a new object. This cycle is ongoing: "Scarcely has the process begun, when the original law of interest begins to operate on this new formation, seizes on the interesting parts and impresses them on the attention to the exclusion of the rest, and the whole process is repeated again with

---

37. John Fiske, "Final Causes in History," *New York World*, December 27, 1869; also in John Fiske, *Outlines of Cosmic Philosophy, Based on the Doctrine of Evolution, with Criticisms on the Positive Philosophy*, 2 vols. (London: Macmillan, 1874), 2:399, 402.

endless variety." Hodgson thus called interest the "secret spring" and "motive power" of spontaneous redintegration.[38]

James was intrigued by the implications of Hodgson's "law of interest" for psychology generally, as it seemed to undermine Spencer's more passive account of the experiencing mind. The mind's active role was even more obvious in voluntary (rather than spontaneous) redintegration. As Hodgson wrote, "it is impossible there to suppose consciousness to be a mere foam, aura, or melody, arising from the brain, but without reaction upon it. The states of consciousness are, in voluntary redintegration, links in the chain of physical events or circumstances in the external world."[39] James thought Hodgson had identified a deficiency in Spencer's view—something it had missed. According to Spencer, feelings of pleasure evolved because "those species survived which came to have emotions of pleasure associated with experiences that were useful to them, whilst others perished." But on this story, said James, "the purely *conscious quale* of the mental event seems to act as a determinant link in the chain of physical causes and effects," which contradicts Spencer's broader claim that "the links of the chain of conscious events" are mere "concomitants of those of the chain of successive physical phenomena." Physical and mental events do not form two independent and parallel chains, as Spencer argued; they are at least sometimes links in the same chain. More specifically, the evolution of pleasure is an example, according to James, of how "quality of consciousness as such, instead of being discontinuous with all the facts of nerve vibration, may influence them in direction or amount."[40]

The foregoing discussion of Hodgson's work appeared in a review of William Carpenter's *Principles of Mental Physiology*. In that same review, James declared that Carpenter's chapters on sensation and perception were "very inadequate" in comparison to research "by German inquirers, among whom we may mention [Wilhelm] Wundt . . . and the immortal [Hermann von] Helmholtz in his Optics."[41] Both of these authors

---

38. Shadworth H. Hodgson, *Time and Space: A Metaphysical Essay* (London: Longman, Green, Longman, Roberts, & Green, 1865), 266–68.

39. Hodgson, *Time and Space*, 280; quoted in William James, "[Draft on Brain Processes and Feelings] 1872," in *Manuscript Essays and Notes* (Cambridge, MA: Harvard University Press, 1988), 249; and in William James, "Carpenter's Principles of Mental Physiology," *North American Review* 119 (1874): 228.

40. James, "Carpenter's Principles of Mental Physiology," 226–28; see Herbert Spencer, *The Principles of Psychology*, 2nd ed., vol. 1 (London: Williams & Norgate, 1870), 280.

41. James, "Carpenter's Principles of Mental Physiology," 228. James had planned to study in 1868 with both Wundt and Helmholtz at the University of Heidelberg but ended up leaving Heidelberg after less than a week, worried about his health: see William James to Henry Pickering

CHAPTER TWO

emphasized the activity of the mind in perception. Helmholtz claimed that "the connection of the sensation with the idea of the object . . . depends in large part on acquired experience, and thus on mental activity," although he argued that most of this activity was unconscious. Wundt went even further in his *Grundzüge der physiologischen Psychologie* (*Principles of Physiological Psychology*):

> [Psychology] has tacitly presupposed that the course of sense-perceptions recapitulates immediately and essentially unchanged the temporal course of external impressions. But this is not so; rather, the way in which external events are pictured in our ideas is co-determined by the qualities of consciousness and attention.

The importance of consciousness and attention in experience would become a prominent theme in James's own psychology and the essence of his polemic against Spencer.[42]

James's 1875 review of *Grundzüge* contained almost all of his characteristic criticisms of Spencer, later presented in his Harvard classes and in a series of journal articles.[43] In the book, Wundt had provided empirical evidence for the importance of attention through a series of reaction-time experiments. James described Wundt's laboratory setup: "A signal is given to the subject who, immediately on its reception, replies by closing an electric key. The instant of the signal and of the closure are chronographically registered, and the time between them ascertained." Wundt explained how he separated this total time into several components:

1. "Transmission from the organ of sense to the brain"
2. "Entrance into the field of vision of consciousness, or perception"
3. "Entrance into the focus of attention, or apperception"

Bowditch, 5 April 1868, and William James to Henry James Sr., 3 July 1868, in James, *Correspondence*, 4:292, 327.

42. Hermann Helmholtz, *Handbuch der physiologischen Optik* (Leipzig, Ger.: Leopold Voss, 1867), §26; Wilhelm Wundt, *Grundzüge der physiologischen Psychologie* (Leipzig, Ger.: Wilhelm Engelmann, 1874), 726.

43. Other authors have also highlighted the importance of this review: Robert J. Richards, *Darwin and the Emergence of Evolutionary Theories of Mind and Behavior* (Chicago: University of Chicago Press, 1987), 433–35; Lucas McGranahan, *Darwinism and Pragmatism: William James on Evolution and Self-Transformation* (New York: Routledge, 2017), chap. 1; David Leary, "Psychology and Philosophy in the Work of William James: Two Good Things," in *The Oxford Handbook of William James*, ed. Alexander Klein (Oxford: Oxford University Press, forthcoming).

4. "The time of willing, which is necessary for the registering motion to be triggered in the central organ"
5. "Transmission of the initiated motor excitation to the muscles"

He called (3) and (4) together "reaction time," and he demonstrated that this time could be altered by manipulating attention. James summarized:

> The experimental circumstances which shorten the time of reaction are mainly those which define beforehand as to its quality, intensity, or time, the signal given to the observer, so that he may accurately expect it before it comes. The focusing of the attention takes place under these circumstances *in advance*.

In other words, if subjects are prepared for the signal, they react more quickly—not surprising perhaps, but now "mathematically" demonstrated.[44]

After describing Wundt's experiments, James launched into a tirade against Spencer and "the *a posteriori* school," repurposing Hodgson's notion of interest:

> The *a posteriori* school, with its anxiety to prove the mind a *product, coûte que coûte* [whatever the cost], keeps pointing to mere "experience" as its source. But it never defines what experience is. *My* experience is only what I agree to attend to. Pure sensation is the vague, a semi-chaos, for the *whole* mass of impressions falling on any individual are chaotic, and become orderly only by selective attention and recognition. These acts postulate *interests* on the part of the subject,—interests which, as ends or purposes set by his emotional constitution, keep interfering with the pure flow of impressions and their association, and causing the vast majority of mere sensations to be ignored. It is amusing to see how Spencer shrinks from explicit recognition of this law, even when he is forced to take it into his hand, so to speak.

James argued that subjective interests lead to selective attention, which in turn has a major role in determining human experience. He thus defended a loose psychological analogue of Immanuel Kant's Copernican turn: the mind is not a mere product but itself shapes and orders experience. James criticized Spencer for ignoring the empirically demonstrated importance of the active mind.[45]

---

44. William James, "Grundzüge der physiologischen Psychologie," *North American Review* 121 (1875): 197–99; Wilhelm Wundt, *Grundzüge der physiologischen Psychologie* (Leipzig, Ger.: Wilhelm Engelmann, 1874), 727.

45. James, "Grundzüge der physiologischen Psychologie," 199–200.

A few months later, citing Wundt's teaching as a model, James proposed a new Harvard undergraduate course in psychology. He suggested in a letter to president Charles William Eliot that the course would take a middle way between philosophy and physiology:

> The principal claim I should make for it is its intrinsic importance at the present day when on every side naturalists & physiologists are publishing extremely crude and pretentious psychological speculations under the name of "Science"; and when professors whose education has been exclusively literary or philosophical, are too apt to show a real inaptitude for estimating the force & bearing of physiological arguments when used to help define the nature of man. A real science of man is now being built up out of the theory of evolution and the facts of archeology, the nervous system and the senses.

James told Eliot that his own varied background ensured that he would not only "realize the force of all the natural history arguments" but also avoid "certain crudities of reasoning which are extremely common in men of the laboratory pure & simple." The course, Natural History 2: Physiological Psychology, was accepted, and James taught it for the first time in 1876–77, using the second edition of Spencer's *Principles of Psychology* as a textbook—representing, of course, the crude naturalistic approach.[46]

James's new course seems to have created some buzz among Harvard students. One complained in September that in James's class "Darwinianism is to be treated metaphysically. That is to say, it is to be treated precisely as Darwin and his followers say it should not be treated." The student would have preferred a "general review of the Darwinian hypothesis" by John Fiske. Then in November, another asked whether James's "very valuable" lectures on evolution "could be repeated for the benefit of those interested in the subject." The following April, yet another praised the transfer of the course to the philosophy department: "now that it has been rightly classified, we may confidently expect that this course will occupy a place equal in favor with that of any philosophical elective." Even his students could appreciate that James was offering a philosophical approach to evolution in the mid-1870s that stood in direct contrast to that of Spencer and Fiske.[47]

---

46. William James to Charles William Eliot, 2 December 1875, in James, *Correspondence*, 4:526–28; *Fifty-Second Annual Report of the President of Harvard College, 1876–77* (Cambridge, MA: John Wilson & Son, 1877), 51.

47. "[Spencerian Philosophy]," *Harvard Crimson*, September 29, 1876; "Evolution," *Harvard Crimson*, November 17, 1876; "[Natural History 2]," *Harvard Crimson*, April 22, 1877. These student

In his lecture notes for the 1876–77 course, again citing "Helmholtz's Optics," James continued to criticize Spencer's account of the perceiving mind:

One may go farther & affirm that it makes no difference at all *what* the sensation is in itself, its function in tho't, its connections with the other ingredients of the mental organism so *determine* it, that if time enough is given it will in all cases contribute to one identical conclusion. Thus blind persons seem to have built up the notion of space as perfectly as those who see it.

"Spencer ignores this whole consideration," James concluded.[48] Becoming more and more confident in his criticism, halfway through the course, James pronounced himself "completely disgusted with the eminent philosopher." After delivering his final lecture, James sent a mocking letter to James Jackson Putnam, a clinical instructor in the medical school who specialized in diseases of the nervous system:

Poor Spencer, reduced to the simple childlike faith of merely timid, receptive uncritical, undiscriminating, worshipful, servile gullible, stupid, idiotic natures like you and Fiske! Would *I* were part of his environment! I'd see if his "intelligence" could establish "relations" that would "correspond" to me in any other way than by giving up the ghost before me! He and all his myrmidons, disciples and parasites! Down with the hell-spawn of 'em! Of all the incoherent, rotten, quackish humbugs & pseudo-philosophasters which the womb of all-inventive time has excreted he is the most infamous and "abgeschmackt" [vulgar]—but even *he* is better than his followers.[49]

Behind James's jokes lay a more serious point: the idea of a correspondence between relations in the mind and relations in the environment, so central to Spencer's psychology, was difficult to interpret. What counted as a better correspondence?

James tackled this question in his first published article, written in the fall of 1877: "Remarks on Spencer's Definition of Mind as Correspondence."[50] For Spencer, as we have seen, the evolution of intelligence was just a

---

comments were first noted in Ralph Barton Perry, *The Thought and Character of William James*, 2 vols. (Boston: Little, Brown, 1935), 1:475–76.

48. "Notes for Natural History 2: Physiological Psychology (1876–1877)," in James, *Manuscript Lectures*, 128; see Hermann Helmholtz, *Handbuch der physiologischen Optik* (Leipzig, Ger.: Leopold Voss, 1867), §33.

49. William James to Thomas Wren Ward, 30 December 1876, and William James to James Jackson Putnam, 26 May 1877, in James, *Correspondence*, 4:552, 4:564.

50. William James to William Torrey Harris, 6 December 1877, in James, *Correspondence*, 4:587. Harris was the editor of the *Journal of Speculative Philosophy*, where the article eventually appeared.

progressively improving correspondence between mind and environment. But how would progress be assessed? According to James, this story had as its implicit "teleological factor" the notion of *interest* (he cited Hodgson's "Spontaneous Redintegration" chapter, discussed earlier). As James had already insisted in the Wundt review, "subjective interests" are "the real *a priori* element in cognition" and "precede the outer relations noticed." His main strategy, as Mathias Girel has shown, was to replace Spencer's two-place relation—organism-environment—with a three-place relation—organism-interests-environment. Only by adding this third element, which provides a kind of norm or end, are we able to judge whether a correspondence has improved, James believed. Spencer's account could not truly avoid this teleology, said James; it just assumed specific interests, "those of physical prosperity or survival."[51]

So James's primary critique of Spencer, first presented in the 1875 review of Wundt and elaborated in the 1876–77 course and the 1878 article, was that he ignored the importance of interests—"the very flour out of which our mental dough is kneaded."[52] When James sent his article to Renouvier, the French philosopher responded by dismissing not only Spencer but evolution in general: "[Spencer's] great renown in Europe arises from his systematization of the theory of evolution. But evolution is a passing fad. It will last 15 or 20 years, and then we'll talk about it the way they talked about Lamarck's system in the age of Cuvier. So it goes."[53] James, in contrast, was happy with the evolutionary-naturalistic picture of the mind; he just thought that Spencer had neglected important phenomena.

James repeated his critique in another early essay, "Brute and Human Intellect":

> Spencer, throughout his work, ignores entirely the reactive spontaneity, both emotional and practical, of the animal. Devoted to his great task of proving that mind from its lowest to its highest forms is a mere product of the environment, he is unwilling, even cursorily, to allude to such notorious facts . . . as the existence of peculiar idiosyncrasies of interest or selective attention on the part of every sentient being.

According to James, Spencer's account of the mind as a "mere product" had absurd implications:

---

51. William James, "Remarks on Spencer's Definition of Mind as Correspondence," *Journal of Speculative Philosophy* 12 (1878): 5, 6n, 10–11; Mathias Girel, "James critique de Spencer: D'une autre source de la maxime pragmatiste," *Philosophie* 64 (2000): 82.

52. James, "Remarks on Spencer's Definition of Mind as Correspondence," 13.

53. Charles Renouvier to William James, 14 May 1878, in James, *Correspondence*, 5:8.

If Spencer's account were true, a race of dogs bred for generations, say in the Vatican, would have characters of visual shape, sculptured in marble, presented to their eyes, in every variety of form and combination. The result of this reiterated "experience" would be to make them dissociate and discriminate before long the finest shades of these particular characters. In a word, they would infallibly become, if time were given, accomplished *connoisseurs* of sculpture. The reader may judge of the probability of this conclusion.

That is, subjective interests—whether innate or acquired—inevitably shape our experience. Whereas we marvel at the sculpted agony of Laocoön, dogs care only for "the odors at the bases of the pedestals." Spencer's theory, said James, could not explain this difference.[54]

But was this really fair to Spencer? It is hard to imagine him denying Auguste Comte's claim, in an 1843 letter to John Stuart Mill, that "it is the organism and not the environment [*milieu*] that makes us men rather than monkeys or dogs."[55] After all, as James might have remembered from Abbot's review of *Principles of Biology*, Spencer had treated the innate features of organisms as "proclivities inherited by them from antecedent organisms, and which past processes of evolution have bequeathed." Fiske's *Outlines of Cosmic Philosophy* had likewise emphasized that human minds are the product of "intercourse with their environment—both their own intercourse and that of ancestral minds." Even Renouvier had noted the point: "Innateness [for Spencer] is nothing but inheritance, an inheritance that one must follow back through the ages along the series of ancestors of each man, and along the longer series of man's animal ancestors, to the first and evanescent origins of life."[56]

This apparent unfairness was probably the reason Wright was so frustrated with James's attack on the "*a posteriori* school." In the Wundt review, James had singled out Wright as one of the few empiricists who

---

54. William James, "Brute and Human Intellect," *Journal of Speculative Philosophy* 12 (1878): 256–57. This example is repeated in William James, *The Principles of Psychology*, 2 vols. (New York: Henry Holt, 1890), 1:403.

55. Émile Littré, *Auguste Comte et la philosophie positive* (Paris: Hachette, 1863), 411; quoted in François Pillon, "L'innéité selon M. Herbert Spencer," *Critique Philosophique* 1 (1872): 211. James seems to have read Littré's book in 1870; he probably also read the Pillon essay in which it was quoted—see *Diary*, vol. 1, seq. 96, Item 4550, James Papers, https://iiif.lib.harvard.edu/manifests/view/drs:45436722.

56. Herbert Spencer, *The Principles of Biology*, vol. 2 (London: Williams & Norgate, 1867), 8; quoted in Francis Ellingwood Abbot, "Philosophical Biology," *North American Review* 107 (1868): 403–4; John Fiske, *Outlines of Cosmic Philosophy* (London: Macmillan, 1874), 1:86; Charles Renouvier, review of *Les premiers principes*, by Herbert Spencer, *Critique Philosophique* 1 (1872): 15.

CHAPTER TWO

actually admitted the importance of the active mind in experience.[57] Nevertheless, Wright was not happy, and told Grace Norton that James had misunderstood empiricism: "In a paragraph in which he distinguishes and compliments me among the 'empiricists' he has so badly misapprehended what the experience philosophy in general holds and teaches, that the compliment to me goes for nothing in mitigation of my resentment." Wright complained to James himself a few days later that the compliment had been "made at the expense of my friends."[58] As we have seen, Wright was no friend of Spencer—so why was he upset? Wright probably thought that philosophers such as Bain and Mill, mentioned alongside Spencer in the review, had already embraced something like James's position. Bain, for example, had built his account of volition around the idea of "spontaneous activity": "Our various organs are liable to be moved by a stimulus flowing out from the nervous centres, in the absence of any impressions from without." Mill, for his part, had clearly acknowledged in *System of Logic*—first published in 1843—that people differed in their susceptibility to certain sensations: "Differences of mental susceptibility in different individuals may be, first, original and ultimate facts, or, secondly, they may be consequences of the previous mental history of those individuals, or thirdly and lastly, they may depend upon varieties of physical organization."[59] According to James, the "*a posteriori* school" was "desperately bent on covering up all tracks of the mind's originality." According to Wright, this claim was simply false: just because Mill thought that "the German school of metaphysical speculation" had erred in failing to attribute mental differences "to the outward causes by which they are for the most part produced," that did not mean that he denied to such differences a role in shaping experience.[60] Mill thought these differences were "for the most part produced" by the environment. Spencer simply extended this line of thought, arguing that even those differences that seem like "original and ultimate facts" could be viewed—from the perspective of evolution—as products of the ancestral environment.

57. William James, "Grundzüge der physiologischen Psychologie," *North American Review* 121 (1875): 199; citing Chauncey Wright, "Evolution of Self-Consciousness," *North American Review* 116 (1873).
58. Chauncey Wright to Grace Norton, 12 July 1875 and 18 July 1875, Items 310–11, Norton Letters.
59. Alexander Bain, *The Senses and the Intellect* (London: John W. Parker, 1855), 289; John Stuart Mill, *Collected Works*, 33 vols. (Toronto: University of Toronto Press, 1963–91), 8:856.
60. William James, "Grundzüge der physiologischen Psychologie," *North American Review* 121 (1875): 200; Mill, *Collected Works*, 8:859. Mill added this passage to the third edition of the *System of Logic* in 1851, and it appeared in all subsequent editions.

Spencer could thus have appealed to evolutionary history to explain variation in interests within and especially across species—something Wright thought James should have acknowledged. But James soon identified a key weakness in this potential explanation: it gave a misleading account of the origin of variation. James claimed to have a better story: variation in interests, and thus in perception and experience, is partly influenced by consciousness.

James believed that consciousness helps determine one's interests. Recall the line from the Wundt review, repeated almost verbatim in later works: "*My* experience is only what I agree to attend to."[61] He was well aware that many critics of the "*a posteriori* school" were theologically motivated: they wanted to preserve consciousness as something linked to the supernatural.[62] But as he had already stressed in the review of Wundt, his own approach to consciousness was strictly scientific. After all, Spencer's own evolutionism implied that consciousness must have a function:

Taking a purely naturalistic view of the matter, it seems reasonable to suppose that, unless consciousness served some useful purpose, it would not have been superadded to life. Assuming hypothetically that this is so, there results an important problem for psycho-physicists to find out, namely, *how* consciousness helps an animal.

Given the "naturalistic" hypothesis that consciousness benefits us in some way, we need to ask how. James's answer was that it may make us more streamlined and efficient in our response to stimuli—that "much complication of machinery may be saved in the nervous centres . . . if consciousness accompany their action":

Might, for example, an animal which regulated its acts by notions and feelings get along with fewer preformed reflex connections and distinct channels for acquired habits in its nervous system than an animal whose varied behavior under varying circumstances was purely and simply the result of the change of course through the nervous reticulations which a minute alteration of stimulus had caused the nervous action to take? In a word, is consciousness an economical *substitute* for mechanism?

---

61. James, "Grundzüge der physiologischen Psychologie," 199; William James, "Brute and Human Intellect," *Journal of Speculative Philosophy* 12 (1878): 256; William James, *The Principles of Psychology*, 2 vols. (New York: Henry Holt, 1890), 1:402.

62. "Notes for Philosophy 4: Psychology (1878–1879)," in James, *Manuscript Lectures*, 136; William James to James Jackson Putnam, 17 January 1879, in James, *Correspondence*, 5:34.

According to James, a mechanical response to a series of individual environmental stimuli might be unwieldy compared with a response regulated by consciousness, and this could be an explanation of why consciousness evolved in the first place. Thus, James accused Spencer of ignoring not only interests and attention but also the function of consciousness in experience.[63]

These three notions came together in the 1879 essay "Are We Automata?"[64] James repeated his earlier points about the active mind: "Whoever studies consciousness, from any point of view whatever, is ultimately brought up against the mystery of *interest* and *selective attention*." Spencer, ignoring these two concepts, seemed to claim that a highly evolved mind would be exquisitely tailored and completely responsive to each and every aspect of its environment—this would be perfect correspondence. James's reply was that in fact "the most perfected parts of the brain are those whose action are least determinate. It is this very vagueness which constitutes their advantage. They allow their possessor to adapt his conduct to the minutest alterations in the environing circumstances, any one of which may be for him a sign." According to James's own evolutionary account, the contribution of consciousness is to, "by its selective emphasis, make amends for the indeterminateness" of what is otherwise "a happy-go-lucky, hit-or-miss affair." Thus the different aspects of James's critique of Spencer were linked: only by emphasizing the discriminating power of consciousness—"the mind's selective industry"—could we explain its evolution.[65]

Spencer could have replied that James's own story of the evolution of consciousness implicitly appealed to the ancestral environment: organisms with consciousness persisted and progressed because they coped with environmental challenges more successfully than those without consciousness. But in "Are We Automata?" James also suggested that ancestral choice—and not only the ancestral environment—was an important factor in evolution:

---

63. William James, "Grundzüge der physiologischen Psychologie," *North American Review* 121 (1875): 201.

64. This essay was based on a November 1 lecture, the last of a series delivered in the autumn of 1878: see "Lowell Lectures on 'The Brain and the Mind' (1878)," in James, *Manuscript Lectures*, 24–30; "The Lowell Institute," *Boston Daily Advertiser*, November 2, 1878.

65. William James, "Are We Automata?" *Mind* 4 (1879): 5, 8, 11, 18. For a more detailed account of this article, see Alexander Klein, "James and Consciousness," in *The Oxford Handbook of William James*, ed. Alexander Klein (Oxford: Oxford University Press, forthcoming).

We may even, by our reasonings, unwind things back to that black and jointless continuity of space and moving clouds of swarming atoms which science calls the only real world. But all the while the world we feel and live in, will be that which our ancestors and we, by slowly cumulative strokes of choice, have extricated out of this, as the sculptor extracts his statue by simply rejecting the other portions of the stone. Other sculptors, other statues from the same stone! Other minds, other worlds from the same chaos! Goethe's world is but one in a million alike embedded, alike real to those who may abstract them. Some such other worlds may exist in the consciousness of ant, crab and cuttle-fish.

James was obviously enamored of this image, as he returned to it several times in later works. It is consistent with his example of the dogs as well: the canine experience of a museum is different because dogs, like crabs and cuttlefish, live in a different world. For James, one's world is the product not only of one's environment and that of one's ancestors but also of a long series of "cumulative strokes of choice."[66]

It is difficult to determine exactly what role James was granting to consciousness and choice in evolution. He was not alone in considering the question. The paleontologist Edward Drinker Cope, a prominent American defender of evolution whom we will meet again in chapter 5, had recently argued that "intelligent choice taking advantage of the successive evolution of physical conditions, may be regarded as the *originator of the fittest*, while natural selection is the tribunal to which all the results of accelerated growth are submitted."[67] The sculpture metaphor, with its emphasis on choice, suggests that James might have agreed with Cope on this point. Like Cope, James thought that consciousness could "immensely shorten the time and labour of natural selection."[68] On the other hand, he was also in the midst of developing a new critique of Spencer that downplayed the importance of direct adaptation to the environment and thus seemed to rule out any cumulative evolutionary effects of intelligent choice. This new critique, at least initially, was directed at Fiske and Spencer's sociology.

---

66. James, "Are We Automata?" 13–14; William James, *The Principles of Psychology*, 2 vols. (New York: Henry Holt, 1890), 1:288–89; William James, *Pragmatism: A New Name for Some Old Ways of Thinking* (New York: Longmans, Green, 1907), 247.
67. Edward Drinker Cope, "The Method of Creation of Organic Forms," *Proceedings of the American Philosophical Society* 12 (1871): 259. Cope's prominence is shown by the inclusion of his essay "On the Hypothesis of Evolution" alongside those of Thomas Henry Huxley and John Tyndall in Noah Porter's *Half Hours with Modern Scientists: Huxley—Barker—Stirling—Cope—Tyndall* (New Haven, CT: Charles C. Chatfield, 1871).
68. William James, "Are We Automata?" *Mind* 4 (1879): 15.

Spencerian Sociology

Just as William James criticized Herbert Spencer's psychology for treating the mind "as absolutely passive clay, upon which 'experience' rains down," he criticized his sociology for treating social change as primarily "due to the environment."[69] But the latter criticism concerned Spencer's whole theory of evolutionary variation, and not just certain underemphasized phenomena. James argued that Spencerians saw the environment as primarily a producer of variation, whereas Darwin (and James) viewed it as primarily a preserver of variation. In this section, I will argue that despite the usefulness of James's distinction between the production and preservation of variation, his attack on Fiske and Spencer's evolutionary sociology was ultimately unsuccessful.

This attack began in the 1878–79 version of James's Natural History 2 class at Harvard, which had been rechristened Philosophy 4. Having again assigned Spencer's *Principles of Psychology*, James accused Spencer of failing to distinguish two independent causal factors in evolution, insisting that "the regulator or preserver of the variation, the environment, is a different part from its producer." James pointed out that Darwin's phrase "spontaneous variation" was meant to capture the fact that variations usually stem from "unknown physiological conditions."[70] In Darwin's view, the environment did not normally directly shape organisms but merely selectively preserved those that happened to possess beneficial variations. Darwin thus argued in *Variation of Animals and Plants* that "in most, perhaps in all cases, the organisation or constitution of the being which is acted on, is a much more important element than the nature of the changed conditions, in determining the nature of the variation." Spencer, in contrast, claimed that "the production of adaptations by direct equilibration"—that is, active adjustment in direct response to the environment—was of primary importance in humans, since we are shielded from natural selection by our social arrangements.[71]

---

69. William James, "Brute and Human Intellect," *Journal of Speculative Philosophy* 12 (1878): 256; William James, "Great Men, Great Thoughts, and the Environment," *Atlantic Monthly* 46 (1880): 442.
70. "Notes for Philosophy 4: Psychology (1878–1879)," in James, *Manuscript Lectures*, 137–38. The textbook for 1877–78 was Hippolyte Taine, *On Intelligence*, trans. T. D. Haye, 2 vols. (New York: Henry Holt, 1875). See *Annual Reports of the President and Treasurer of Harvard College, 1877–78* (Cambridge, MA: John Wilson & Son, 1879), 59; *Annual Reports of the President and Treasurer of Harvard College, 1878–79* (Cambridge, MA: John Wilson & Son, 1880), 60.
71. Charles Darwin, *The Variation of Animals and Plants under Domestication*, 2 vols. (London: John Murray, 1868), 2:291; Herbert Spencer, *The Principles of Biology*, vol. 1 (London: Williams &

Although Philosophy 4 was billed as a psychology class, James spent part of it arguing that the direct adaptation view also vitiated Spencerian sociology. In the early 1870s, Spencer had attacked "the great man theory of history" for being primitive and unscientific:

> If, not stopping at the explanation of social progress as due to the great man, we go back a step and ask whence comes the great man, we find that the theory breaks down completely. . . . He must be classed with all other phenomena in the society that gave him birth, as a product of its antecedents. Along with the whole generation of which he forms a minute part—along with its institutions, language, knowledge, manners, and its multitudinous arts and appliances, he is a resultant of an enormous aggregate of causes that have been cooperating for ages.[72]

This view of the "great man" as a mere "resultant" of his society, said James, again failed to distinguish between producing and preserving causes: Spencer, "in falling back on the environment as implicating the causes of the great man," was "clumping into one term things which it is useful to have distinguished." The sociologist, according to James, should treat the "great man" as a datum, analogous to the zoologist's "spontaneous variation" (as illustrated in figure 7). James did admit a disanalogy between the cases: in social change, the environment preserves not the variation itself but the entire social organism, fundamentally altered by a Martin Luther or a Muhammad. But he maintained that, whatever causes produced Muhammad, they should be kept strictly separate from those causes that allowed Islamic society to flourish.[73]

Despite the quotation from Spencer, James's proximal target was Grant Allen, a Canadian popularizer of evolution whom James had previously accused of falling "into the vice which is the curse of the Spencerian school—the vice of an illusory simplicity gained only by leaving

---

Norgate, 1864), 468–69. Spencer's term for *natural selection* was "indirect equilibration." The idea that humans are to some extent shielded from natural selection was also promoted in Alfred Russel Wallace, "The Origin of Human Races and the Antiquity of Man Deduced from the Theory of Natural Selection," *Journal of the Anthropological Society of London* 2 (1864), which was reviewed in William James, "Wallace's Origin of Human Races," *North American Review* 101 (1865), and cited in Herbert Spencer, *The Principles of Biology*, vol. 1 (London: Williams & Norgate, 1864), 469n.

72. Herbert Spencer, "Is There a Social Science?" *Contemporary Review* 19 (1872): 708; also in Herbert Spencer, *The Study of Sociology* (New York: D. Appleton, 1873), 33; cited in "Notes for Philosophy 4: Psychology (1878–1879)," in James, *Manuscript Lectures*, 140; William James, "Great Men, Great Thoughts, and the Environment," *Atlantic Monthly* 46 (1880): 448–49.

73. James, "Notes for Philosophy 4: Psychology (1878–1879)," in James, *Manuscript Lectures*, 139–40.

CHAPTER TWO

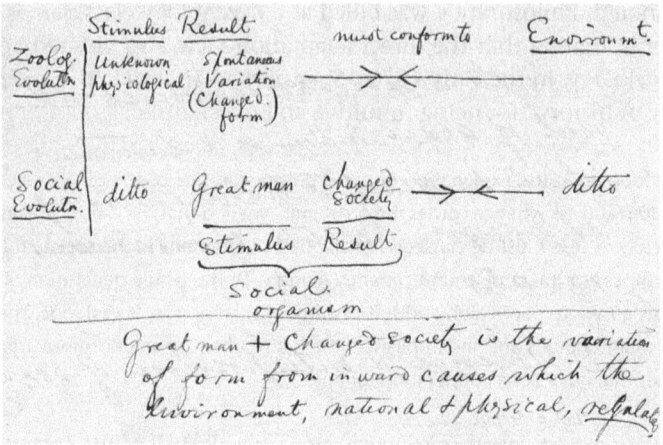

Figure 7 William James's analogy between zoological and social evolution. From William James's lecture notes for his Philosophy 4 class at Harvard in 1878–79: "Spencer's 'Law of Intelligence,'" MS Am 1092.9 (4493), William James Papers, Houghton Library, Harvard University.

out essential data."[74] Allen, in an 1878 article on "Nation-Making," had attacked the English journalist Walter Bagehot's application of Darwin's theories to social change. Bagehot had begun his own essay of the same title, later collected in *Physics and Politics*, by dismissing "the direct effect of climate." He described how zoologists such as Wallace had shown that regions with a similar climate could contain different forms of life. Analogously, said Bagehot, "climate is clearly not *the* force which makes nations, for it does not always make them, and they are often made without it."[75] Bagehot argued instead that it was "great men" who produce and alter national characters: "Some one attractive type catches the eye, so to speak, of the nation"; this is followed by "unconscious imitation and encouragement of [this] appreciated character," combined with "unconscious shrinking from and persecution" of other characters.

---

74. William James, "Allen's Physiological Aesthetics," *Nation*, September 20, 1877, 186.

75. Grant Allen, "Nation-Making: A Theory of National Characters," *Gentleman's Magazine* 243 (1878). Allen was attacking Walter Bagehot, "Nation Making," *Fortnightly Review* 10 (1869): 59–60; also in Walter Bagehot, *Physics and Politics; Or, Thoughts on the Application of the Principles of "Natural Selection" and "Inheritance" to Political Society* (London: K. Paul, Trench, Trübner, 1872), 84–86. The incident was cited in William James, "Great Men, Great Thoughts, and the Environment," *Atlantic Monthly* 46 (1880): 452. Bagehot and James both quoted Alfred Russel Wallace, *The Malay Archipelago: The Land of the Orang-Utan, and the Bird of Paradise; A Narrative of Travel, with Studies of Man and Nature*, 2 vols. (London: Macmillan, 1869), 1:16.

Thus, nations are made by "the confluence of congenial attractions and accordant detestations," a process paralleling the unconscious selection of favored types that Darwin had described in the *Origin*.[76]

Echoing a standard criticism of Darwin that I will discuss in chapter 5, Allen attacked Bagehot for failing to explain the origin of his models for imitation:

> *How* are these spontaneous variations set up? Apparently, in Mr. Bagehot's view, by mere causeless accident. . . . Of course Mr. Bagehot would answer (were he here amongst us to do so) that these minor variations were set up by surrounding circumstances. So far, good. But then he seems to regard those surrounding circumstances as of little importance, mere fugitive collocations of petty causes, varying from moment to moment, and only worthy of note because of the effects which they conspire remotely to produce.[77]

Bagehot and Allen agreed that Greek culture was the origin of free and progressive civilization, but Allen was unsatisfied with Bagehot's admission that "we cannot in the least explain why the incipient type of curious characters broke out, if I may so say, in one place rather than in another."[78] Allen asked how Bagehot could still be treating variations "as results of unknown laws" when Wallace, the very naturalist Bagehot cited, had recently explained the origin of color variation in birds. According to Wallace, "the endless processes of growth and change during the development of feathers, and the enormous extent of this delicately-organized surface, must have been highly favorable to the production of varied colour-effects."[79]

Like Spencer, Allen wanted an explanation of the "great man" as a "resultant." Like the classical and modern writers criticized by Bagehot, he found it in a nation's environment:

---

76. Bagehot, "Nation Making," 62, 66, 70; also in Bagehot, *Physics and Politics*, 90, 97, 105. On unconscious selection, see Charles Darwin, *On the Origin of Species by Means of Natural Selection, or the Preservation of Favoured Races in the Struggle for Life* (London: John Murray, 1859), 34–38.

77. Grant Allen, "Nation-Making: A Theory of National Characters," *Gentleman's Magazine* 243 (1878): 585. For examples of this criticism of Darwin, see the Duke of Argyll, *The Reign of Law* (London: Alexander Strahan, 1867), 229; Edward Drinker Cope, "The Method of Creation of Organic Forms," *Proceedings of the American Philosophical Society* 12 (1871): 230.

78. Grant Allen, "Hellas and Civilisation," *Gentleman's Magazine* 243 (1878); Walter Bagehot, *Physics and Politics; Or, Thoughts on the Application of the Principles of "Natural Selection" and "Inheritance" to Political Society* (London: K. Paul, Trench, Trübner, 1872), 183.

79. Alfred Russel Wallace, *Tropical Nature, and Other Essays* (London: Macmillan, 1878), 198; this book is cited in Grant Allen, "Nation-Making: A Theory of National Characters," *Gentleman's Magazine* 243 (1878): 586n.

To me it seems rather that the differentiating agency must be sought in the great permanent geographical features of land and sea, and that these have necessarily and inevitably moulded the characters and the histories of every nation upon earth.... We cannot regard any nation as an active agent in differentiating itself. Only the surrounding circumstances can have any effect in such a direction.[80]

After attacking this view in his 1878–79 lecture notes, James repeated his criticisms in an 1880 essay titled "Great Men, Great Thoughts, and the Environment":

I affirm that the relation of the visible environment to the great man is in the main exactly what it is to the "variation" in the Darwinian philosophy. It chiefly adopts or rejects, preserves or destroys, in short *selects* him.... The mutations of societies, then, from generation to generation, are in the main due directly or indirectly to the acts or the example of individuals whose genius was so adapted to the receptivities of the moment, or whose accidental position of authority was so critical, that they became ferments.[81]

That is, James—following Bagehot—attributed social change to the social environment as preserver at two levels: the "great man," to make or change a nation, must be adopted as a model within the narrower environment of that nation, and a nation changed by a "great man" is then able to flourish within the broader international environment. He also claimed that the idea of the environment as producer was a scientific dead end: sociologists could no more explain the origin of a "great man" than zoologists could explain the origin of a new variant—both were the result of unknown conditions.

Those targeted by this attack immediately realized that although James described his essay as demonstrating that "the Spencerian 'philosophy' of social and intellectual progress is an obsolete anachronism, reverting to a pre-Darwinian type of thought," it really only amounted to an argument about which phenomena the sociologist should emphasize. James's rhetoric concealed substantial agreement: all parties admitted that variation was both produced and preserved by the environment, whether or not it was appropriate to use the term *environment* to denote the agent of both causal processes, and everyone agreed that social change was the proximate result of individual actions. Fiske was mystified by James's at-

---

80. Allen, "Nation-Making," 585, 589; quoted in William James, "Great Men, Great Thoughts, and the Environment," *Atlantic Monthly* 46 (1880): 450.

81. James, "Great Men, Great Thoughts, and the Environment," 445–46.

tack, and he cited Spencer's claim—in the very book referenced by James—that growth and change in society are "brought about by the mutual actions of individuals." Fiske suggested that the approaches of James and Allen were complementary:

> One writer may turn his attention chiefly to the consideration of those individual variations in opinion and conduct which, in our ignorance concerning their complex modes of genesis, we call spontaneous variations. Another writer may be more deeply interested in pointing out such circumstances as those of geographical position, of commercial intercourse, of political cohesiveness, by which the broad outlines of history have been more or less determined.[82]

Allen likewise said that "individual men are the units whose movements make up social changes," though he still maintained that these individuals "are wholly created by their external circumstances." In a reply to the replies of Fiske and Allen, written at the time but not published until a decade later, James granted that the debate was really "a disagreement about which of many universally admitted factors is the one upon which *emphasis* should be placed." Back in 1880, he had already admitted as much in a letter to Fiske: "Perhaps I laid it too strong on the individual's share in my polemic passage, as [Spencer] on the 'Conditions' in his polemic passage."[83]

Lucas McGranahan has suggested that James's distinction between the production and preservation of variation allowed him to provide "not merely an *externalist* story of how the social environment shapes the individual, but rather an *interactionist* story of how societies and individuals shape one another."[84] But this contrast is unfair to James's opponents: not only were they themselves telling an interactionist story, but that story featured a rich description of the relevant environment. Although Allen, despite his reference to "the various other countries and races with which the nation under consideration is brought into relations," comes off as a naive geographical determinist, the sociological approach of Fiske and Spencer was much more nuanced.

---

82. John Fiske, "Sociology and Hero-Worship," *Atlantic Monthly* 47 (1881): 77–78. The "mutual actions" line is in Herbert Spencer, *The Study of Sociology* (New York: D. Appleton, 1873), 53; this book is cited in James, "Great Men, Great Thoughts, and the Environment," 449.

83. Grant Allen, "The Genesis of Genius," *Atlantic Monthly* 47 (1881): 372; William James, "The Importance of Individuals," *Open Court*, August 7, 1890, 2438; William James to John Fiske, 19 December 1880, in James, *Correspondence*, 5:145–46.

84. Lucas McGranahan, "William James's Social Evolutionism in Focus," *Pluralist* 6 (2011): 86.

CHAPTER TWO

Because Spencer's sociological writings did not appear until the 1870s, it was actually Fiske who first applied the philosophy of evolution to social change. In several of his Positive Philosophy lectures at Harvard in 1869, later incorporated into *Outlines of Cosmic Philosophy*, Fiske provided an analysis of social evolution that was almost completely ignored by James in his 1880 critique. Fiske began by noting that the relevant environment for social evolution was both physical and social:

The prime factors of progress are the community and its environment. The environment of a community comprises all the circumstances, adjacent or remote, to which the community may be in any way obliged to conform its actions. It comprises not only the climate of the country, its soil, its flora and fauna, its perpendicular elevation, its relation to mountain-chains, the length of its coast-line, the character of its scenery, and its geographical position with reference to other countries; but it includes also the ideas, feelings, customs, and observances of past times, so far as they are preserved by literature, traditions, or monuments; as well as foreign contemporary manners and opinions, so far as they are known and regarded by the community in question.[85]

With this extended notion of the environment in hand, Fiske presented his main thesis: "the heterogeneity of the environment is the chief determining cause of social progress." The cause of this environmental heterogeneity, said Fiske, was "the integration or growing interdependence of communities that were originally isolated." Fiske thought that social progress, and especially moral progress, had been the result of the expansion of social interaction both within and across social groups—that is, increased integration within the group and increased communication and trade with other groups.[86]

In contrast to his treatment of psychology, Fiske was careful to stress that he was not an externalist in sociology. Social progress, he noted, is not simply the result of physical or geographical heterogeneity; communities can change their environment and pass those changes on to subsequent generations:

Every city that is built, every generalization that is reached, every invention that is made, every new principle of action that is suggested, alters in some degree the social environment—alters the sum-total of external relations to which the community must

---

85. John Fiske, "The Factors of Progress," *New York World*, December 31, 1869; also in John Fiske, *Outlines of Cosmic Philosophy, Based on the Doctrine of Evolution* (London: Macmillan, 1874), 2:197.
86. John Fiske, "The Law of Progress," *New York World*, January 1, 1870; also in Fiske, *Outlines of Cosmic Philosophy*, 2:203–4, 213, 215.

adjust itself by instituting new internal relations. The entire organized experience of each generation, so far as it is perpetuated by literature or oral tradition, adds an item to the environment of the next succeeding generation; so that the sum-total of the circumstances to which each generation is required to conform itself, is somewhat different from the sum-total of circumstances to which the immediately preceding generation was required to conform itself. Thus the community, by the inevitable results of its own psychical activity, is continually modifying the environment; and to the environment, as thus continually modified, the community must reciprocally conform itself.[87]

According to Fiske, then, social evolution proceeds by reciprocal interaction between communities and their environments. Although he held that environmental heterogeneity was of primary importance, he acknowledged that, since the relevant heterogeneity was social, communities played a key role in producing it.

After reading these Positive Philosophy lectures in 1870, Spencer wrote to Fiske praising them:

In several of the sociological propositions you set forth, you have to some extent forestalled me in the elaboration of the doctrine of Evolution under its sociological aspects. I refer to the dominance you have given to the influence of the sociological environment, and to the conception of social life as having its actions adjusted to actions in the environment, which you have presented in a more distinct way than I have as yet had the opportunity of doing.[88]

Spencer thus endorsed Fiske's account of social evolution as properly Spencerian.

Although Spencer's phrase "the influence of the sociological environment" might suggest that he favored a purely externalist approach, his later work demonstrated that he was as much an interactionist as Fiske when it came to social evolution. In the first volume of *The Principles of Sociology*, published in 1877, Spencer distinguished between a society's "intrinsic factors" (the physical, emotional, and intellectual character of its individuals) and its "extrinsic factors" (climate, surface geography, flora and fauna, etc.). More important, he presented a list of "secondary

---

87. Fiske, "Law of Progress"; also in Fiske, *Outlines of Cosmic Philosophy*, 202–3.
88. Herbert Spencer to John Fiske, 2 February 1870, HM 13722, "Herbert Spencer and John Fiske," Huntington Library; published in John Spencer Clark, *The Life and Letters of John Fiske*, 2 vols. (Boston: Houghton Mifflin, 1917), 1:366–67.

or derived sets of factors" arising from the interaction of these two primary sets:

1. "Progressive modifications of the environment, inorganic and organic, which the actions of societies effect"
2. "The increasing size of the social aggregate, accompanied, generally, by increasing density"
3. "The reciprocal influence of the society and its units"
4. "The influence of the super-organic environment—the action and reaction between a society and neighboring societies"
5. "Accumulation of super-organic products" —that is, "material appliances," "development of knowledge," "systems of laws," etc.

This last set of secondary factors, according to Spencer, becomes in effect a new social environment:

All these various orders of super-organic products . . . are ever modifying individuals and modifying society, while being modified by both. They gradually form what we may consider either as a non-vital part of the society itself, or else as an additional environment, which eventually becomes even more important than the original environments.[89]

It is hard to imagine a more complete statement of the interactionist position in sociology.

James's portrayal of Spencerian sociology as naive externalism was thus unfair and misleading. Even if he had not read Spencer's 1877 volume (which he did not cite), he had certainly encountered the same idea in Fiske's *Outlines of Cosmic Philosophy*. Spencer had even made similar points—albeit in passing—in *Principles of Psychology*, the book James assigned to his classes in 1876–77 and 1878–79: "The conditions to which we must be re-adapted are themselves changing. Each further modification of human nature makes possible a further social modification. The environment alters along with alteration of the constitution. Hence there is required re-adjustment upon re-adjustment."[90] James thus either overlooked or deliberately ignored the explicitly interactionist orientation of Spencerian sociology, focusing on the crude

---

89. Herbert Spencer, *The Principles of Sociology*, vol. 1 (London: Williams & Norgate, 1877), 10–15. According to Spencer's preface, this installment of the book was issued to subscribers in June 1874.

90. Herbert Spencer, *The Principles of Psychology*, 2nd ed., vol. 1 (London: Williams & Norgate, 1870), 284.

determinism of Allen rather than the more sophisticated approach of Spencer and Fiske.

Since James could not fairly peg them as simple-minded geographical determinists, his disagreement with the Spencerian sociologists boiled down to the relative importance not of intrinsic factors but of "great men": Fiske said that sociology was "primarily concerned with *institutions* rather than with *individuals*," and James replied that he would then rather be a "hero-worshipper" than a sociologist.[91] Allen replied in turn that James's personal preferences were beside the point: "As to the emphasis question, I don't know that it is quite enough to say that we *do* feel interested in such and such a question, and that that gives the question its importance. . . . The real question is, what *ought* we to interest ourselves in; what, to a philosophic mind, *ought* to seem the most important." James was engaged, as he himself put it, "in picking out from history our heroes and communing with their kindred spirits"; Allen and Fiske merely pointed out that if history and sociology were to become scientific and predictive, they needed to develop "general propositions" about when and where "great men" are likely to appear.[92]

Although James's distinction between the production and preservation of variation was useful and clarifying, it did not undermine Spencerian sociology. Nor did it threaten Spencer's general approach to evolution, since Spencer could admit the distinction but still insist that we ought to study the origin of variation: after all, unknown causes need not remain so. In taking this position, the Spencerians had the naturalists on their side: Darwin had written his own two-volume work on variation, which James reviewed; Cope was searching "for the causes of the origin of the fittest"; and Wallace, as noted by Allen, claimed to have identified the source of color variation in birds and butterflies.[93] As we will see in more detail in chapter 5, the origin of variation was a central problem in biology at the time; variation could not be set to one side as merely "spontaneous" or "fortuitous." Characterizing James's 1878–81 critique as an attack on Spencer's "Lamarckism" does not help his case: all of the points James made against direct adaptation were

---

91. John Fiske, "Sociology and Hero-Worship," *Atlantic Monthly* 47 (1881): 82; William James, "The Importance of Individuals," *Open Court*, August 7, 1890, 2439.
92. Grant Allen to William James, 6 April 1881, in James, *Correspondence*, 5:159; James, "Importance of Individuals," 2439; John Fiske, "Sociology and Hero-Worship," *Atlantic Monthly* 47 (1881): 82. James sent his then-unpublished reply to Allen in 1881, prompting Allen's letter.
93. Charles Darwin, *The Variation of Animals and Plants under Domestication*, 2 vols. (London: John Murray, 1868); Edward Drinker Cope, "The Method of Creation of Organic Forms," *Proceedings of the American Philosophical Society* 12 (1871): 230; Alfred Russel Wallace, *Tropical Nature, and Other Essays* (London: Macmillan, 1878), 158–220.

consistent with the inheritance of acquired characters, which most naturalists (including Darwin) accepted at the time. The Spencerian sociologists were in fact defending the more plausible view: when it comes to social change, it really is geography, history, culture, and institutions that matter.[94]

Several years later, James merged his criticisms of Spencer's psychology and sociology in the final chapter of his own *Principles of Psychology*, published in 1890. James argued in this chapter, drafted in 1885, that "the experience of the race can no more account for our necessary or *a priori* judgments than the experience of the individual can." To support this thesis, James needed his earlier distinction between the production and preservation of variation. He began with a concession: Spencer's environmentalist account of the mind is true, but only for the special case of "time- and space-relations":

> Here the mind is passive and tributary, a servile copy, fatally and unresistingly fashioned from without. . . . The degree of cohesion of our inner relations, is, in this part of our thinking, proportionate, in Mr. Spencer's phrase, to the degree of cohesion of the outer relations; the causes and the objects of our thought are one; and we are, in so far forth, what the materialistic evolutionists would have us altogether, mere offshoots and creatures of our environment, and naught besides.

James claimed that the evolution of this aspect of human intelligence, which we have in common with other animals, was equivalent to a steady improvement in the correspondence between relations in the environment and relations in our minds. That is, the "experience of the race" can account for judgments of this general sort.[95]

However, James thought Spencer's approach inadequate when it came to abstraction, classification, logic, aesthetic appreciation, and other more advanced forms of judgment. Referring to his recent critique of Spencer, James argued that the external environment was not the direct cause of these judgments. Instead, proposed James, they may "be pure *idiosyncrasies*, spontaneous variations, fitted by good luck (those of them which have survived) to take cognizance of objects (that is, to steer us in our active dealings with them), without being in any intelligible sense immedi-

---

94. John Luke Gallup, Jeffrey D. Sachs, and Andrew D. Mellinger, "Geography and Economic Development," *International Regional Science Review* 22 (1999); Daron Acemoglu, Simon Johnson, and James A. Robinson, "The Colonial Origins of Comparative Development: An Empirical Investigation," *American Economic Review* 91 (2001).

95. William James, *The Principles of Psychology*, 2 vols. (New York: Henry Holt, 1890), 2:617–18, 632, 686.

ate derivatives from them." Sometimes, said James, experience directly teaches the mind by impressing its order upon it, as in the case of time and space relations. But more often, he suggested, what does the work are "indirect causes of mental modification—causes of which we are not immediately conscious as such, and which are not the direct *objects* of the effects they produce." Most of the interesting aspects of the human mind, according to James, stem from the latter kind of process: "Our higher aesthetic, moral, and intellectual life seems made up of affections of this collateral and incidental sort, which have entered the mind by the back stairs, as it were, or rather have not entered the mind at all, but got surreptitiously born in the house." For James, the higher mental life of human beings is the product of spontaneous variations, only some of which have been preserved.[96]

But what determines which variations survive? The environment, at least in part: James implied that if variations do not "steer us in our active dealings with [objects]," they will not persist. Thus "natural selection" of helpful thoughts should produce a rough correspondence between mind and environment, even without direct adaptation. However, as in his earlier critique of Spencer's psychology, James claimed that our subjective interests—especially our need for system—also play a key role, undermining the correspondence:

The popular notion that "Science" is forced on the mind *ab extra* [from outside], and that our interests have nothing to do with its constructions, is utterly absurd. The craving to believe that the things of the world belong to kinds which are related by inward rationality together, is the parent of Science as well as of sentimental philosophy.

That is, James insisted that the "rational order of comparison" is part of the selection process; thus, it is not adequate to say that our knowledge is shaped by the environment, whether that shaping is direct or indirect.[97]

Although Alexander Klein has interpreted this argument as a critique of Spencer's "Lamarckism," I think this interpretation neglects the structure of James's chapter.[98] It is divided into two parts: the first, as just discussed, argued that "the *theoretic* part of our organic mental structure ... can be due neither to our own nor to our ancestor's experience";

---

96. James, *Principles of Psychology*, 2:627, 631.
97. James, *Principles of Psychology*, 2:667, 676.
98. Alexander Klein, "Was James Psychologistic?" *Journal for the History of Analytical Philosophy* 4, no. 5 (2016): 11–15.

CHAPTER TWO

the second argued the same for the "practical parts of our organic mental structure"—that is, for our instinctive behavior. As I will discuss in chapter 5, the second part was clearly an argument against Spencer's Lamarckian theory of instinct, according to which instincts are "'secondarily-automatic' habits."[99] The argument of the first part, however, did not depend on a critique of Lamarckism. It claimed that direct adaptation to ancestral environments could not have been the cause of the more advanced forms of human judgment, not because such acquired forms of judgment could not have been inherited or because mental variations must be treated as spontaneous, but rather because conscious choice and a "rational order of comparison" were essential to the evolution of these forms. James did not oppose Spencer because his philosophy of evolution was Lamarckian; he opposed him because his theory of evolutionary variation neglected the importance of human interests and human reason. Thus, one could even argue that James's theory of mental evolution, despite its rejection of the environment as producer and its purported commitment to spontaneity, involved a kind of interest-directed variation, analogous to that endorsed—as we will see in chapter 5—by American neo-Lamarckians such as Cope and Peirce.

Everyone associated with the Metaphysical Club, apart from Peirce, engaged closely with Spencer's philosophy of evolution in the 1860s and '70s. Although they disagreed in their evaluation of the English philosopher, they all attempted to criticize him on his own terms—from a naturalistic point of view. This approach was obvious for Wright, Abbot, and Fiske, who were all positivists in the broad sense.[100] But even James focused on the scientific aspects of the philosophy of evolution and admitted that his own research in psychology had been shaped by Spencer's naturalistic framework. He wrote to his publisher Henry Holt in the midst of his late 1870s attacks on Spencer:

So far am I from leaving out the environment, that I shall call my text-book "Psychology, as a Natural Science," and have already in the introduction explained that the constitution of our mind is incomprehensible without reference to the external circumstances in the midst of which it grew up. My quarrel with Spencer is not that he makes

---

99. William James, *The Principles of Psychology*, 2 vols. (New York: Henry Holt, 1890), 2:681; see Herbert Spencer, *The Principles of Psychology*, 2nd ed., vol. 1 (London: Williams & Norgate, 1870), 432–43.

100. Trevor Pearce, "'Science Organized': Positivism and the Metaphysical Club, 1865–1875," *Journal of the History of Ideas* 76 (2015).

much of the environment but that he makes *nothing* of the glaring and patent fact of subjective interests which cooperate with the environment in moulding intelligence.

James made good on his promise: in his chapter 1 as published, he contrasted the fertility of Spencer's naturalistic approach with that of traditional psychology:

> On the whole, few recent formulas have done more real service of a rough sort in psychology than the Spencerian one that the essence of mental life and of bodily life are one, namely, "the adjustment of inner to outer relations." Such a formula is vagueness incarnate; but because it takes into account the fact that minds inhabit environments which act on them and on which they in turn react; because, in short, it takes mind in the midst of all its concrete relations, it is immensely more fertile than the old-fashioned "rational psychology," which treated the soul as a detached existent, sufficient unto itself, and assumed to consider only its nature and properties.

James's critique of Spencer's psychology, as we have seen, was not that he was too naturalistic or scientific, but rather that he neglected certain facts about subjective interests, mental activity, and consciousness.[101]

To highlight his strategic adoption of what he would eventually call the "natural history point of view," James often explicitly bracketed his own metaphysical stance. Discussing his account of the evolution of consciousness, for example, he insisted that "free-will is in short, no necessary corollary of giving causality to consciousness. My phrase about choosing one's own character is perfectly consistent with fatalism." In sum, although James was criticizing the *"a posteriori* school," he was doing it with respectably naturalistic arguments: "The antithesis between inner and outer may subsist on a purely natural plane and a Philosophy accentuating the inner element be true without in any sense being a supernatural Philosophy." Thus, like Wright and Abbot, he ultimately opposed Spencer's evolutionism from the viewpoint of a broader evolutionary naturalism.[102]

Chapters 1 and 2 of this book have described how the first cohort of philosophers later associated with pragmatism—Wright, Peirce, Abbot, James, and Fiske—reacted to the work of the two most famous late

---

101. William James to Henry Holt, 22 November 1878, in James, *Correspondence*, 5:24–25; William James, *The Principles of Psychology*, 2 vols. (New York: Henry Holt, 1890), 1:6.

102. William James to James Jackson Putnam, 17 January 1879, in James, *Correspondence*, 5:34; "Notes for Philosophy 4: Psychology (1878–1879)," in James, *Manuscript Lectures*, 136. For the "natural history point of view," see William James, *The Will to Believe, and Other Essays in Popular Philosophy* (New York: Longmans, Green, 1897), 116, 128, 187.

nineteenth-century evolutionists, Darwin and Spencer. The positivists Wright, Fiske, and Abbot, as we saw in chapter 1, immediately embraced Darwin's ideas about the natural world and wrote extensively about evolution in the 1860s and early 1870s. Peirce and James, on the other hand, had close ties to Agassiz and did not explicitly endorse Darwin until later. During the same period, Wright, Abbot, and Fiske engaged with Spencer's philosophy of evolution, reviewing his various books in the 1860s. Fiske and Wright, however, were split on Spencer: Fiske became one of his key American disciples, whereas Wright criticized his view of evolution as overly progressive. James began teaching and writing about Spencer in the mid-1870s, several years after the first meetings of the Metaphysical Club. He criticized Spencer's psychology for ignoring the mental phenomena of interests and selective attention and attacked his sociology—along with that of his followers—for overemphasizing the role of the environment in directing social change.

This first cohort graduated from college in the 1850s and '60s, and its members were thus involved in the initial debates over the evolutionary views of Darwin and Spencer. These views still loomed large for the second cohort, whose members are the subject of the next two chapters, but their experience was quite different: by the time they got to college in the 1870s and '80s, evolution was literally in the textbooks. In chapter 3, I will explore how Royce, Dewey, Addams, Mead, and Du Bois were exposed to evolutionary ideas in college and graduate school, and how they later deployed these ideas in their early teaching.

THREE

Evolution at School: Educating a New Generation

When Christine Ladd-Franklin (née Ladd) attended Vassar College in the 1860s, students were required to take several natural history courses, including zoology and geology. She developed an unexpected interest in the latter: as she wrote in her diary in 1866, "it is a daily wonder to me how I came to study Geology. I have less taste for it than for any of the other sciences, and I never intended to know more than enough to enable me to graduate, yet now I am pursuing independent investigations and really getting up some enthusiasm in the subject." Ladd's geology class was taught by Sanborn Tenney, who was a student of Louis Agassiz, the staunch opponent of Charles Darwin whom we met in chapter 1. It should thus come as no surprise that the assigned textbooks were explicitly opposed to evolution: James Dwight Dana's *Manual of Geology* stated that geology "has brought to light no facts sustaining a theory that derives species from others, either by a system of evolution, or by a system of variations of living individuals, and bears strongly against both hypotheses"; Tenney's own *Geology* flatly declared, "there is no such thing as the development of lower species into higher."[1]

1. Christine Ladd-Franklin, "Diary, 1866–1873," p. 03a, Box 135, Vassar College Student Materials Collection, Archives and Special Collections, Vassar College, https://digitallibrary.vassar.edu/islandora/object/vassar%3A2708; *Second Annual Catalogue of the Officers and Students of Vassar College, Poughkeepsie, N. Y., 1866–67* (New York: John A. Gray & Green, 1867), 5, 8, 28, 30; James D. Dana, *Manual*

CHAPTER THREE

As we saw in chapter 1, the other members of the first cohort of pragmatists heard a similar story in the 1850s at Harvard, where Agassiz dominated. But things were different for the second cohort: by the time they attended college in the 1870s and '80s, evolution had been accepted by the majority of naturalists, including Dana (the author of Ladd's textbook) and even many of Agassiz's own students. Whereas for the first cohort of pragmatists, evolution was something about which one could ask, as William James did in 1865, "whether it may not be destined eventually to prevail," for the second cohort it was the standard scientific account of the history of life, despite continuing controversy over its causes (the topic of chapter 5).[2]

In the first part of this chapter, I will demonstrate that the older members of the second cohort of pragmatists—specifically Josiah Royce, John Dewey, Jane Addams, George Herbert Mead, and W. E. B. Du Bois—were not only exposed to evolutionary ideas in college but were also interested in these ideas. Although they went on to graduate work in philosophy and the social sciences rather than biology, they continued to encounter debates about the significance of evolution, as I will show in the second part of the chapter. Finally, all of them went on to teach their own students about biology and evolution, ensuring that the next generation of philosophers and social scientists would take natural history seriously.

Evolution in College

American students today can major in engineering, business, or philosophy without taking classes in biology or paleontology, but this was not the case in the nineteenth century. The pragmatists were obligated to take

---

of Geology: Treating of the Principles of the Science, with Special Reference to American Geological History, for the Use of Colleges, Academies, and Schools of Science, rev. ed. (Philadelphia: Theodore Bliss, 1866), 602; Sanborn Tenney, *Geology: For Teachers, Classes, and Private Students* (Philadelphia: E. H. Butler, 1860), 279. For Tenney's study with Agassiz, see *A Catalogue of the Officers and Students of Harvard University, for the Academical Year 1857–58, Second Term* (Cambridge, MA: John Bartlett, 1858), 68; *A Catalogue of the Officers and Students of Harvard University, for the Academical Year 1858–59, First Term* (Cambridge, MA: John Bartlett, 1858), 68.

2. William James, "Huxley's Comparative Anatomy," *North American Review* 100 (1865): 291; Ronald L. Numbers, *Darwinism Comes to America* (Cambridge, MA: Harvard University Press, 1998), 24; Jon H. Roberts, "Louis Agassiz on Scientific Method, Polygenism, and Transmutation: A Reassessment," *Almagest* 2 (2011): 96; William F. Sanford, "Dana and Darwinism," *Journal of the History of Ideas* 26 (1965): 537; Edward Lurie, *Louis Agassiz: A Life in Science* (Chicago: University of Chicago Press, 1960), chap. 8; Mary P. Winsor, *Reading the Shape of Nature: Comparative Zoology at the Agassiz Museum* (Chicago: University of Chicago Press, 1991), 37–42.

a set curriculum as undergraduates, with few if any electives. This curriculum included classical languages and literature (Greek, Latin); modern languages and literature (English, French, German); mathematics, physics, astronomy, chemistry, physiology, and natural history (botany, zoology, mineralogy, geology); history and political economy; philosophy (logic, psychology, morals); and religion (the Bible and Christian evidences). So even though Royce, Dewey, Addams, Mead, and Du Bois attended very different institutions, the content of their college education was roughly the same.[3]

Royce and the rest were thus required to take a series of natural history classes. These probably provided their first introduction to evolutionary ideas, which were discussed by their textbooks and defended by their teachers. There were four subjects in natural history: botany, zoology, mineralogy, and geology. None of their botany textbooks—neither Asa Gray's various texts (used by Royce, Dewey, and Mead) nor Alphonso Wood's *Class Book of Botany* (used by Addams and Du Bois) addressed evolution.[4] This is somewhat surprising in light of Gray's support for Darwin's ideas, noted in chapter 1. However, the original editions of the Gray and Wood textbooks were written before Darwin's *Origin of Species* appeared, and neither author added evolution into later editions. For example, both the 1857 and 1868 editions of Gray's *Lessons in Botany* contained the following declaration: "the Creator established a definite number of species at the beginning."[5]

Their textbooks in zoology and geology, on the other hand, did discuss evolution, and sometimes in detail. Royce's zoology class at California still used Agassiz's textbook, which was firmly opposed to evolution (as we saw in chapter 1) and only revised once, in 1851. Dewey and Du Bois were assigned the more up-to-date *Elements of Zoology*, published in 1875 by Sanborn Tenney (Ladd's teacher at Vassar). Although Tenney

---

3. For courses and assigned textbooks, see the *Register of the University of California* from 1871 to 1875 (Royce); the *Catalogue of the Officers and Students of the University of Vermont* from 1875 to 1879 (Dewey); the *Annual Catalogue of the Officers and Students of Rockford Seminary* from 1877 to 1881 (Addams); the *Catalogue of the Officers and Students of Oberlin College* from 1879 to 1883 (Mead); and the *Catalogue of the Officers and Students of Fisk University* from 1885 to 1888 (Du Bois). Information on when Addams took which classes was taken from her academic transcript, held in the Jane Addams Collection at the Rockford University Archives. Information on Du Bois's classes and textbooks was cross-checked with the four-page list accompanying W. E. B. Du Bois to Secretary Harvard College, 29 October 1887, Du Bois Folder.
4. Botany was an optional class at California, but Royce elected to take it: see "[Science and other course notes, Profs. Joseph and John LeConte and Prof. Carr]," Folder 5, Box 114, Royce Papers.
5. Asa Gray, *First Lessons in Botany and Vegetable Physiology* (New York: Ivison & Phinney, 1857), 173; Asa Gray, *Gray's Lessons in Botany and Vegetable Physiology* (New York: Ivison, Blakeman, Taylor, 1868), 173.

had remained silent about evolution in the revised edition of his earlier zoology textbook, published only three years before, he was led by changing scientific opinion to include the following footnote in *Elements of Zoology* despite his continued opposition to the theory:

> The doctrine held by many distinguished naturalists—that in a certain sense *all* animals have had a common origin, that is, that our present "species" have been evolved from other and earlier species, . . . cannot be dwelt upon here. For the views of these naturalists, as to the "Origin of Species," and as to the doctrine of "Evolution," see the writings of Darwin, Huxley, Mivart, Cope, [Alpheus] Hyatt, [Theodore] Gill, and others.

Mead was assigned Henry Alleyne Nicholson's *Manual of Zoology*, which was also a recommended reference (along with Tenney's book) for Dewey's classes. Whereas Tenney consigned evolution to the footnotes, Nicholson devoted an entire section of his book to the "Origin of Species." Although he noted—writing in 1869—that "the opinions of scientific men are still divided upon this subject," he provided a step-by-step description of "the doctrine of the development of species by variation and natural selection—propounded by Darwin, and commonly known as the Darwinian theory."[6]

Although there was no zoology class in the college curriculum at Rockford, Addams would almost certainly have learned about evolution in geology, which she and the other pragmatists took during their final year of college. This class was often restricted to seniors because of its potential to conflict with the biblical narrative in Genesis, including the Creation and the Flood. Geology was divided at the time into three main areas: structural geology, dynamical geology, and historical geology. The latter subject, covered by all textbooks, was equivalent to paleontology—the study of the fossil record. Royce, Dewey, Addams, and Du Bois were all assigned one of Dana's various textbooks, which had been updated in the 1870s to reflect his belated acceptance of evolution. Thus, the 1866 edition of Dana's *Manual of Geology* (used by Ladd at Vassar) declared that geology contained "no facts sustaining a theory that derives species

---

6. Louis Agassiz and Augustus Addison Gould, *Principles of Zoölogy: Touching the Structure, Development, Distribution, and Natural Arrangement of the Races of Animals, Living and Extinct, with Numerous Illustrations*, rev. ed. (Boston: Gould & Lincoln, 1851); Sanborn Tenney, *Natural History: A Manual of Zoology for Schools, Colleges, and the General Reader*, rev. ed. (New York: American, 1872); Sanborn Tenney, *Elements of Zoology: A Text-Book* (New York: Scribner, Armstrong, 1875), 21n; Henry Alleyne Nicholson, *A Manual of Zoology for the Use of Students, with a General Introduction on the Principles of Zoology*, 2nd ed. (New York: D. Appleton, 1876), 39–40; also in Henry Alleyne Nicholson, *A Manual of Zoology for the Use of Students, with a General Introduction on the Principles of Zoology* (Edinburgh: William Blackwood & Sons, 1870), 37–39.

from others," but the 1874 edition stated instead that "the evolution of the system of life went forward through the derivation of species from species, according to natural methods not yet clearly understood, and with few occasions for supernatural intervention."[7]

Although Dewey recalled using Dana's *Manual*, the college catalog indicates that he was assigned Joseph LeConte's recently published *Elements of Geology*, which was also used by Mead. According to LeConte, evolution was "the central idea" of historical geology, and he defined *geology* more broadly as "the history of the evolution of the earth and its inhabitants." LeConte was keenly interested in the dynamics of evolutionary change. He argued that the primary mode of evolution was "a gradual, *extremely slow* evolution of organic forms under the operation of all the forces and factors of evolution known and unknown." Gradual evolution takes "different directions in different places and under different physical conditions," and so it tends to increase geographical diversity. But this increase is tempered in critical periods by "*physical changes* and consequent *migrations*," which lead to more rapid evolution caused by "*severer pressures of external conditions*, . . . *severer struggle for life*, . . . and the more active operation of *other factors of change*, which we do not yet understand." Thus, although LeConte highlighted human ignorance about many of the factors of evolution, he was committed to an evolutionary account of historical geology.[8]

Most of the early second-cohort pragmatists also took physiology, which was not considered part of natural history. Dewey recalled that his physiology class at Vermont led him "to desire a world that had the same properties as had the human organism." But unlike those assigned in zoology and geology, physiology textbooks did not normally discuss evolution. Despite Thomas Henry Huxley's role as "Darwin's bulldog," his textbook (used by Dewey) did not even mention the topic; nor did that of Joseph Chrisman Hutchison (used by Addams). Henry Newell Martin's textbook (used by Mead and Du Bois) did occasionally explain how "those who accept the doctrine of evolution" would account for

---

7. John Dewey and Jane M. Dewey, "Biography of John Dewey," in *The Philosophy of John Dewey*, ed. Paul Arthur Schilpp (Evanston, IL: Northwestern University Press, 1939), 10; William F. Sanford, "Dana and Darwinism," *Journal of the History of Ideas* 26 (1965): 537; James D. Dana, *Manual of Geology: Treating of the Principles of the Science, with Special Reference to American Geological History, for the Use of Colleges, Academies, and Schools of Science*, rev. ed. (Philadelphia: Theodore Bliss, 1866), 602; James D. Dana, *Manual of Geology: Treating of the Principles of the Science, with Special Reference to American Geological History*, 2nd ed. (New York: Ivison, Blakeman, Taylor, 1874), 603–4; also in James D. Dana, *New Text-Book of Geology*, 4th ed. (New York: American, 1883), 393.

8. Dewey and Dewey, "Biography of John Dewey," 10; Joseph LeConte, *Elements of Geology: A Text-Book for Colleges and for the General Reader* (New York: D. Appleton, 1878), iv, 1, 396, 553–54.

some particular physiological feature, but Martin did not describe this doctrine any further.[9]

Textbooks are not always a sure guide to the perspective of a course: as discussed in chapter 2, William James used to assign Herbert Spencer's books in order to attack them. However, the natural history teachers of Royce, Dewey, Addams, Mead, and Du Bois all appear to have supported evolution—as we might expect, given its general acceptance among naturalists by the mid-1870s. Royce's teacher at California was Joseph LeConte, author of Mead's geology textbook. Although LeConte had been a student of Agassiz, by the time he taught zoology and geology to Royce—from 1872 to 1875—he had accepted evolution. As Royce later recalled, "[LeConte's] transition to the acceptance of the general doctrine of evolution seems to have occurred early in his California period [circa 1870]. Within a few years it had determined the whole character of his lectures. I myself heard him speak as a convinced evolutionist." Royce thus learned his natural history from an evolutionist in the early 1870s, and twenty-first-century scholarship has shown that LeConte—who owned a slave plantation in Georgia before the Civil War—also influenced Royce's later discussion of "race questions."[10]

Dewey was taught natural history by George Henry Perkins, who also supported evolution. Perkins's work cut across the standard divisions of the subject: after writing a dissertation on mollusks under Addison Emery Verrill—another student of Agassiz who eventually endorsed evolution—Perkins studied everything from flowering plants to archaeology. His specialty during Dewey's years at Vermont was agricultural biology, and he

---

9. John Dewey, "From Absolutism to Experimentalism," in *Contemporary American Philosophy: Personal Statements*, ed. George Plimpton Adams and William Pepperell Montague, vol. 2 (New York: Russell & Russell, 1930), 13; Thomas Henry Huxley and William Jay Youmans, *The Elements of Physiology and Hygiene*, rev. ed. (New York: D. Appleton, 1873); Joseph Chrisman Hutchison, *A Treatise on Physiology and Hygiene for Educational Institutions and General Readers* (New York: Clark & Maynard, 1875); Henry Newell Martin, *The Human Body: An Account of Its Structure and Activities and the Conditions of Its Healthy Working* (New York: Henry Holt, 1881), 77, 462; Frank Fanning Jewett and Frances Gulick Jewett, "The Chemical Department of Oberlin College from 1833 to 1912," *Oberlin Alumni Magazine* 18 (1922): 7. For "Darwin's bulldog," see Henry Fairfield Osborn, "Memorial Tribute to Professor Thomas H. Huxley," *Transactions of the New York Academy of Sciences* 15 (1896): 47.

10. Josiah Royce, "Joseph Le Conte," *International Monthly* 4 (1901): 332; see also Josiah Royce, "Words of Professor Royce at the Walton Hotel at Philadelphia, December 29, 1915," *Philosophical Review* 25 (1916): 509; Lester D. Stephens, *Joseph LeConte: Gentle Prophet of Evolution* (Baton Rouge: Louisiana State University Press, 1982), 69, 80–81, 94, 113. For LeConte and Royce on race, see Joseph LeConte, "The Race Problem in the South," in *Man and the State: Studies in Applied Sociology* (New York: D. Appleton, 1892); Josiah Royce, "Race Questions and Prejudices," *International Journal of Ethics* 16 (1906); Marilyn Fischer, "Locating Royce's Reasoning on Race," *Pluralist* 7 (2012): 111–18; Tommy J. Curry, *Another White Man's Burden: Josiah Royce's Quest for a Philosophy of White Racial Empire* (Albany: State University of New York Press, 2018), chap. 4.

gave a series of talks to farmers in the 1870s on insect pests and the parasites of farm animals.[11] Although most of Perkins's writings were silent on the topic, Dewey later recalled that Perkins, in his zoology and geology classes, "ordered his presentation of material on the theory of evolution." This recollection is corroborated by a lecture Perkins delivered to medical students a few years after Dewey left:

The modern theories of evolution have done great things for medicine, and will do far more in the future. They have put in action forces that may revolutionize medical science. Evolution has shown as nothing else could, how profoundly animals are affected by their environment, their food, habits, climate, etc., and, by showing how inevitable is the modification of structure in other animals, has called attention to the same facts in man's existence.

Perkins thus introduced Dewey to the study of biology and natural history from an evolutionary point of view.[12]

Addams's natural history teacher at Rockford was Mary Emilie Holmes, who later earned her doctorate in paleontology at the University of Michigan and founded Mary Holmes Seminary, a Presbyterian school (named after her mother) for young black women in Mississippi. Holmes was the first American woman to earn a doctoral degree in any area of geology. Aspiring women scientists faced serious resistance at the time: for example, although Holmes defended her thesis in 1886, she was initially only granted an MA; her PhD was conferred belatedly in 1888, a year after the thesis was published. Although there are no direct indications of Holmes's position on evolution during Addams's time at Rockford (1877–81), Addams later recalled, speaking of "the theory of evolution," that "our science teacher had accepted this theory."[13]

---

11. Edward Lurie, *Louis Agassiz: A Life in Science* (Chicago: University of Chicago Press), 358; Wesley R. Coe, "Biographical Memoir of Addison Emery Verrill," *Biographical Memoirs of the National Academy of Sciences* 14 (1929): 39; George Henry Perkins, "Molluscan Fauna of New Haven," *Proceedings of the Boston Society of Natural History* 13 (1869); George Henry Perkins, "Certain Internal Parasites of Domestic Animals," in *Eighth Annual Report of the Secretary of the Vermont Dairymen's Association* (Montpelier, VT: Poland, 1877); George Henry Perkins, "On Some of the Injurious Insects of Vermont," in *Fifth Report of the Vermont Board of Agriculture* (Montpelier, VT: Poland, 1878); George Henry Perkins, "General Remarks upon the Archaeology of Vermont," *Proceedings of the American Association for the Advancement of Science* 27 (1878); George Henry Perkins, *A General Catalogue of the Flora of Vermont* (Montpelier, VT: Freeman, 1882).

12. John Dewey and Jane M. Dewey, "Biography of John Dewey," in *The Philosophy of John Dewey*, ed. Paul Arthur Schilpp (Evanston, IL: Northwestern University Press, 1939), 10; George Henry Perkins, "The Physician of the Future," *Popular Science Monthly* 21 (1882): 639.

13. *Order of Examinations for Higher Degrees and for the Bachelors' Degrees on the University System* ([Ann Arbor]: University of Michigan, 1886), 6; *Calendar of the University of Michigan for 1886–87* (Ann Arbor: University of Michigan, 1887), 157; Mary Emilie Holmes, *The Morphology of the*

Holmes did employ evolutionary arguments in her doctoral thesis, which focused on an obscure morphological feature of the Rugosa, an extinct order of corals. The rugose coral polyp—like that of modern stony corals—was contained in a hard skeletal cup (or corallite). Solitary rugose corallites were shaped like drinking horns—thus their common name, "horn corals." As they grew, rugose polyps would slowly lengthen their corallite, creating the rings visible along the "horn" in figure 8. These corallites, like those of other orders, were divided into vertical compartments by radial plates called *septa*, the top edges of which are visible at the open end of the horn in figure 8, looking something like the spokes of a wheel. Many species also had *carinae*, a series of vertical flanges protruding from the face of each septum. In transverse sections of the corallite (i.e., circular slices through the horn), carinae are visible as hash marks perpendicular to the "spokes" of the septa (marked "a" in figure 8).[14] Assuming an evolutionary perspective, Holmes claimed in her thesis that the carinae—as a morphological character present in many rugose coral species across several geological periods—must have had some "functional value," since "in the survival of the fittest, any structure merely decorative would not have perpetuated itself." She suggested that the primary function of the carinae was "furnishing a support" to the soft body of the living polyp, a view shared by modern paleontologists. Thus, although Holmes's evolutionary views during her tenure at Rockford are uncertain, she had definitely embraced evolution a few years later and may even have chosen to study at Michigan because of her familiarity with the popular evolutionary writings of Alexander Winchell, the geologist (and racist pre-Adamite) who became her PhD supervisor.[15]

Carinae upon the Septa of Rugose Corals (Boston: Bradlee Whidden, 1887); *Proceedings of the Board of Regents, 1886–91* ([Ann Arbor]: [University of Michigan]), 203, https://hdl.handle.net/2027/mdp.35112204232435; *Calendar of the University of Michigan for 1888–89* (Ann Arbor: University of Michigan, 1889), 160; Margaret W. Rossiter, "Geology in Nineteenth-Century Women's Education in the United States," *Journal of Geological Education* 29 (1981): 231; Mary Emilie Holmes, "Mary Holmes Seminary," *Church at Home and Abroad* 11 (1892); *27th Annual Report of the Board of Missions for Freedmen of the Presbyterian Church in the United States of America: Presented to the General Assembly, May, 1892* (Pittsburgh: James McMillin), 97; "News from the Classes," *Michigan Alumnus* 5 (1899): 396; Jane Addams, *Twenty Years at Hull-House, with Autobiographical Notes* (New York: Macmillan, 1910), 62.

14. Holmes, *Morphology of the Carinae*, 7–8; Dorothy Hill, *Coelenterata: Rugosa and Tabulata*, 2 vols., Part F, Supplement 1 of *Treatise on Invertebrate Paleontology* (Boulder, CO: Geological Society of America, 1981), 1:6–36.

15. Holmes, *Morphology of the Carinae*, 12, 22; cf. William A. Oliver Jr. and James E. Sorauf, "The Genus *Heliophyllum* (Anthozoa, Rugosa) in the Upper Middle Devonian (Givetian) of New York," *Bulletins of American Paleontology* 362 (2002): 5, 11; Alexander Winchell, *The Doctrine of Evolution: Its Data, Its Principles, Its Speculations, and Its Theistic Bearings* (New York: Harper & Brothers, 1874);

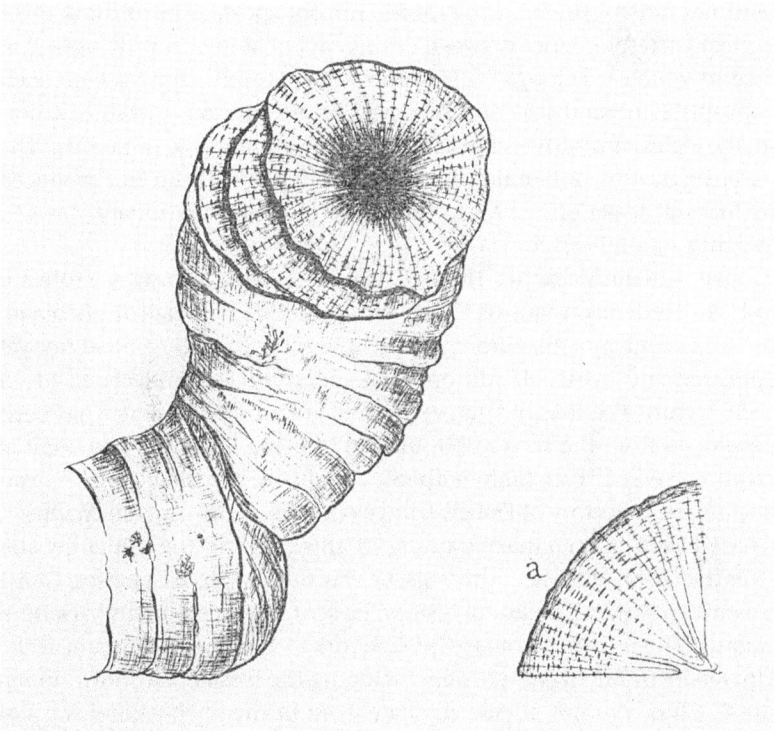

Figure 8 *Heliophyllum halli*, a rugose coral fossil, including part of a transverse section (marked "a").
Detail from Plate 3 of Mary Emilie Holmes, *The Morphology of the Carinae upon the Septa of Rugose Corals* (Boston: Bradlee Whidden, 1887). Reproduced courtesy of the University of Chicago Library.

Mead learned natural history from Albert Allen Wright, also a geologist. Wright's endorsement of evolution was unambiguous: in a lecture at Oberlin after Darwin's death in 1882, he praised his "inductive method" of developing theories from a great number of facts and spoke of the "almost universal acceptance . . . by working naturalists" of his theories. Wright's classes also contained some discussion of evolution. Notes taken in Mead's zoology class by his friend Henry Northrup Castle confirm that it covered the topic: in his second lecture, Wright presented Spencer's law of evolution as well as Ernst Haeckel's biogenetic law: "Simple preceeds [sic] complex or homogeneous by differentiation

---

David N. Livingstone, "The Preadamite Theory and the Marriage of Science and Religion," *Transactions of the American Philosophical Society* 82, no. 3 (1992): 40–52.

becomes heterogeneous. . . . Progress in zoology leads us to think there is great correspondence between ontog[eny] of animal & phyl[ogeny] of race to which it belongs."[16] That the students were familiar with such concepts as natural selection is proved by a humorous sketch of a normally legless primitive chordate in Castle's notebook (figure 9). The parenthetical remark under the sketch reads "N.B. These legs produced by 'natural selection.'"[17] Mead thus encountered evolutionary views in Wright's natural history classes.

Like Addams's teacher Holmes, Du Bois's natural history professor at Fisk—Frederick Augustus Chase—studied with Winchell at Michigan, but as an undergraduate from 1857 to 1859. Darwin's *Origin* had not yet appeared and Winchell still opposed evolution. As he declared in an 1858 lecture, the theory "that every grade of animal existence has been developed from the next lower, instead of being a distinct and original creation," was a "baneful hypothesis, . . . grossly and fatally false."[18] Paralleling the trajectory of Francis Ellingwood Abbot, discussed in chapter 1, Chase began his seminary training in 1860, during the initial debates over the *Origin of Species*. Although he was studying to be a pastor, Chase was still interested in natural history: he sent Devonian fossils back from Auburn Theological Seminary in New York to Winchell's museum at the University of Michigan, possibly including the fossil coral shown in figure 8. Chase did not pursue graduate work in the sciences and was not an active researcher; hence, it is difficult to tell what he thought about evolution. But if he took the approach of Edwin Hall, his theology professor at Auburn, he probably remained open-minded and followed the lead of religiously inclined naturalists such as Dana. As Hall wrote in 1866, citing Dana's *Manual* and addressing the conflict between geology and Genesis, "it is probable that we have not yet learned the exact interpretation of either book, the Earth or the Word. Instead of feeling ourselves bound to adopt either theory, we are rather privileged to wait for more light." Given its widespread support among naturalists (including

---

16. Henry Northrup Castle, "Notes on Zoology," p. 2, Folder 6, Box 1, Castle Papers; see Herbert Spencer, *First Principles of a New System of Philosophy*, 4th ed. (New York: D. Appleton, 1880), 396; Ernst Haeckel, *The History of Creation: Or the Development of the Earth and its Inhabitants by the Action of Natural Causes*, trans. Edwin Ray Lankester, 2 vols. (New York: D. Appleton, 1876), 1:212.

17. Castle, "Notes on Zoology," p. 45.

18. *Catalogue of the Officers and Students of the University of Michigan: 1858* (Ann Arbor: University of Michigan, 1858), 6, 19; *Catalogue of the Officers and Students for 1859* (Ann Arbor: University of Michigan, 1859), 6, 16; "Obituary [of Frederick Augustus Chase]," *American Missionary* 58 (1904): 152; Alexander Winchell, *Creation the Work of One Intelligence, and Not the Product of Physical Forces, Being the Closing Lecture of a Course upon Geology and Natural History, Delivered before the Young Men's Literary Association of Ann Arbor* (Ann Arbor: E. B. Pond, 1858), 15.

Figure 9 Sketch of Amphioxus, with legs added by Henry Castle.
From "Notes on Zoology," p. 45, Folder 6, Box 1, Castle Papers. Reproduced by permission of the Special Collections Research Center, University of Chicago Library.

Dana) by the mid-1870s, Chase would almost certainly have supported evolution in 1885–88, when Du Bois attended Fisk.[19]

After graduating, Du Bois—like Mead the year before—began studying for a second bachelor's degree at Harvard, taking a plurality of his courses in philosophy. But unlike Mead, Du Bois also took several courses in the sciences, including two geology classes in 1888–89. The professor of geology at Harvard was Nathaniel Southgate Shaler, another Agassiz student who (like his classmates Hyatt and Verrill) ultimately embraced evolution.[20] Harvard's geology curriculum began with two general courses, Natural History 4 and Natural History 8. Du Bois was excused from the lecture portion of the former because of his studies at Fisk, but he did the associated "laboratory and field exercises." He also enrolled in the latter, which was taught by Shaler (along with two assistant instructors) and included both lectures and fieldwork.[21]

Shaler was a racist and followed Agassiz in claiming that although "black children are surprisingly quick, ... with the maturing of the body

---

19. Alexander Winchell, *Report, Historical and Statistical, on the Collections in Geology, Zoölogy and Botany in the Museum of the University of Michigan, Made to the Board of Regents, Oct. 2d, 1863* (Ann Arbor: University of Michigan, 1864), 13 ("Fossils from the Hamilton Group, N. Y."); Mary Emilie Holmes, *The Morphology of the Carinae upon the Septa of Rugose Corals* (Boston: Bradlee Whidden, 1887), 15 ("Heliophyllum Halli E. and H., of the Hamilton"); William A. Oliver Jr. and James E. Sorauf, "The Genus *Heliophyllum* (Anthozoa, Rugosa) in the Upper Middle Devonian (Givetian) of New York," *Bulletins of American Paleontology* 362 (2002): 6–10; Edwin Hall, *Digest of Studies and Lectures in Theology* (Auburn, NY: Wm. J. Moses, 1866), 39.

20. Du Bois Transcript; Mead Transcript; *Quinquennial Catalogue of the Officers and Graduates of Harvard University, 1636–1900* (Cambridge, MA: Harvard University, 1900), 401–2; Ralph W. Dexter, "Three Young Naturalists Afield: The First Expedition of Hyatt, Shaler, and Verrill," *Scientific Monthly* 79 (1954): 51; David N. Livingstone, *Nathaniel Southgate Shaler and the Culture of American Science* (Tuscaloosa: University of Alabama Press, 1987), 58; Mary P. Winsor, *Reading the Shape of Nature: Comparative Zoology at the Agassiz Museum* (Chicago: University of Chicago Press, 1991), 37–42.

21. Du Bois Transcript; *The Harvard University Catalogue, 1888–89* (Cambridge, MA: Harvard University, 1888), 125.

the animal nature generally settles down like a cloud on that promise." He had a paternalistic attitude toward African Americans, arguing that they should be "thoughtfully cared for"—that is, "put under such conditions of training as shall open to the abler members of the race higher places in life than they now have a chance to fill." This was a "gentlemanly" racism, as was common among intellectuals at the time: Du Bois later recalled that Shaler "invited a Southerner, who objected to sitting by me, out of class."[22] Despite his racism, including the view that "in most cases the result of miscegenation is a feebler man than the unmixed descendants of the primitive stocks," Shaler was impressed by Du Bois. He expressed this opinion in an 1891 letter which, although filled with racist and backhanded compliments, supported one of Du Bois's scholarship applications:

I am inclined to believe he has Arab as well as negro "blood in his veins." His type of mind is rather Shemitic than Hamitic [see Genesis 10]: he is a person of very fair ability and may perhaps be regarded as distinctly promising. He seems to me decidedly the best specimen of his race we have had in our classes. . . . He is a sober minded, industrious and discreet person: singularly free from the self assertion so often found in educated negros. I think it would be well to give him a thorough training. Were I myself a man of fortune I should make the essay.

As this assessment suggests, Du Bois performed very well in Natural History 8, and we even have direct evidence that he learned about evolution in the class. In one of the last lectures of the 1888–89 academic year, as attested by Du Bois's own notes, Shaler discussed both the Lamarckian and Darwinian theories of evolution, concluding that there is "a measure of truth in both hypotheses" despite their respective shortcomings.[23]

Debates about evolution featured even more prominently in William James's class on Recent English Contributions to Theistic Ethics, which

---

22. Nathaniel Southgate Shaler, "The Negro Problem," *Atlantic Monthly* 54 (1884): 700, 708; Nathaniel Southgate Shaler, "Race Prejudices," *Atlantic Monthly* 58 (1886): 518; W. E. B. Du Bois, *Dusk of Dawn: An Essay toward an Autobiography of a Race Concept* (New York: Harcourt, 1940), 38. On Shaler's "gentlemanly" racism, see Joel Williamson, *The Crucible of Race: Black/White Relations in the American South since Emancipation* (New York: Oxford University Press, 1984), 119–21.
23. Nathaniel Southgate Shaler, "The African Element in America," *Arena* 2 (1890): 672; Nathaniel Southgate Shaler to Rutherford Birchard Hayes, 2 March 1891, Folder 28, Box 43, Series 1, Daniel Coit Gilman Papers (MS 0001), Special Collections, Johns Hopkins University; W. E. B. Du Bois, "Philosophy IV Notebook," Series 10, Du Bois Papers. Although most of this notebook contains notes from the second semester (February to May 1889) of Philosophy 4, the last three pages are notes from a late session (8 May 1889) of Natural History 8.

Du Bois also took in 1888–89. The second semester of the class focused on James Martineau's two-volume *Study of Religion*, which opposed positivism and attempted to reconcile scientific and religious thought. Martineau argued that "the Teleological interpretation of nature" was consistent with modern evolutionary ideas, criticizing John Fiske's dismissal of William Paley's famous design argument:

> If marks of Thought were truly found before, they have now become marks of larger and sublime thought; all that was detached having passed into coherence, so that one intellectual organism embraces the whole, from the animalcule in a dewdrop to the birth and death of worlds. I see no reason to doubt that Paley would have welcomed the new theory of organic life upon the globe, as a magnificent expansion of his idea.[24]

In a section titled "Place of Teleology," Martineau claimed that the fundamental marks of thought can indeed be found in nature, despite the denials of some evolutionists. As Du Bois's notes indicate, James covered this section in two lectures. Martineau argued that "accidental variation & the survival of the fittest" could not explain adaptation because "successful variations should be in the minority instead of [the] majority," implying that divine intelligence must play some role in evolution. James agreed that variation was a problem for Darwin but insisted that God's will could not be "deduce[d] from the phenomena logically." According to James, foreshadowing his later views, "Will" and "Chance" are options "we vote for . . . with our hearts"—effectively "synonymous terms" with respect to "all the facts of the world." Following up on these points at the end of the next lecture, James insisted that the idea of an "antithesis between God & mechanism . . . was the older view from which we are cut off by modern thought." Referring to mechanistic evolution, he concluded, "if we can conceive a God a Universe who uses this means, all right." Thus, Du Bois's natural history professors at both Fisk and Harvard supported evolution, and he also encountered the topic in one of James's classes.[25]

Like Martineau and James, the pragmatists' natural history professors were committed to the harmony of science and religion. Most of them endorsed Dana's view, expressed in the closing paragraph of his *Manual*

---

24. James Martineau, *A Study of Religion: Its Sources and Contents*, 2 vols. (New York: Macmillan, 1888), 1:xiii–xiv.
25. W. E. B. Du Bois, "Philosophy IV Notebook," Series 10, Du Bois Papers; covering Martineau, *A Study of Religion*, 1:254–302. For James's summary of the class, see James, *Manuscript Lectures*, 182–86.

of Geology (in all its editions): "There can be no real conflict between the two Books of the Great Author. Both are revelations made by him to man." This issue was salient for the young pragmatists: for example, Royce praised the attempt of his teacher LeConte to portray evolution as "not only reconcilable with, but an aid in, the interpretation of the world of man's spiritual nature." LeConte employed this approach in an article Dewey may have read in college, interpreting all "the forces of nature as an effluence from the Divine Person." Similarly, in a class at Oberlin designed as an "Answer to Modern Forms of Skepticism," Mead learned that the "various theories of Evolution do not explain the Universe without a God. Evolution is not a substitute for Creation but only a different mode of Creation & is not necessarily Atheistic." A year later, he and his friend Castle highlighted this perspective in an editorial responding to their teacher's obituary for Darwin: "The lecture given some weeks ago on Charles Darwin has impressed us more favorably than any. It is pleasing to observe how rapidly the religious craze against evolutionary theories is dying out, and theologians are beginning to discover that science may not after all be their most deadly foe." Along similar lines, Shaler told Du Bois's geology class that because physical geography is "critically adjusted to life[,] . . . . the possibility of human life must either depend on Design or 1 chance in ∞." In short, all of the pragmatists received a clear message in college: evolution is the correct account of the history of life and it does not conflict with religious belief.[26]

But they were not mere passive recipients of this message: they were actively enthusiastic about developments in biology and their philosophical implications. In 1881, Royce recalled his feelings upon abandoning the simple Christian faith of his childhood: "I remember the failing at heart when I first had to throw overboard my little old creed, and felt that I must for example accept the modern theory of evolution as the real truth of nature, against which a poor mortal with his blind hope of immortality might struggle in vain."[27] Thus, although he was later committed to

---

26. James D. Dana, *Manual of Geology: Treating of the Principles of the Science, with Special Reference to American Geological History*, 2nd ed. (New York: Ivison, Blakeman, Taylor, 1874), 770; Josiah Royce, "Joseph Le Conte," *International Monthly* 4 (1901): 333; Joseph LeConte, "Man's Place in Nature," *Princeton Review* 55 (1878): 794; Lewis S. Feuer, "John Dewey's Reading at College," *Journal of the History of Ideas* 19 (1958): 420; "Evidences of Christianity," p. 3, Folder 6, Box 1, Castle Papers; Henry Northrup Castle, "[Thursday Lectures]," *Oberlin Review*, November 18, 1882, 55; Du Bois, "Philosophy IV Notebook." Castle was editor in chief of the *Oberlin Review* for 1882–83 and was "assisted in the Editorial department by Mr. G. H. Mead": see *Oberlin Review*, September 23, 1882, 6.

27. Josiah Royce to George Buchanan Coale, 5 December 1881, in Josiah Royce, *The Letters of Josiah Royce*, ed. John Clendenning (Chicago: University of Chicago Press, 1970), 104.

an evolutionary and religious idealism, Royce—like William James—was initially attracted to a strictly empirical account of human existence. During his junior year at California (1873–74), he was a keen reader of *Popular Science Monthly*, founded in 1872. Its editor, Edward Livingston Youmans, was one of Herbert Spencer's most fervent American supporters, and the magazine, which published much of Spencer's sociological work, was intended to promote the "immense extension of the conception of science" into realms of "intellect, feeling, human action, language, education, history, morals, religion, law, commerce, and all social relations and activities." As this list indicates, the topics covered were diverse: in just the first half of the December 1873 issue, Royce encountered the concluding chapter of Spencer's *Study of Sociology* and an article by LeConte defending the "Correlation of Vital with Chemical and Physical Forces," along with others on animal furs; heredity and education; the simplest forms of animal life; the experimental localization of brain functions; and the geography and atmosphere of Mars.[28]

Royce was particularly interested in Spencer, biology, and evolution. Over just a few months in 1873–74, for example, he mentioned an article in the *British Quarterly Review* that "attacks Spencer's philosophy from an empirical stand-point," spotted a veiled reference to Spencer in George Eliot's *Middlemarch*, and noted Spencer's "Replies to Criticisms," published in *Popular Science Monthly*. One of Spencer's arguments in this last article was that his philosophy was a kind of evolutionary empiricism, and thus more Lockean than Kantian:

The Evolution-view is completely experiential. It differs from the original view of the experientialists by containing a great extension of it. With the relatively-small effects of individual experiences, it joins the relatively-vast effects of the experiences of antecedent individuals. But the view of Kant is avowedly and absolutely unexperiential. Surely this makes the predominance of kinship manifest.[29]

Around the same time, Royce claimed that Spencer was "generally acknowledged to be the greatest thinker now living," calling him "the

---

28. Edward Livingston Youmans, "Purpose and Plan of Our Enterprise," *Popular Science Monthly* 1 (1872): 113–14; *Popular Science Monthly* 4 (1873): 129–93. Royce mentioned one of the articles from the December 1873 issue in "Notebook 2 [General and miscellaneous]," p. 21, Folder 4, Box 114, Royce Papers.

29. Royce, "Notebook 2 [General and miscellaneous]," pp. 21–23, 26; "Herbert Spencer," *British Quarterly Review* 58 (1873): 308; George Eliot, *Middlemarch: A Study of Provincial Life* (New York: Harper & Brothers, 1873), 93; Herbert Spencer, "Replies to Criticisms," *Popular Science Monthly* 4 (1874): 308.

living expounder of Evolution" and a "far-reaching grasper of scientific truth under every form." Royce was also influenced by the work of Spencer's follower John Fiske: he cited Fiske's "Primeval Ghost-World," which drew on the evolutionary anthropology of Spencer and Edward Burnett Tylor, as well as Fiske's discussion of spontaneous generation in the *Atlantic Monthly*, which argued that naturalists should gather "as extensive a mass of facts as possible, and see which theory, that of germs or that of spontaneous molecular reconstruction, is, on the whole, the simplest and easiest 'fit' for them all." But although Royce praised Spencer and Fiske, he did not want to become their disciple:

That Spencer is wholly unsurpassed by any of his contemporaries in his mastery of the highest Scientific generalizations, we do believe. But while we have every desire to study him and to be to a degree formed by him, we are not and hope we never will be worshippers of him. . . . The Philosophical, the Metaphysical side of Spencer's system we have a suspicion of.

Even as a senior in college, Royce already had a clear sense of Spencer's shortcomings.[30]

Agassiz died in 1873 while Royce was in college. In a January 1874 address that Royce almost certainly read, his teacher LeConte claimed that Agassiz had—despite appearances—been "the great apostle of evolution": "It was only the present theories of evolution, or evolution by transmutation, which he rejected. His was an evolution, . . . not by transmutation of species, but by substitution of one species for another." LeConte also credited Agassiz with anticipating Spencer's view that "society, too, passes by evolution from lower to higher, from simpler to more complex, from general to special, by a process of successive differentiation."[31] In his reading notebook a few months later, Royce mentioned some "Recollections of Agassiz" in which the "writer takes the opportunity to strike at Positivism," as well as a *Popular Science Monthly* editorial discussing "the false use that has been made of his (Agassiz's) name with regard to evolution." In the latter, Youmans

---

30. "Mill and Spencer," Document 15, Box 53, Royce Papers, which is an unpublished sequel to Josiah Royce, "Literary Education," *Berkeleyan* 1, no. 5 (1874); Royce, "Notebook 2 [General and miscellaneous]," pp. 16–18, 21; John Fiske, "The Primeval Ghost-World," *Atlantic Monthly* 30 (1872); John Fiske, "Science," *Atlantic Monthly* 32 (1873): 760; Josiah Royce, "Notes on Exchanges," *Berkeleyan* 2, no. 5 (1875): 13.

31. Joseph LeConte, "Agassiz and the Basis of His Scientific Reputation," *Berkeleyan* 1, no. 2 (1874): 5. An article by Royce appeared in the same issue of the *Berkeleyan*, directly after LeConte's.

argued that there was no necessary opposition between Christianity and evolution:

> When we are told that Agassiz was a Christian *because of his opposition to Darwinism,* we decidedly object. Prof. Agassiz was a Theist, who ascribed the universe to a Divine Mind; Darwinians [may] do the same. . . . A literature of reconciliation is springing up, and we are beginning to hear of Christian evolutionists, as we have long heard of Christian astronomers and Christian geologists.

Youmans pointed out that even Andrew Preston Peabody—a fierce critic of positivism whose work drew responses in the 1860s from both John Fiske and Chauncey Wright—had admitted at Agassiz's funeral that accepting evolution would not affect his Christian faith. Royce's reading in college thus demonstrates that Spencer, evolution, and positivism were among his chief interests.[32]

Dewey was also excited by science and evolution as a college student. Discussions of the implications of evolutionist, empiricist, and materialist ideas filled the pages of the magazines Dewey read at the time. Looking back, he emphasized the importance of this reading for his intellectual development:

> The University library subscribed to English periodicals which were discussing the new ideas which centered about the theory of evolution. . . . These periodicals discussed far more than this particular subject, however, for the controversy about evolution was but the forefront of the rising interest in the relation between the natural sciences and traditional beliefs. English periodicals which reflected the new ferment were the chief intellectual stimulus of John Dewey at the time and affected him more deeply than his regular courses in philosophy.[33]

Dewey's library charging records confirm this recollection. Journals he checked out contained, for example, an idealist attack on Spencer's evolutionary empiricism; a discussion of the application of Darwinism to the history of languages; the story of Agassiz's visit to the Galapagos

---

32. Josiah Royce, "Notebook 2 [General and miscellaneous]," pp. 21–23, 26, Folder 4, Box 114, Royce Papers; Theodore Lyman, "Recollections of Agassiz," *Atlantic Monthly* 33 (1874): 228–29; Edward Livingston Youmans, "[Agassiz and Evolution]," *Popular Science Monthly* 4 (1874): 499. On Fiske, Wright, and Peabody, see Trevor Pearce, "'Science Organized': Positivism and the Metaphysical Club, 1865–1875," *Journal of the History of Ideas* 76 (2015): 451–53.

33. John Dewey and Jane M. Dewey, "Biography of John Dewey," in *The Philosophy of John Dewey*, ed. Paul Arthur Schilpp (Evanston, IL: Northwestern University Press, 1939), 10–11. This biography was coauthored by John Dewey, despite referring to him in the third person.

Islands, including comments on the origin of species; William Kingdon Clifford's views on the relation between mind and body; speculation by Arthur Balfour about whether evolution can explain true beliefs; a series of discussions of evolution, design, materialism, and the mind-brain relation in the *Princeton Review*; essays on the connections between ethics, conscience, and evolution in the *Nineteenth Century*; a translated chapter of Eduard von Hartmann's *Truth and Falsity in Darwinism*; and reviews by Chauncey Wright, William James, Joseph Bangs Warner, and others of recent books such as John Stuart Mill's *Examination of Sir William Hamilton's Philosophy*, William Benjamin Carpenter's *Principles of Mental Physiology*, John Fiske's *Outlines of Cosmic Philosophy*, George Henry Lewes's *Problems of Life and Mind*, and Friedrich Albert Lange's *History of Materialism*. Dewey was not exaggerating when he spoke of a "new ferment" in the late 1870s: the addition of evolution into existing philosophical and religious debates had created an enticing mixture for the precocious college student.[34]

Dewey was particularly interested, beginning in his junior year, in physiological psychology and what it implied about the mind-body relationship. He later recalled that he "got great stimulation from the study" of Huxley's human physiology textbook, assigned by Perkins. It is likely no coincidence that in the fall of 1877 and again a year later he borrowed journal volumes that contained Huxley's infamous 1874 address, "On the Hypothesis that Animals are Automata," the target of the 1879

---

34. Lewis S. Feuer, "John Dewey's Reading at College," *Journal of the History of Ideas* 19 (1958): 416–20; William Torrey Harris, "Herbert Spencer," *Journal of Speculative Philosophy* 1 (1867); William Dwight Whitney, "Darwinism and Language," *North American Review* 119 (1874); Elizabeth Cabot Agassiz, "Cruise through the Galapagos," *Atlantic Monthly* 31 (1873); William Kingdon Clifford, "Body and Mind," *Fortnightly Review* 22 (1874); Arthur Balfour, "A Speculation on Evolution," *Fortnightly Review* 28 (1877); John T. Duffield, "Evolutionism Respecting Man, and the Bible," *Princeton Review* 54 (1878); Paul Ansel Chadbourne, "Design in Nature," *Princeton Review* 54 (1878); Francis Bowen, "Dualism, Materialism, or Idealism" *Princeton Review* 54 (1878); Laurens P. Hickok, "Evolution from Mechanical Force," *Princeton Review* 54 (1878); James McCosh, "Contemporary Philosophy: Mind and Brain," *Princeton Review* 54 (1878); John William Dawson, "Evolution and the Apparition of Animal Forms," *Princeton Review* 54 (1878); Lionel S. Beale, "The Materialist Revival," *Princeton Review* 55 (1878); Thomas Welbank Fowle, "The Place of Conscience in Evolution," *Nineteenth Century* 4 (1878); William Angus Knight, "Ethical Philosophy and Evolution," *Nineteenth Century* 4 (1878); Edward von Hartman[n], "The True and the False in Darwinism," *Journal of Speculative Philosophy* 12 (1878); Chauncey Wright, "Mill on Hamilton," *North American Review* 103 (1866); "Fiske's Outlines of Cosmic Philosophy," *North American Review* 120 (1875); Joseph B. Warner, "Fiske's Cosmic Philosophy," *Atlantic Monthly* 35 (1875); William James, "Carpenter's Principles of Mental Physiology," *North American Review* 119 (1874); Frederic Harrison, "Mr. Lewes's Problems of Life and Mind," *Fortnightly Review* 22 (1874); William James, "Lewes's Problems of Life and Mind," *Atlantic Monthly* 36 (1875); "History of Materialism," *Contemporary Review* 30 (1877).

article by William James discussed in chapter 2.[35] Huxley argued that *"the living body is a mechanism"* was the "fundamental proposition of the whole doctrine of scientific Physiology." He rejected René Descartes's notion that animals are machines without consciousness, as evolution suggests a continuity between humans and animals. He claimed instead that both should be seen as conscious automata, for "states of consciousness . . . are immediately caused by molecular changes of the brain-substance." Huxley ended the paper with a denial that materialism was implied by these strictly scientific facts. Nevertheless, he realized that many observers would accuse him of materialism and antireligious sentiment regardless of such protests—after all, he was famously an agnostic about anything supernatural. For most readers, the links between physiological psychology, the mind-body problem, and materialism would have been apparent. Dewey was clearly fascinated by such questions, as he also checked out the "Nervous System" volume of Austin Flint's *Physiology of Man* and Alexander Bain's *Mind and Body* in the fall of 1878.[36]

But the most famous empiricist account of the mind at the time was that of Herbert Spencer. In his senior year (1878–79), Dewey borrowed the first volume of Spencer's *Principles of Psychology* no fewer than four times, more than any other book during his time at Vermont. In this book, as discussed in chapter 2, Spencer introduced his theory of "the correspondence between the organism and its environment" and argued that mind was just a special form of life.[37] Dewey probably also encountered James's critique of Spencerian psychology. A few weeks before checking out the first volume of Spencer one last time, he borrowed a journal volume that contained two of James's early articles: "Spencer's

---

35. John Dewey, "From Absolutism to Experimentalism," in *Contemporary American Philosophy: Personal Statements*, ed. George Plimpton Adams and William Pepperell Montague, vol. 2 (New York: Russell & Russell, 1930), 13; Thomas Henry Huxley and William Jay Youmans, *The Elements of Physiology and Hygiene*, rev. ed. (New York: D. Appleton, 1873); Feuer, "John Dewey's Reading at College," 418–19; Thomas Henry Huxley, "On the Hypothesis that Animals are Automata, and its History," *Fortnightly Review* 22 (1874); Thomas Henry Huxley, "On the Hypothesis that Animals are Automata, and its History," *Littell's Living Age* 123 (1874); William James, "Are We Automata?" *Mind* 4 (1879).

36. Huxley, "On the Hypothesis that Animals are Automata," 555, 574, 577–78, italics added; Richard Holt Hutton, "Pope Huxley," *Spectator*, January 29, 1870; Bernard Lightman, "Huxley and Scientific Agnosticism: The Strange History of a Failed Rhetorical Strategy," *British Journal for the History of Science* 35 (2002); Feuer, "John Dewey's Reading at College," 419; Austin Flint, *The Physiology of Man: Designed to Represent the Existing State of Physiological Science as Applied to the Functions of the Human Body*, 2nd ed., vol. 4 (New York: D. Appleton, 1875); Alexander Bain, *Mind and Body: The Theories of their Relation* (New York: D. Appleton, 1873).

37. Feuer, "John Dewey's Reading at College," *Journal of the History of Ideas* 19 (1958): 419–20; Herbert Spencer, *The Principles of Psychology*, 2nd ed., 2 vols. (New York: D. Appleton, 1873), 1:294.

CHAPTER THREE

Definition of Mind as Correspondence" and "Brute and Human Intellect." As we saw in chapter 2, James criticized Spencer for regarding the organism "as absolutely passive clay, upon which 'experience' rains down," and insisted that "there belongs to mind, from its birth upward, a spontaneity, a vote. It is in the game, and not a mere looker-on." Dewey's reading of Spencer and James would have taught him that the organism-environment relation was at the heart of debates over mind, matter, and evolution.[38]

Addams was also very interested in science and natural history in college, as the pioneering research of Barbara Bair has shown. Toward the end of her sophomore year at Rockford, she became a founding member of a new Scientific Association at the seminary. Holmes, their natural history teacher, was the fledgling association's president from spring 1878 to fall 1879, with Addams serving as her vice president beginning early in 1879.[39] A few months after the founding of the association, Addams gave a talk to its members about a rock fissure full of animal skeletons that she had found near her hometown of Cedarville, Illinois. Perhaps aided by Holmes, Addams showed impressive geological knowledge, identifying the relevant rocks as part of "the Trenton Group of the lower Silurian age." As Bair has suggested, Holmes served for Addams "as a role model of the possibilities of science as a pathway for the professional woman." She even instructed her in the art of taxidermy: Addams proudly told her mother in 1880 that after receiving a live hawk in the mail (apparently from someone familiar with her interests), she and Holmes had "killed it with chloroform and [will] have it all nicely cleaned and tanned ready to stuff to-morrow morning."[40]

Addams was familiar at the time with the work of evolutionists such as Darwin, Spencer, and Huxley. Recalling her college days in 1910, she implied that she had accepted "the theory of evolution," which "even thirty years after the publication of Darwin's 'Origin of Species' had about

---

38. Feuer, "John Dewey's Reading at College," *Journal of the History of Ideas* 19 (1958): 420; William James, "Brute and Human Intellect," *Journal of Speculative Philosophy* 12 (1878): 256; William James, "Remarks on Spencer's Definition of Mind as Correspondence," *Journal of Speculative Philosophy* 12 (1878): 17.

39. Addams, *Selected Papers*, 1:209, 252. The editorial material relating to Addams's education at Rockford was researched and written by Barbara Bair.

40. Jane Addams, "[Address on Illinois Geography]," in Addams, *Addams Papers* 45:1781; on the Trenton Group, see James D. Dana, *Manual of Geology: Treating of the Principles of the Science, with Special Reference to American Geological History*, 2nd ed. (New York: Ivison, Blakeman, Taylor, 1874), 194–210 (these strata are also known as the Galena Group and are now classified as part of the Late Ordovician Period); Addams, *Selected Papers*, 1:173; Jane Addams to Anna Hostetter Haldeman Addams, 7 March 1880, in Addams, *Selected Papers*, 1:347–48.

it a touch of intellectual adventure."[41] Evolutionary ideas were not restricted to zoology and geology: Spencer and others had popularized the *nebular hypothesis*, a theory of cosmological evolution according to which our solar system was gradually formed by the condensation of nebulae. Addams wrote an essay on this topic for her astronomy class in 1880, noting the naturalistic tendency of the time and alluding to Spencer's idea of the persistence of force as the physical basis of all evolution:

> In this age of intellectual force and intensive living, man spurns the Egyptian creation of the great brooding mother to whom time & knowledge were as nothing, and claims, in his birth-right—and as the parent to whom he can lay the most natural right—the dull heated mass extending throughout the universe in constant & frightful motion, from which by the very persistence of force man himself was finally evolved.[42]

A month later, in another draft essay, Addams argued that the "mechanical and material" civilization of the nineteenth century had forced intellectuals to recast simple truths as overarching ideals. She praised the literary ideal of Matthew Arnold and the scientific ideal of Thomas Henry Huxley, asking, "if men are not working for one of those two ends, what *are* they working for?" According to Addams, Huxley believed in "learning thoroughly the physical and moral laws which govern man, in seeing things exactly as they are, and improving our capacities as thinking beings to the uttermost." The students at Rockford were struck by all of these scientific ideas. As Addams wrote in the college magazine, which she edited, they were jointly discovering "how hard it is to become natural in their mode of thinking." She reported the thoughts of one "perplexed maiden," struggling to reconcile the doctrines of science with her traditional beliefs: "If I didn't know anything and didn't believe anything I read in the *Popular Science Monthly* I could get along." As this comment suggests, Youmans's magazine was required reading for the young women of the Scientific Association at Rockford. Addams, like Royce and Dewey, eagerly followed the latest scientific developments and reflected on their implications for religion and philosophy.[43]

---

41. Jane Addams, *Twenty Years at Hull-House, with Autobiographical Notes* (New York: Macmillan, 1910), 62; see also James Weber Linn, *Jane Addams: A Biography* (New York: D. Appleton-Century, 1935), 60.
42. Jane Addams, "The Nebular Hypothesis" (28 January 1880), *Addams Papers* 46:157, published in Addams, *Selected Papers*, 1:325; see Herbert Spencer, *First Principles of a New System of Philosophy*, 2nd ed. (New York: D. Appleton, 1869), 308–9, 555.
43. Jane Addams, "Resolved. The Civilization of the 19th cent. tends to fetter intellectual life and Expression" (18 February 1880), in Addams, *Addams Papers* 46:181–83; Jane Addams, "Literary Notes," *Rockford Seminary Magazine* 8 (1880): 50; Addams, *Selected Papers*, 1:253n2.

CHAPTER THREE

These implications weighed especially heavily on Mead, whose father was professor of sacred rhetoric and pastoral theology at Oberlin.[44] Although we know little about Mead's personal reading at college, the interests of his best friend Castle confirm the pattern we have seen: evolution, psychology, and materialism were unavoidable topics for philosophically minded students. In 1882, Castle boasted to his family of a reading regimen that paralleled Dewey's: the *Origin of Species*, Carpenter's *Principles of Mental Physiology*, Bain's *Mind and Body*, and Lange's *History of Materialism*, as well as works by Huxley, Haeckel, Lewes, and LeConte.[45] Mead and Castle were sometimes taught rhetoric by their philosophy professor, John Millot Ellis, which explains why Castle's senior assignment in the class was "to present the argument of materialism as fairly, as fully, and as strongly as I can." In his essay, Castle—like Spencer and Huxley—argued against claims of a gap between life and nonlife, as well as between lower and higher forms of life: "Life became self conscious by steps as slow as those of the dawn when its light faintly illuminates the eastern skies." Nevertheless, he reassured his parents: "I shall never be a materialist. I have a comfortable door open, just like Huxley. Only my door is not that of wretched agnosticism. I can always take refuge in Idealism, and say that we know nothing of matter except through the agency of mind, so that instead of saying that there is nothing but matter, I shall say that there is nothing but mind." Mead was thus familiar with the evolutionary and religious ideas of Huxley and others.[46]

Castle's "comfortable door" was not open to all, however: during four years of teaching and tutoring after graduation, Mead continuously struggled with agnosticism stemming from a loss of religious faith. In a letter to Castle written the year after graduation, he lamented that he had "to[o] feminine a nature to ever become a philosopher. My sentiments . . . are too large a part of my life to admit of that mental abstraction which becomes a lover of truth." He thought that his sentimentality better suited him to missionary work, but this was not possible because of his loss of faith. In another letter he picked up the same theme: "Perhaps I am utterly wrong in my doubts and they are only

44. *Catalogue of the Officers and Students of Oberlin College, for the College Year 1879–80* (Oberlin, OH: Mattison's Steam Printing House, 1879), 4.
45. Henry Northrup Castle to Caroline Castle, 15 March 1882; Henry Northrup Castle to Family, 24 May 1882; Henry Northrup Castle to Samuel and Mary Castle, 13 October 1882; Henry Northrup Castle to Samuel and Mary Castle, 4 November 1882, in Castle, *Letters*, 158, 167, 198, 212.
46. Henry Northrup Castle, "The Materialistic Argument," Folder 7, Box 1, Castle Papers; Castle to Samuel and Mary Castle, 13 October 1882, in Castle, *Letters*, 198–200.

supterfuges [sic] and I long to throw them all aside and leap with my eyes shut and heart open in Christian work. But I cannot do that." Back in Oberlin a few weeks later, he moaned, "I am wallowing in the depths of Agnosticism." He raised his doubts with James Harris Fairchild, Oberlin's president and ethics teacher:

I mentioned [to Fairchild] the fact that I saw no reason why the mind might not be a material evolution[,] a later quality of matter. He wanted me to start with the mind à la Spencer. He said that we knew the mind only at first and the not me we knew only in its resistance to us, but what this resistance was we could not know and no one could disprove that this external not me was the will of God giving certain qualities of resistance to matte to positions in Space.

This letter indicates that at least one source of Mead's agnosticism was the materialist-evolutionary account of the mind provided by philosophers such as Spencer, an account that seemed to leave no room for God. Despite Fairchild's reassurance, Mead continued to doubt: "My creed is dark and agnostic."[47]

Mead was also interested "in popularizing metaphysics among the common people," and he mentioned in this context Alexander Wilford Hall's the *Microcosm*.[48] Mead thought Hall's philosophy ridiculous, but his magazine—"devoted to the discoveries, theories, and investigations of modern science, and their bearings upon the religious thought of the age"—does give a sense of what troubled Mead. Most contributors to the *Microcosm*, like its editor, saw a clear conflict between evolutionary ideas on the one hand and religion and morality on the other. For example, Fletcher Hamlin was worried that evolution would lead to skepticism:

Who has not observed that multitudes of the young men of America are being unsettled in their theological views by the fact that some so-called great men are skeptics? We must all admit that "No man who thoroughly accepts a principle in the philosophy of Nature which he feels to be inconsistent with a doctrine of religion, can help having his belief in that doctrine shaken and undermined." Now that the Doctrines of Development [i.e., evolution] and spontaneous generation have this tendency is evident not only from the rejoicing of infidelity at their first announcement, and the clearly logical argument of Haeckel based upon them in favor of Atheism, but also from

---

47. George Herbert Mead to Henry Northrup Castle, n.d., 23 February 1884, 7 March 1884, and 16 March 1884, Folder 16, Box 1, Mead Papers. The first of these letters (n.d.) was sent from Berlin Heights and thus dates from between September 1883 and February 1884.
48. Mead to Castle, 16 March 1884.

the almost universal skepticism which immediately follows the espousal of any type of either theory.

Even those who grudgingly accepted some of the facts of evolution, such as Allan Conant Ferrin, worried about its implications for faith: "Darwin has been in natural science what Kant was in mental science. He destroyed dogmatism by introducing a critical study of Nature; but by confining himself too strictly to physical phenomena, and by confounding the physical with the spiritual, he ran into agnosticism." Mead had been unsettled in just this way; he had, like Darwin, run into agnosticism.[49]

Like the other future pragmatists, Du Bois was fascinated by evolutionary empiricism and its implications. This fascination is evidenced by notes taken during his senior year at Fisk (1887–88): two pages on John Jay Elmendorf's *Outlines of Lectures on the History of Philosophy* and four pages relating to his Mental Science class.[50] Du Bois copied the following two lines from Elmendorf's introduction: (a) "Comte's three eras of progress[:] Theological, Metaphysical, Positive" and (b) "Teleology is impossible," the latter being a key commitment of what Elmendorf called "Philosophical Atheism." Then in his notes on the "Scholasticism" chapter, Du Bois highlighted Elmendorf's claim that the historical controversies over nominalism and Averroism were echoed in recent discussions of evolution, copying down citations to Max Müller, St. George Mivart, and Herbert Spencer. If he tracked down these references, Du Bois would have encountered Spencer's claim that the explanation of adaptation given by "the teleologist" is at best "an obverse to the truth," as well as Mivart's appeal to "an innate force and tendency" in the production of new species, discussed in chapter 1.[51]

---

49. Fletcher Hamlin, "Science and the Clergy," *Microcosm* 3 (1883): 100–101; internal quotation from Duke of Argyll, *The Reign of Law* (London: Alexander Strahan, 1867), 57; Allan Conant Ferrin, "The Limits of Development.—A Plea for Theistic Evolution," *Microcosm* 3 (1884): 168. For more on this period in Mead's life, see Trevor Pearce, "Naturalism and Despair: George Herbert Mead and Evolution in the 1880s," in *The Timeliness of George Herbert Mead*, ed. Hans Joas and Daniel R. Huebner (Chicago: University of Chicago Press, 2016), 124–27.

50. W. E. B. Du Bois, "Mental Science Notes," Series 10, Du Bois Papers. The archival description is incorrect: Mental Science was a class at Fisk, not Harvard, and one of the notes is dated 17 January 1888.

51. Du Bois, "Mental Science Notes"; John J. Elmendorf, *Outlines of Lectures on the History of Philosophy* (New York: G. P. Putnam's Sons, 1876), 3–4, 104, 113; citing Max Müller, "Lectures on Mr. Darwin's Philosophy of Language: Second Lecture," *Littell's Living Age* 118 (1873); St. George Mivart, "Likenesses; Or, Philosophical Anatomy," *Contemporary Review* 26 (1875); Herbert Spencer, *The Principles of Biology*, vol. 1 (London: Williams & Norgate, 1864), 234; St. George Mivart, *On the Genesis of Species* (London: Macmillan, 1871), 264. The relevant issue of *Littell's Living Age* is incorrectly cited by Elmendorf as "No. 1578" instead of "No. 1518."

Du Bois's focus on evolution is even more obvious in the Mental Science notes. They are without context and seem to be Du Bois's own musings, although they could conceivably be based on statements made by his teacher Erastus Milo Cravath. Here is one series:

The watch (as proof of maker) an evolution
There is no God—Jesus lover of my soul
My life is lain on the altar of Truth. If T[ruth] lead me to dethrone the Infinite, I will do it. If not I will throw around him the royal purple of a soul's adventure.
Systematic observation of savages, of children[,] of likenesses of knowledge.
Truth, deep, Echoless!
Experiments in compulsory evolution.
Study human lives.
Idealism Realism Materialism the same
If thought *is* force what's the difference[?]

The last two notes seem to suggest that idealism, which reduces everything to thought, and materialism, which reduces everything to matter and force, could be reconciled if thought and force were seen as equivalent. This position was at odds with the textbook assigned in the class, John Bascom's *The Science of Mind*. Du Bois may have been alluding instead to Spencer, who argued in *First Principles* that "the law of metamorphosis, which holds among the physical forces, holds equally between them and the mental forces." Spencer concluded his book with the claim that "the establishment of correlation and equivalence between the forces of the outer and the inner worlds, may be used to assimilate either to the other," meaning that materialism and spiritualism were just alternative interpretations of the same data. Even if Du Bois were not familiar with Spencer's work at the time, he certainly shared the English philosopher's interest in evolutionary progress, as shown by two more notebook entries: "Evolution might not imply a lower type growing into a higher, but (?)"; "Grades of Evolution are but steps to Infinity. From gorilla to Man, a mighty stride." These notes from his senior year demonstrate that Du Bois was particularly excited by evolutionary ideas and their broader consequences while at Fisk.[52]

Du Bois may also have been inspired by Alfred Oscar Coffin, who in 1889 became the first African American to be awarded a doctorate in

---

52. Du Bois, "Mental Science Notes"; John Bascom, *The Science of Mind* (New York: G. P. Putnam's Sons, 1881), 423, 454; Herbert Spencer, *First Principles of a New System of Philosophy*, 4th ed. (New York: D. Appleton, 1880), 217, 559.

biology.⁵³ Coffin was a recent graduate of Fisk whose work appeared frequently in the *Fisk Herald*, where Du Bois was on staff for all three of his years in Tennessee (1885–88). As the paper—then edited by Coffin's brother Samuel Allen Coffin—reported in January 1886, Coffin had "matriculated in Illinois Wesleyan University for a non-resident, postgraduate course" and was "deep in the mysteries of biology." The article concluded by predicting that in the coming decades Fisk might have the honor of training "a colored Agassiz commanding the respect of the world." A few months later, it was announced that Coffin was "working hard for the degree of Ph.D." (Coffin's graduate research was probably supervised by James Branch Taylor, an Illinois Wesleyan graduate who traveled with John Wesley Powell and then studied in Germany before earning his MD at Columbia in 1882.)⁵⁴ In February 1888, by which time Du Bois had become editor in chief, Coffin published an essay in the *Fisk Herald* titled "Evolution; Pro and Con." It was more pro than con: "Whatever objections may be raised against Darwin's Theory, he has so fortified himself with practical scientific investigations that no one can gainsay his premises, and those who might bitterly oppose him find themselves instinctively drawn to the conviction and acceptance of his views." Coffin, after presenting some of the embryological evidence for evolution, noted that although evolutionists are often saddled with the idea that "man is descended from an ape," they in fact "argue no such thing," claiming only "that since [man and ape] are so closely connected, they probably had a parentage of like similarities, presum-

---

53. *Thirty-Second Annual Catalogue of the Illinois Wesleyan University, Bloomington, Ill.* (Bloomington: Illinois Wesleyan University, 1889), 8; "'What Do Your Graduates Do?'" *Fisk Herald* 6, no. 8 (1889); Harry Washington Greene, *Holders of Doctorates among American Negroes: An Educational and Social Study of Negroes Who Have Earned Doctoral Degrees in Course, 1876–1943* (Boston: Meador, 1946), 182. Coffin's dissertation, which we would now classify as anthropology rather than biology, was published as Alfred Oscar Coffin, *The Origin of the Mound Builders* (Cincinnati: Elm Street Printing Company, 1889).

54. *Twenty-Eighth Annual Catalogue of the Illinois Wesleyan University, Bloomington, Illinois* (Bloomington: Illinois Wesleyan University, 1886), 56; "A New Field," *Fisk Herald* 3, no. 5 (1886): 5; "Personals," *Fisk Herald* 3, no. 9 (1886): 8. Du Bois was one of several associate editors of the *Fisk Herald* in 1885–86, the literary editor in 1886–87, and the editor in chief in 1887–88: see *Fisk Herald* 3, no. 3 (November 1885): 6; *Fisk Herald* 4, no. 4 (December 1886): 6; *Fisk Herald* 5, no. 3 (November 1887): 8. On Taylor, see *Thirty-First Annual Catalogue of the Illinois Wesleyan University, Bloomington, Illinois* (Bloomington: Illinois Wesleyan University, 1888), 60; William H. Wilder, ed. *An Historical Sketch of the Illinois Wesleyan University, Together with a Record of the Alumni, 1857–1895* (Bloomington: Illinois Wesleyan University Press, 1895), 97; Elmo Scott Watson, ed. *The Professor Goes West: Illinois Wesleyan University—Reports of Major John Wesley Powell's Explorations: 1867–1874* (Bloomington: Illinois Wesleyan University Press, 1954), 26–27; *Catalogue of the Officers and Graduates of Columbia College (Originally King's College) in the City of New York, 1754–1888* (New York: Columbia College, 1888), 181.

ably lower than the present form."⁵⁵ Coffin's account added detail to Du Bois's picture of evolution, and the older man—a Fisk alumnus pursuing his doctorate—was a likely role model for Du Bois, who also planned on graduate work.

Evolution in Graduate School

The early second-cohort pragmatists also engaged with biological ideas during their graduate studies. Josiah Royce spent 1875–76 in Germany, with a semester in Leipzig and another in Göttingen. He was studying philosophy, but because of Immanuel Kant's legacy—specifically the antinomy between mechanism and teleology featured in the Third Critique—his courses in logic and metaphysics would have included discussion of biology.⁵⁶ Royce took two classes with Wilhelm Wundt at Leipzig: Logic and Theory of Method, with Special Consideration of the Method of Natural Science; and Anthropology (Natural History and Prehistory of Man). The second of these could hardly have avoided evolution, and the first probably also dealt with the topic: when Wundt published his *Logic* in the early 1880s, the second volume (*Theory of Method*) included a long chapter on "The Logic of Biology" that demonstrated a sophisticated understanding of recent developments in the life sciences. Wundt criticized Haeckel's position, according to which Darwin had replaced "purposively acting living forces" with "a 'causal explanation' of living forms through the laws of heredity, adaptation, and the struggle for existence." He argued that these laws in fact "have a purely teleological character. For the significance of [Darwin's] theory consists precisely in this, that it has replaced a barren with a presumably more fruitful teleology, in that the teleological principles established by the theory offer better prospects for future causal explanation than the living forces of the older biology." Hence, it is likely that the classes Royce took with Wundt in Leipzig included some analysis of biology and evolution.⁵⁷

---

55. Alfred Oscar Coffin, "Evolution; Pro and Con," *Fisk Herald* 5, no. 6 (1888): 10, 12.
56. Immanuel Kant, *Critique of the Power of Judgment*, trans. Paul Guyer and Eric Matthews (Cambridge: Cambridge University Press, 2000), §70. On Kant's legacy for the life sciences in Germany, see Timothy Lenoir, *The Strategy of Life: Teleology and Mechanics in Nineteenth Century German Biology* (Dordrecht, Neth.: D. Reidel, 1982); Robert J. Richards, *The Romantic Conception of Life: Science and Philosophy in the Age of Goethe* (Chicago: University of Chicago Press, 2002); John H. Zammito, *The Gestation of German Biology: Philosophy and Physiology from Stahl to Schelling* (Chicago: University of Chicago Press, 2018).
57. John Clendenning, *The Life and Thought of Josiah Royce*, 2nd ed. (Nashville, TN: Vanderbilt University Press, 1999), 70; Josiah Royce to Daniel Coit Gilman, 2 February 1876, in Josiah Royce,

Royce then went to Göttingen to study with Rudolf Hermann Lotze, whose philosophy he described as "more traditionally German than Wundt's, and certainly with a more Idealistic tendency in it." Royce's classes at Göttingen included one with Lotze's former student Johann Eduard Rehnisch called Social Statistics, with Special Consideration of the Controversy over the Relation of the Results of Moral Statistics to Free Will. His two courses with Lotze himself were Practical Philosophy and Metaphysics.[58] Although Lotze, like Wundt, was originally trained in medicine, he had developed his philosophical system in the 1840s and '50s and so engaged only superficially with more recent developments in biology. Nevertheless, he also argued that there was no necessary conflict between teleological principles and the mechanistic outlook of modern biology. The following passage from his *Metaphysics* (published three years after Royce's course) gives the flavor of his view:

> Every organic development seems to take place step by step through reciprocal effects made necessary by the constant nature of the connected elements, and life need never escape the mechanical conception of its formation. But we also never have a right to speak of a merely mechanical development of life, as if nothing lay behind it; in fact, behind it always lies, as the real activity which assumes this form of appearance, that unifying vivacity of the absolute.

Unlike Wundt, however, Lotze was opposed to evolution: although he praised "the remarkable natural-historical facts that Darwin's tireless art of observation has discovered," he retreated "with complete contempt from his ambitious and misguided theories." He did at least admit that philosophy should not be in the business of contradicting what "empirical verification" teaches us, noting that even if humans were descended from apes, this would not undermine our humanity: what matters is who we are, not where we came from. Wundt and Lotze thus provided

---

The Letters of Josiah Royce, ed. John Clendenning (Chicago: University of Chicago Press, 1970), 49; *Verzeichniss der im Winter-Halbjahre 1875/76 auf der Universität Leipzig zu haltenden Vorlesungen* (Leipzig, Ger.: Alexander Edelmann, 1875), 14–15, http://histvv.uni-leipzig.de/vv/1875w.html; Wilhelm Wundt, *Logik: Eine Untersuchung der Principien der Erkenntniss und der Methoden wissenschaftlicher Forschung*, vol. 2, *Methodenlehre* (Stuttgart, Ger.: Ferdinand Enke, 1883), 436–37.

58. Royce to Gilman, 2 February 1876; Clendenning, *Life and Thought of Josiah Royce*, 72; *Verzeichniss der Vorlesungen auf der Georg-Augusts-Universität zu Göttingen während des Sommerhalbjahrs 1876* (Göttingen, Ger.: Dieterich, 1876), 7, http://resolver.sub.uni-goettingen.de/purl?PPN654655340 _1876_SS. Clendenning mistakenly claims that Rehnisch's course concerned Herbert Spencer's *Social Statics*.

Royce with examples of how philosophers could critically evaluate recent developments in biology.[59]

A few years later, Addams also learned more about these developments: after graduating from Rockford, she attended the Women's Medical College of Pennsylvania in Philadelphia for a semester (October 1881 to March 1882) before leaving due to health and family problems.[60] Her professor of physiology there was Frances Emily White, who was more philosophically radical than any of the pragmatists' other biology teachers. In 1877, for example, White openly supported materialism in *Popular Science Monthly*:

> It has been conclusively shown, by experimental methods . . . , that emotion and thought are correlated with heat and electricity; and the correlation between *thought* and *mass motion* through the action of nerve and muscle, is constantly exhibited in the human body. It must, then, be admitted that these forces (thought, etc.), like those with which they are correlated, are manifestations of matter.

The next year, she applied the same view to consciousness. "This manifestation of matter," said White, "differs from other kinds of force only in the experience of the conscious subject."[61]

White gave the lecture that opened the 1881–82 academic year at Women's Medical College, with Addams in attendance. She chose as her subject protoplasm—"the universal life-substance from which all organisms, whether vegetable or animal, originate, and modifications of which constitute even the most complex tissues of the highest animal forms." Her account of the properties of protoplasm followed that of Michael Foster, who had opened his physiology textbook (recommended to her students at the college) by introducing the amoeba—"hardly anything more than a minute lump of protoplasm"—and declaring that "the higher animals . . . may be regarded as groups of amoebae peculiarly associated together." White also echoed her Philadelphia colleague Edward Drinker Cope (who had recently named a fossil mammal species after her), claiming that "in the life of protoplasm we behold the dawning

---

59. Hermann Lotze, *System der Philosophie*, vol. 2 of *Drei Bücher der Metaphysik* (Leipzig, Ger.: S. Hirzel, 1879), 455, 465.

60. *Thirty-Third Annual Announcement of the Woman's Medical College of Pennsylvania, Session of 1882–83* (Philadelphia: Grant, Faires, & Rodgers, 1882), 17; Addams, *Selected Papers*, 2:3–18.

61. *Thirty-Second Annual Announcement of the Woman's Medical College of Pennsylvania, Session of 1881–82* (Philadelphia: Grant, Faires, & Rodgers, 1881), 3; Frances Emily White, "Matter and Mind," *Popular Science Monthly* 11 (1877): 183; Frances Emily White, "The Doctrine of Persistence of Individual Consciousness," *Penn Monthly* 9 (1878): 618.

of voluntary motion—of those spontaneous movements especially characteristic of animals." At the climax of her lecture, White made protoplasm the key to our evolutionary history:

> The history of the growth and development of every animal—whether moner, mollusk, or man—is a history of cell-multiplication and cell-differentiation; and the most highly endowed individual of them all possesses no property, no faculty, no power, which is not at last foreshadowed in the formless, structureless, protoplasmic cell from which they are all alike derived. . . . From the beginning of his career, as a microscopic speck of living matter, to its close, although he figures as the most highly endowed and transcendent of beings, man, biologically considered, is protoplasm, protoplasm, only protoplasm; and, whatever his perfections, regarded as a member of the animal series, he has the high privilege of knowing if not of feeling himself the brother of all living things.

Addams was excited by evolutionary ideas in college, but she had never heard their materialistic basis stated so explicitly: from the point of view of biology, said White, we are nothing but protoplasm.[62]

Royce, Dewey, and Addams were connected in various ways to Johns Hopkins University in the 1870s and '80s: Royce (1876–78) and Dewey (1882–84) both did their doctoral work at Hopkins, and Addams lived in Baltimore for a year (1886) while her stepbrother was a graduate student there. Beyond the link between G. W. F. Hegel and evolution, which I will discuss in chapter 4, Royce's studies over his first three semesters at Hopkins made little contact with the sciences. This changed early in 1878, when William James and George Sylvester Morris—both of whom had strong views on the relationship between philosophy and biology—were visiting lecturers.[63] At the time, as we saw in chapter 2, James was developing his critique of Spencer's psychology. He thus opened his Hopkins lectures by announcing that "geology, zoology, astronomy, & human history all seem to be coalescing into a vast system called the theory of evolution," and alluded to Spencer's view that "evolution accounts for mind as a product." Although few notes from these lectures survive, we know that James at least introduced his claim that our

---

62. *Thirty-Second Annual Announcement of the Woman's Medical College of Pennsylvania, Session of 1881–82*, 15; Frances Emily White, "Protoplasm," *Popular Science Monthly* 21 (1882): 362, 364, 366–67; Michael Foster, *A Text Book of Physiology*, 4th ed. (New York: Macmillan, 1880), 1, 4; Edward Drinker Cope, "Consciousness in Evolution," *Penn Monthly* 6 (1875): 565; Edward Drinker Cope, "Contributions to the History of the Vertebrata of the Lower Eocene of Wyoming and New Mexico, Made During 1881," *Proceedings of the American Philosophical Society* 20 (1881): 164.

63. John Clendenning, *The Life and Thought of Josiah Royce*, 2nd ed. (Nashville, TN: Vanderbilt University Press, 1999), 73–77.

"senses [are] organs of *selection*," one of the main points of his 1879 essay "Are We Automata?" and part of his attack on Spencer.⁶⁴

Morris lectured on the history of philosophy, and his background suggests that his course probably included discussion of more recent German philosophers—who were often in conversation with the sciences, as we have just seen. He probably assigned Friedrich Ueberweg's *History of Philosophy*, which he had translated in the early 1870s. The English version of this book was more than a translation: Morris added an overview of the work of his teacher Friedrich Adolf Trendelenburg, who "philosophized with constant reference to the positive sciences." As he wrote in a longer article on Trendelenburg, "recognizing fully the necessity of experience for all concrete knowledge, respecting the various positive sciences as sovereign within their respective spheres, he sought in philosophy the common band which should unite the sciences, and not the speculative principle which should produce them *a priori.*"⁶⁵ Building on Trendelenburg's treatment of biology in *Logical Investigations*, Morris had also recently argued that purpose or "final cause" was a constitutive principle of both thought and nature, criticizing the materialist view that "nature is a complex of unconscious forces." He and other "intelligent teleologists" granted "that the formation of every organism is a complex mechanical problem" and that "nothing is to be accomplished in the world except on the basis of 'mechanical' conditions." But they treated even "the apparent blunders and impotences of the 'Idea' which would realize itself in nature"—monstrous births, bad design, etc.—as "simply incidents in a process by which, as matter of fact, *the Idea is, after all, realized.*" Even Darwin could not avoid "the language of teleology," wrote Morris; "the facts speak for themselves in language too loud to be mistaken" and "they cannot be fully apprehended or described without reference to the adaptations and purposes manifested in them." Morris—who was Royce's doctoral examiner—thus argued that Darwinism was consistent with a teleological view of the natural world, reiterating the position of Lotze, Trendelenburg, and Wundt.⁶⁶

---

64. James, *Manuscript Lectures*, 4; cf. James, "Are We Automata?" *Mind* 4 (1879): 9.

65. Friedrich Ueberweg, *A History of Philosophy, from Thales to the Present Time*, trans. George Sylvester Morris, 2 vols. (New York: Scribner, Armstrong, 1872–74); 2:330; George Sylvester Morris, "Friedrich Adolf Trendelenburg," *New Englander* 33 (1874): 297; cf. Adolf Trendelenburg, *Logische Untersuchungen*, 3rd ed., 2 vols. (Leipzig, Ger.: S. Hirzel, 1870), chap. 1.

66. George Sylvester Morris, "The Final Cause as Principle of Cognition and Principle in Nature," *Journal of the Transactions of the Victoria Institute, or, Philosophical Society of Great Britain* 9 (1876): 184, 187–88, 193; cf. Trendelenburg, *Logische Untersuchungen*, chap. 9. On Royce's doctoral thesis and examination, see Josiah Royce to Charles Rockwell Lanman, 11 June 1878, and Josiah Royce to William James, 11 June 1878, in Josiah Royce, *The Letters of Josiah Royce*, ed. John Clendenning

Morris was also Dewey's doctoral supervisor, but Dewey was able to take a wider range of courses at Hopkins because by 1882 there were three men competing for the professorship of philosophy there: Morris, Granville Stanley Hall, and Charles Sanders Peirce. The course of study that Dewey proposed in his graduate application indicated their respective specialties: "History of Philosophy, Psychology and Logic." Each was employed part-time, with Morris teaching in the fall, Hall teaching in the spring, and Peirce teaching a half load all year. Despite his later status as the founder of pragmatism, Peirce had little influence on Dewey, who dismissed his logic classes as too mathematical; it was Morris and Hall who were Dewey's graduate mentors.[67]

Morris helped incline Dewey toward idealism. One of the first classes Dewey took at Hopkins was Morris's "History of Philosophy in Great Britain," which Dewey described as a long attack on the empiricist school: "[Morris's] lectures upon Brit. Phil will be rather of a critical character than descriptive, & tend to show the inadequacies, contradictions &c of the Sensationalism & Agnosticism." These lectures, like Morris's book *British Thought and Thinkers*, culminated in an attack on Spencer's evolutionary philosophy. One of Morris's strategies was to distinguish between philosophical and scientific questions, placing debates over evolution among the latter:

Strictly speaking, therefore, the phrase "Philosophy of Evolution" is an egregious misnomer. Evolution is no more philosophy than gravitation is. It has no other *kind* of philosophical significance than that which may be indirectly connected with any other scientific law. Conceding that the law of evolution has been established, the nature and the wording of philosophical problems have not been changed one whit. . . . The so-called "*philosophy* of evolution" is an extra-scientific accretion of philosophical convictions, for the most part negative, wholly dogmatic, amusingly oracular, and thoroughly irrelevant, about a scientific law of phenomena.

Morris insisted that it was not some new philosophy of evolution that determined Spencer's views but only "the old sensational theory of knowledge which has prevailed in British thought since Bacon's time." What

---

(Chicago: University of Chicago Press, 1970), 52–56; *Third Annual Report of the Johns Hopkins University, Baltimore, Maryland, 1878* (Baltimore: John Murphy, 1878), 39.

67. John Dewey to Johns Hopkins University, 20 September 1882, in Dewey, *Correspondence*, no. 00419; Christopher D. Green, "Johns Hopkins's First Professorship in Philosophy: A Critical Pivot Point in the History of American Psychology," *American Journal of Psychology* 120 (2007): 308–10. For Dewey's comments on Peirce's logic, see John Dewey to H. A. P. Torrey, 5 October 1882, and John Dewey to W. T. Harris, 17 January 1884, in Dewey, *Correspondence*, nos. 00415 and 00429.

was missing in Spencer—as in the work of all empiricists, according to Morris—was the notion of mind as a "dynamic principle which effectuates or is active underneath phenomena." Through Morris, Dewey was introduced to the idealist-metaphysical version of James's critique of Spencer: mind as not only actively involved in experience but as a principle behind phenomena.[68]

Dewey heard a different story the following spring in his physiology and psychology classes. Hall, although he had some sympathy for Hegel, was convinced that the future of philosophy depended on the new psychology. Hall wrote, in an essay Dewey may have read in college:

It is because physiological psychology . . . puts the same old question of philosophy in such new, tangible terms, and with such a divine soul of curiosity, that we love its spirit, and hope much from its methods. Nothing, since the phenomenology, which seems to us to contain the immortal soul of Hegelism, is so fully inspired with the true philosophic motive.

Hall had been converted by the experimental approach to psychology, the basis of the textbook Dewey was assigned in his advanced psychology class: Wundt's *Grundzüge*, discussed in chapter 2. In the introduction to the second edition, published in 1880, Wundt stated:

Psychological introspection goes hand in hand with the methods of experimental physiology, and by the application of the latter to the former, psychophysical methods have developed as a distinct branch of experimental research. Placing the emphasis on the peculiarity of the method, we might call our science *experimental psychology* to distinguish it from the standard theory of the mind based on introspection.

The new psychology thus relied not only on the results of physiology but also on its own experimental techniques. Dewey was "engaged . . . in observation and experiment" with Hall in the spring of 1883 and learned at least some of these techniques: he apparently participated in "some simple experiments in attention," a topic that Wundt had made famous. Although Dewey did not do any laboratory work in physiology proper, he was also enrolled in Henry Newell Martin's "Animal Physiology" class that semester. These experiences apparently made an

---

68. "Enumeration of Classes, First Half-Year, 1882–83," *Johns Hopkins University Circulars* 2 (1882): 18; Dewey to Torrey, 5 October 1882; George Sylvester Morris, *British Thought and Thinkers: Introductory Studies, Critical, Biographical and Philosophical* (Chicago: S. C. Griggs, 1880), 346–47, 361, 364.

CHAPTER THREE

impression on Dewey: he began by thinking that psychology would "furnish grist for the mill, if nothing else," but a year later after several more classes with Hall, he was arguing that "the rise of this physiological psychology has produced a revolution in psychology" by providing "a new instrument, . . . a new *method*,—that of experiment."[69]

In addition to his classes with Hall and Morris, Dewey participated in the university's Metaphysical Club, founded by Peirce in 1879 and not to be confused with the earlier Cambridge club of the same name, discussed in chapter 1. It was here that idealism and the perspective of the empirical sciences clashed, with differing professors in the same room together. Both students and teachers presented on a variety of topics, from physiology to philosophy. Dewey's first presentation to the group, likely stemming from Morris's class on British philosophy, was a critique of the idea that knowledge is relative to the knowing subject, an implication of Spencer's evolutionary account of the mind. Continuing in this antiempirical vein the following semester, Dewey gave presentations on both Hegel and Thomas Hill Green. Other talks were more scientific in nature, with Hall speaking on reaction-time experiments and the biologists Newell Martin and William Thompson Sedgwick presenting on the evolution of vision and the nature of reflex action, respectively. The philosophical analyses of science that Royce had encountered in Germany were also featured: James McKeen Cattell discussed Joseph Cook's misinterpretation of Lotze's "defence of the mechanism of organic bodies," and Ira Remsen reviewed Wundt's "Logic of Chemistry."[70]

The tension between spiritualism and materialism became a focal topic during Dewey's second year in the club (1883–84), with a series of presenters focusing on issues such as materialism and the boundary between the animate and the inanimate. The first talk of the academic year was Morris on the concept of life, in which he argued that attempting to explain life in terms of matter, motion, and force is fruitless because of its active, teleological nature. Both Dewey and Peirce participated in the discussion following this presentation; Peirce even formally replied

---

69. Granville Stanley Hall, "Notes on Hegel and his Critics," *Journal of Speculative Philosophy* 12 (1878): 102; Lewis S. Feuer, "John Dewey's Reading at College," *Journal of the History of Ideas* 19 (1958): 420; "Philosophy, Ethics, Psychology, and Logic," *Johns Hopkins University Circulars* 1 (1882): 233; Wilhelm Wundt, *Grundzüge der physiologischen Psychologie*, 2nd ed., 2 vols. (Leipzig, Ger.: Wilhelm Engelmann, 1880), 1:2–3; "Enumeration of Classes, Second Half-Year, 1882–3," *Johns Hopkins University Circulars* 2 (1883): 90, 93; John Dewey to H. A. P. Torrey, 14 February 1883, in Dewey, *Correspondence*, no. 00422; John Dewey, "The New Psychology," *Andover Review* 2 (1884): 282.

70. "24th Meeting" (17 October 1882) to "31st Meeting" (8 May 1883), in Metaphysical Club Records; James McKeen Cattell, "Hermann Lotze," *Nation*, March 15, 1883, 232; Wilhelm Wundt, "Die Logik der Chemie: Eine methodologische Betrachtung," *Philosophische Studien* 1 (1883).

to Morris at the next meeting, emphasizing "the contrast between Materialism and Spiritualism." These topics carried over to the third meeting of the semester, with Joseph Jastrow speaking on "Materialism, Spiritualism, and the Scientific Spirit" and Dewey presenting "remarks on [Joseph] Delboeuf on Living and Dead Matter." The records do not include any more details about the discussions, but Dewey almost certainly sided with Morris against materialism. The essay by Delboeuf that was the subject of Dewey's remarks argued that "non-living matter cannot generate life, nor, in consequence, sensibility and thought," and that "organisms are not combinations comparable to those of dead matter." It is not clear whether any of the participants would have argued in favor of materialism: Jastrow was a crusader against spiritualism later in his career, though he was not a professed materialist; and Peirce, in his next presentation to the club, lauded the work of the Society for Psychical Research for engaging "in the careful examination of all kinds of phenomena which suggest the possibility of the relation between body and soul being different from what ordinary experience leads us to conceive it." Although Dewey left Hopkins in 1884 as an idealist, his courses with Hall as well as the Metaphysical Club discussions had given him a new respect for experimental psychology, including the claim that the "idea of environment is a necessity to the idea of organism."[71]

Addams arrived in Baltimore a year and a half after Dewey's departure, joining her stepmother as well as her stepbrother George Bowman Haldeman, who had been studying biology at Hopkins since 1883 and was then doing research under the guidance of William Keith Brooks. Although biology was not Addams's main interest at the time, she must have had many conversations with Haldeman about his graduate work. Addams went with Haldeman in February 1886 to Brooks's reading group on Burt Wilder's *The Life of Agassiz*, and Brooks's wife—Amelia Katherine Schultz Brooks—became one of her closest friends in Baltimore.[72] During the summers of 1885 and 1886, Haldeman did research in Beaufort,

---

71. "32nd Meeting" (9 October 1883) to "39th Meeting" (13 May 1884), in Metaphysical Club Records; George Sylvester Morris, "The Philosophical Conception of Life [Abstract]," *Johns Hopkins University Circulars* 3 (1883): 12; Joseph Delboeuf, "La matière brute et la matière vivante," *Revue Philosophique de la France et de l'Étranger* 16 (1883): 339; Joseph Jastrow, *Fact and Fable in Psychology* (Boston: Hougton Mifflin, 1900); "[Design and Chance]," in Peirce, *Writings*, 4:544; Dewey, "The New Psychology," *Andover Review* 2 (1884): 285.

72. "List of Officers and Students, 1883–84," *Johns Hopkins University Circulars* 3 (1883): 19; "Check List of Students," *Johns Hopkins University Circulars* 4 (1884): 8; "Preliminary Register of Officers and Students," *Johns Hopkins University Circulars* 5 (1885): 17; Jane Addams to Sarah Alice Addams Haldeman, 1 February 1886, quoted in Addams, *Selected Papers*, 2:404n28; Addams, *Selected Papers*, 2:388–98.

CHAPTER THREE

North Carolina, as a member of the marine zoological laboratory directed by Brooks. The English biologist William Bateson had recently argued, based in part on research at this laboratory in 1884, that chordates (the phylum to which humans and other vertebrates belong) were more closely related to tunicates and acorn worms than to annelids—a view now accepted. More specifically, Bateson claimed that in acorn worms such as *Balanoglossus*, "the three features which alone distinguish Chordata from other animals are present, and associated from an early period in development." Haldeman's 1886 research was designed to extend and complement this work. Brooks reported in December 1886:

The observations which were made [at Beaufort] two years ago by Mr. Bateson upon the development of a species which has no Tornaria [i.e., larval] stage has thrown new light upon the vexed question of the affinities of these animals and has rendered a complete history of the Tornaria larva peculiarly desirable. Mr. Haldeman has this summer traced the metamorphosis of the Tornaria into the young Balanoglossus.

Haldeman was aware of the evolutionary context, underlining in his own brief report the importance of determining whether the morphological features shared between the larvae of acorn worms and starfish were true homologies or merely "the result of secondary adaptations to the same surroundings," a question biologists are still asking today. Haldeman almost certainly discussed his research with Addams, since he wrote this report while she was living with him and his mother in Baltimore.[73]

Her interest in George's research is unknown, but we do know that they attended several scientific lectures together at the Peabody Institute that December, including one by Brooks on "The Oyster" and at least two by Alfred Russel Wallace on "The Theory of Development and the Origin and Uses of Color in Animals and Plants." She described Wallace as "a firm upholder of [Darwin's] theory" but regretted that his first lecture, presumably on the theory of evolution in general, "was unfortunately

---

73. William Bateson, "Abstract of Observations on the Development of Balanoglossus," *Johns Hopkins University Circulars* 3 (1883); William Keith Brooks, "Chesapeake Zoölogical Laboratory: Report of the Director for the Year 1884," *Johns Hopkins University Circulars* 4 (1884): 14; William Bateson, "The Ancestry of the Chordata," *Quarterly Journal of Microscopical Science* 26 (1886): 537–38, 553; cf. Oleg Simakov et al., "Hemichordate Genomes and Deuterostome Origins," *Nature* 527, no. 7579 (2015); William Keith Brooks, "The Zoölogical Work of the Johns Hopkins University, 1878–86," *Johns Hopkins University Circulars* 6 (1886): 38–39; George Bowman Haldeman, "Notes on Tornaria and Balanoglossus," *Johns Hopkins University Circulars* 6 (1886): 45; cf. Leonid P. Nezlin and Vladimir V. Yushin, "Structure of the Nervous System in the Tornaria Larva of *Balanoglossus proterogonius* (Hemichordata: Enteropneusta) and Its Phylogenetic Implications," *Zoomorphology* 123 (2004).

technical and a little above the average intelligence." A few days later, she attended "a very interesting illustrated lecture on 'protective coloring in animals,'" the third of Wallace's four lectures in Baltimore.[74] He probably gave his audience a preview of the chapters on coloration in *Darwinism*, which argued that it was "millions of years" of natural selection—rather than divine design or the direct influence of environment—that was responsible for protective coloration:

> To many persons it will seem impossible that such beautiful and detailed resemblances as those now described . . . can have been brought about by the preservation of accidental useful variations. But this will not seem so surprising if we keep in mind . . . the rapid multiplication, the severe struggle for existence, and the constant variability of these and all other organisms. And, further, we must remember . . . that we now see the small percentage of successes among the myriads of failures.[75]

Thus, during her year in Baltimore, Addams participated in popular scientific discussions and was also only one step removed from cutting edge research in evolutionary embryology.

Like Royce, both Mead and Du Bois did much of their graduate work in Germany. Mead followed in Royce's footsteps to Leipzig in 1888–89, enrolling in Wundt's Fundamentals of Metaphysics course. Wundt's metaphysics, like his logic, was closely tied to the sciences. As he wrote in the preface to *System of Philosophy*, published in 1889,

> I consider metaphysics neither "concept-poetry" nor a system of reason constructed from a priori valid conditions by means of specific methods. . . . Metaphysics does not have to build from scratch, but can begin with the hypothetical elements presented to it by the individual sciences; these it must logically examine, bringing them into agreement with one another and thus uniting them in a consistent whole.

In his metaphysical treatment of the concept of purpose, Wundt repeated the view expressed in his *Logic*: the theory of evolution does not require "a complete displacement of the teleological explanation of

---

74. Jane Addams to Laura Shoemaker Addams, 1 December 1886; Jane Addams to Sarah Alice Addams Haldeman, 8 December 1886; and Jane Addams to Sarah Alice Addams Haldeman, 15 December 1886, in Addams, *Addams Papers*, 2:377, 383, and Addams, *Selected Papers*, 2:403n23; *Twentieth Annual Report of the Provost to the Trustees of the Peabody Institute of the City of Baltimore, June 1, 1887* (Baltimore: Isaac Friedenwald, 1887), 11. Thanks to Marilyn Fischer for bringing these letters to my attention.

75. Alfred Russel Wallace, *Darwinism: An Exposition of the Theory of Natural Selection with Some of Its Applications* (London: Macmillan, 1889), 205.

CHAPTER THREE

organic life by its causal explanation." He then went further, arguing that the integrated and automatic systems of the higher animals, with each system adapted to the purpose of the whole, are the culmination of an evolutionary process that began with the completely voluntary movements of the single-celled protozoan, "a being acting in all its parts according to volitional impulses." According to Wundt, the "division of labor" in multicellular plants and animals is therefore a product of selective automation, or "the mechanization of countless purposive actions," with the original voluntary "struggle for existence" still playing a guiding role (Cope's similar view will be examined in chapter 5). Mead's class at Leipzig probably included at least some discussion of these ideas, given their prominence in Wundt's metaphysics.[76]

Mead also took a class called The Relationship of German Philosophy to Christianity since Kant with Georg Karl Rudolf Seydel, who was interested not only in philosophy and religion but also in their complicated relationship with the sciences. In the preface to his collection *Religion and Science*, published in 1887, Seydel declared himself "in search of a philosophy that is at the same time religious and scientific, standing firmly on the ground of the experience of nature but also incorporating the truth content of Christian theology." His class on German philosophy and Christianity could hardly have avoided the topic of evolution and its implications for discussions of teleology and mechanism. Seydel had suggested in "Toward Reconciliation with Darwinism," also in the 1887 collection, that if a role could be preserved for the will, "all the objections to Darwinism justly raised by ethical and religious groups would vanish." Taking an approach similar to Wundt's, Seydel argued that in amoebae and other simple animals, "which the Darwinists aptly call 'organisms without organs,'"

> the use of organs and the building of organs are one and the same: the same life activity that [in the case of higher animals] we had to treat as voluntary action or goal-directedness, in this case produces organs in the moment of their use, and so their use

---

76. Hans Joas, *G. H. Mead, A Contemporary Re-examination of His Thought* (Cambridge, MA: MIT Press, 1985), 18; *Verzeichniss der im Winter-Halbjahre 1888/89 auf der Universität Leipzig zu haltenden Vorlesungen* (Leipzig, Ger.: Alexander Edelmann, 1888), 16, http://histvv.uni-leipzig.de/vv/1888w.html; Wilhelm Wundt, *System der Philosophie* (Leipzig, Ger.: Wilhelm Engelmann, 1889), v–vi, 326, 334–36; see also Wilhelm Wundt, *Essays* (Leipzig, Ger.: Wilhelm Engelmann, 1885), 197–98. On metaphysics as "concept-poetry" (*Begriffsdichtung*), see Friedrich Albert Lange, *Geschichte des Materialismus und Kritik seiner Bedeutung in der Gegenwart*, 2nd ed., 2 vols. (Iserlohn, Ger.: J. Baedeker, 1873–75), 2:540.

is itself nothing but their production.... We are compelled to believe as good Darwinists that organs originally emerge through voluntary action and with the tendency to accomplish the definite goals of the living being.

Seydel would thus have conveyed to his students the view he shared with most other German philosophers: the doctrine of evolution does not force us to abandon the concepts of purpose and will.[77]

After one semester in Leipzig, Mead moved to Berlin and stayed there until 1891, when he was hired by Dewey at the University of Michigan. In Berlin, as Daniel Huebner has shown, Mead worked in the experimental psychology laboratory of Hermann Ebbinghaus. Despite this training in the sciences, which also included classes in anatomy and physiology, Mead gravitated toward philosophy: Friedrich Paulsen (a student of Trendelenburg) was arguably his most important teacher, and he planned to write his (ultimately unfinished) dissertation under Wilhelm Dilthey. Mead took four lecture courses with Paulsen—Psychology and Anthropology, History of Modern Philosophy with Consideration of the Entire Development of Modern Culture, Pedagogy, and Anthropology and Psychology—along with tutorials (*Übungen*) on Arthur Schopenhauer's *The World as Will and Representation* and Kant's *Critique of Pure Reason*.[78] As his course titles indicate, Paulsen sought to bring philosophy into contact with the sciences and the broader culture. In the preface to *Introduction to Philosophy*, published in 1892, Paulsen declared that "the moving factor in the entire development of modern philosophy" was "to reconcile the religious worldview with the scientific explanation of nature." He listed Lotze and Wundt among those attempting this project and claimed that "the philosophy of the present" was characterized in part by its "evolutionary-teleological approach."[79]

Paulsen invoked Lotze specifically, arguing that although the mechanistic explanations of the sciences can in principle account for everything in the natural world, they can never answer questions about its

---

77. Joas, *G. H. Mead*, 18; *Verzeichniss der . . . Vorlesungen* , 16; Rudolf Seydel, *Religion und Wissenschaft: Gesammelte Reden und Abhandlungen* (Breslau, Ger.: S. Schottlaender, 1887), Vorwort [foreword], 252–53, 264.

78. Daniel R. Huebner, *Becoming Mead: The Social Process of Academic Knowledge* (Chicago: University of Chicago Press, 2014), 43–48; Joas, *G. H. Mead*, 18–19, 218n15. For course listings, see the *Verzeichniss der Vorlesungen, welche auf der Friedrich-Wilhelms-Universität zu Berlin . . .* from Summer 1889 to Summer 1891, https://www.digi-hub.de/.

79. Friedrich Paulsen, *Einleitung in die Philosophie* (Berlin: Wilhelm Hertz, 1892), VI–IX; translation adapted from Friedrich Paulsen, *Introduction to Philosophy*, trans. Frank Thilly (New York: Henry Holt, 1895), xii–xv.

CHAPTER THREE

significance: "If astronomy had completely explained the cosmical processes by physical laws, if biology had completely revealed to us the origin and mechanism of organic vital processes, the question would remain, what is the meaning of this entire play of forces, what is that which meets us in the thousand forms and movements of the corporeal world?" He also provided a detailed description of Darwin's theories, which he described as having "a significant influence on our entire worldview, and above all on the historical sciences, including politics and morals." Although Paulsen was critical of "the teleological argument which tries to pick flaws in the physical explanation of nature in order thus to prove the need of assuming non-physical causes," he endorsed Wundt's previously discussed theory about protozoans and the role of the will in evolution: "The organism is, as it were, congealed voluntary action. Of course, the result was not represented beforehand as a purpose; the will was at any given moment directed only toward some particular activity. But the effects transcended the immediate goal—a relation which we still find at the highest stage of evolution, in mental-historical life." Since evolution was a likely topic in Paulsen's class on philosophy and modern culture as well as in his psychology and anthropology classes, Mead would have encountered this particular viewpoint in both Leipzig and Berlin. Dilthey's lectures covered similar topics, including a critique of the evolutionary ethics of Spencer and others.[80]

Although Lotze had died in 1881, shortly after moving to the University of Berlin, his ideas cast a long shadow. While studying there, Mead read the English translation of Lotze's *Mikrokosmus*, which was framed by the conflict "between spiritual needs [*Bedürfnissen des Gemüthes*] and the results of human science," or in other words between "the Philosophy of the Feelings [*Weltansicht des Gemüthes*]" and "the mechanical view of Nature." Mead told Castle that the book was "easy reading and very attractive and uplifting." His one criticism concerned the treatment of evolution, about which Lotze (as we saw earlier) was dismissive. Whatever explanation modern biologists give of the history of life, said Lo-

80. Paulsen, *Einleitung in die Philosophie*, 165–67, 187, 195; translation adapted from Paulsen, *Introduction to Philosophy*, 160–61, 182, 190; Wilhelm Dilthey, *Selected Works*, 6 vols., ed. Rudolf A. Makkreel and Frithjof Rodi (Princeton, NJ: Princeton University Press, 1991–2019), 6:47–48, 66–67. Dilthey's "System of Ethics" corresponds to the Summer 1890 lectures that Mead attended: see Wilhelm Dilthey, *Gesammelte Schriften*, 26 vols. (Stuttgart: Teubner, 1957–), 10:9; Sigrid Schulenburg, ed., *Briefwechsel zwischen Wilhelm Dilthey und dem Grafen Paul Yorck v. Wartenburg, 1877–1897* (Halle, Ger.: Max Niemeyer, 1923), 90; *Verzeichniss der Vorlesungen, welche auf der Friedrich-Wilhelms-Universität zu Berlin im Sommer-Semester vom 16. April bis 15. August 1890 gehalten werden* (n.p., 1890), 14.

tze, it would never "explain the wondrous drama as a whole more adequately than that modest belief which sees nothing but the immediate creative will of God from which the races of living beings can have been derived." Mead was not impressed:

> [Lotze] underestimates Evolution very decidedly—development—both in point of fact and in point of metaphysical importance. There is no more valuable or fruitful way of thinking his god behind and in all than along the lines of evolution[,] and however absurd it may be to find the value of life in mere forms and categories[,] even those that run out into [the] future, it is no less absurd to rob development of all its content and make it out only a form.

Like his Leipzig and Berlin professors, Mead believed that modern evolutionary thought had something to contribute to philosophy and that Lotze had neglected its conceptual resources.[81]

Du Bois also did much of his graduate work in Berlin. Although he and Mead had a few professors in common, Du Bois's classes—mostly in history and political economy—were not as obviously connected to biological ideas. But these ideas still lay in the background of work by his two major professors, Adolph Wagner and Gustav Friedrich Schmoller. Du Bois took Wagner's class on General or Theoretical Political Economy in 1892–93 and was probably assigned the new edition of his *Foundations of Political Economy*. In the first part of this textbook, Wagner described the position of his Historical School as follows: "The main viewpoint of 'historical political economy' . . . is that of 'relativity,' the avoidance of the 'absolutism of solutions' in practical, political-economic questions. Indeed, the conception underlying this viewpoint contains the true core of the 'theory of evolution' in its application also to human, societal, economic life." That is, as we will see in chapter 6, Wagner and his colleagues argued that economic policies should be adapted to particular social situations: what worked in one country might not work in another. Given this approach, Wagner believed that the sociological writings of Auguste Comte and Herbert Spencer, along with the results of

---

81. Hermann Lotze, *Microcosmus: An Essay Concerning Man and His Relation to the World*, trans. Elizabeth Hamilton and E. E. Constance Jones, 2 vols. (Edinburgh: T. & T. Clark, 1885), 1:vii, xv, 527; translated from Hermann Lotze, *Mikrokosmus: Ideen zur Naturgeschichte und Geschichte der Menschheit*, 3rd ed., 3 vols. (Leipzig, Ger.: S. Hirzel, 1876–80), 1:V, XIV–XV, 2:138; George Herbert Mead to Henry Northrup Castle, 29 September 1890 and 24 October 1890, Folder 18, Box 1, Mead Papers. For Lotze's view of God, see Lotze, *Microcosmus*, 1:291.

evolutionary anthropology, could "offer a great deal of very stimulating and valuable material to political and social economists."[82]

Although famously critical of laissez-faire economics, Schmoller—in a handbook chapter based on a seminar Du Bois attended in 1893–94—expressed similar gratitude for "Spencer's collections of material." He also went further, praising the unifying tendency of Spencer's evolutionary philosophy. Social scientists, said Schmoller, must comprehend their objects within

> a universal, historical-philosophical, and sociological thought-formation [*Geistesbildung*]. . . . This is even more true for our own field because, as H. Spencer proves so convincingly, all mental causes are intertwined and inseparable from one another, and because all social phenomena, from social drives to economic and political institutions, are inseparably connected, have uniform causes.

Spencer was an amateur compared with modern German economists, but Schmoller still saw value in his work. "Herbert Spencer and the evolutionary theorists," he wrote, "believe they have found 'the law of evolution'"; and although they remain in "the realm of philosophy of history, teleology, hopes, and prophecies, . . . such bold syntheses will always be necessary for practical action." Du Bois's German teachers, although they were often critical of the superficial biological analogies of Spencer and his German counterparts, were thus sympathetic to evolutionary accounts of economic and social change.[83]

All of the early second-cohort pragmatists, from Royce to Du Bois, encountered evolutionary ideas in college and graduate school and met them with great enthusiasm, following and sometimes participating in

---

82. Adolph Wagner, *Grundlegung der politischen Oekonomie: Erster Theil, Grundlegung der Volkswirthschaft (Erster Halbband)*, 3rd ed. (Leipzig, Ger.: C. F. Winter, 1892), 12, 66; W. E. B. Du Bois to the John F. Slater Fund, 10 March 1893, in Du Bois, *Correspondence*, 1:24; *Verzeichniss der Vorlesungen, welche auf der Friedrich-Wilhelms-Universität zu Berlin im Winter-Semester vom 16. October 1892 bis 15. März 1893 gehalten werden* (n.p., 1892), 22; W. E. B. Du Bois, *The Autobiography of W. E. B. Du Bois: A Soliloquy on Viewing My Life from the Last Decade of Its First Century* (New York: International, 1968), 166.

83. Gustav Schmoller, "Volkswirtschaft, Volkswirtschaftslehre und -methode," in *Handwörterbuch der Staatswissenschaften*, ed. Johannes Conrad et al., vol. 6 (Jena, Ger.: Gustav Fischer, 1894), 541, 551, 560; see also Gustav Schmoller, *Zur Litteraturgeschichte der Staats- und Sozialwissenschaften* (Leipzig, Ger.: Dunker & Humblot, 1888), 232. For notes from part of Schmoller's class, see "On Method, Schmoller in Seminar, Winter Semester 93–94," in "Lecture Notebook," Series 10, Du Bois Papers. On Schmoller and Spencer, see also Heino Heinrich Nau, "Gustav Schmoller's *Historico-Ethical Political Economy*: Ethics, Politics and Economics in the Younger German Historical School, 1860–1917," *European Journal of the History of Economic Thought* 7 (2000): 515, 523n30.

debates over their metaphysical and social consequences. As they began their professional careers, the pragmatists' engagement with evolution continued, as shown by the courses they chose to teach.

Teaching Evolution

Evolutionary approaches to philosophy had already gained some institutional ground by the time Josiah Royce and the other pragmatists began their teaching careers in the 1880s and '90s. As discussed in previous chapters, John Fiske gave a special course of lectures at Harvard on Herbert Spencer and positivism in 1869, and William James was assigning Spencer's books to his Harvard classes by the late 1870s. Philosophy departments began to think of Spencer's philosophy of evolution—and his related idea of a correspondence between organism and environment—as worthy of discussion. The pragmatists participated in this shift: all of them engaged with biological ideas in the classroom.

Royce taught English language and literature at the University of California from 1878 to 1882 before being hired at Harvard as a philosophy instructor. (Although most of his teaching at Harvard was in philosophy, he taught writing there as well, including Du Bois's class on Argumentative Composition in 1888–89.) When Royce arrived in the early 1880s, Spencer's philosophy was already in the catalog at Harvard, and not just in psychology: one of the two assigned texts in Francis Greenwood Peabody's class on Ethics in Its Relation to Religion was Spencer's recently published *The Data of Ethics*, and James's Philosophy of Evolution class assigned the same book as well as *First Principles*. Following his colleagues' lead, Royce assigned *Data of Ethics* in his 1883–84 course Introduction to the History of Ethics. This book was also a prominent target in Royce's *The Religious Aspect of Philosophy*, prepared through public lectures at Harvard and Johns Hopkins that same year and published in 1885. Royce accused Spencer's evolutionary ethics of tacitly presuming Kant's categorical imperative:

If this maxim is essential to the foundation of a moral system, then how poor the pretense that the law of evolution gives us any foundation for ethics at all. The facts of evolution stand there, mere dead realities, wholly without value as moral guides, until the individual assumes his own moral principle. . . . Grant that principle, and you have an ideal aim for action. Then a knowledge of the course of evolution will be useful, just as a knowledge of astronomy is useful to a navigator. But astronomy does not tell us why we are to sail on the water, but only how to find our way.

CHAPTER THREE

According to Royce, it was Kant's maxim—and not the facts of evolution—that was doing the real work in Spencer's moral philosophy: evolution might provide a means but never an end.[84]

Royce was promoted to assistant professor in 1885–86 and immediately began teaching a new course, Philosophy 13: Modern Discussion of the Philosophy of Nature. As Royce put it, in a formulation reminiscent of Lotze, "the main problem of this course is in fact the relation between the postulates of the scientific explanation of nature, and that ethical interpretation of the external world which has in past given rise to so many forms of teleology." The course had an immediate impact on several other pragmatists: Mead was enrolled in 1887–88, and Du Bois heard Royce discuss the course topic (teleology vs. mechanism) at the Harvard Philosophical Club in 1889. Royce structured his course as a comparison of seventeenth- and nineteenth-century views of nature, represented by Baruch Spinoza's *Ethics* and Spencer's *First Principles*, respectively; the dean of the College Faculty thus reported that "the philosophy of Evolution, heretofore taught in a half-course by Professor James, will be provided for in Professor Royce's new course on the Philosophy of Nature." In an essay that contained the substance of the course, Royce characterized the seventeenth century as committed to "the substantial, objective, mathematically perfect unity of nature," a view attributed both to Spinoza's philosophy and to the "new mechanical science" culminating in the deterministic "world-formula" of Pierre-Simon Laplace. The nineteenth century, in contrast, was purportedly obsessed with history and evolution, exemplified most recently by the work of Darwin and Spencer.[85]

---

84. *The Harvard University Catalogue, 1882–83* (Cambridge, MA: Charles W. Sever, 1882), 36; *Annual Reports of the President and Treasurer of Harvard College, 1882–83* (Cambridge, MA: Harvard University Press, 1883), 66, 74–75; *Annual Reports of the President and Treasurer of Harvard College, 1883–84* (Cambridge, MA: Harvard University, 1885), 70; *The Harvard University Catalogue, 1883–84* (Cambridge, MA: Harvard University, 1883), 98; "Public Lectures," *Johns Hopkins University Circulars* 3 (1884): 64; Josiah Royce, *The Religious Aspect of Philosophy* (Boston: Houghton, Mifflin, 1885), 83–84. For Royce's teaching at Berkeley, see the *Register of the University of California* from 1878 to 1882. For Du Bois's English class with Royce, see Du Bois Transcript; *Annual Reports of the President and Treasurer of Harvard College, 1888–89* (Cambridge, MA: Harvard University, 1890), 60.

85. Josiah Royce, "Courses in Ethics at Harvard College," *Ethical Record* 2 (1889): 141; *Annual Reports of the President and Treasurer of Harvard College, 1884–85* (Cambridge, MA: Harvard University, 1886), 96–97; *The Harvard University Catalogue, 1885–86* (Cambridge, MA: Harvard University, 1885), 32, 95; *Annual Reports of the President and Treasurer of Harvard College, 1885–86* (Cambridge, MA: Harvard University, 1887), 50; Josiah Royce, "Is There a Philosophy of Evolution?" *Unitarian Review* 32 (1889): 11–12. For Mead's attendance, see Mead Transcript; Henry Northrup Castle to Helen Castle, 9 October 1887, in Castle, *Letters*, 496. On Royce's talk and Du Bois's membership in the Harvard Philosophical Club, see "Fact and Rumor," *Harvard Crimson*, October 26, 1889; "Fact and Rumor," *Harvard Crimson*, November 14, 1889; "Harvard Philosophical Club," *Harvard Index* 16 (1889): 52.

Royce argued in his 1886–87 lecture notes for the class that the "modern period" was characterized by a tension between these two philosophies of nature: on the one hand, "the clear formulation of universal mechanical postulates in the great doctrine of the Conservation of Energy," and on the other, "the accompanying growth of the historical sense & the tendency to explain by the origins," grandly summarized in the "Doctrine of Evolution." He described the tension as follows:

> Is nature a mechanical sum total of energy, whose forms vary with conditions? If so, evolution is an inessential fact, & the mechanical view returns upon our hands, apparently in 17th century form. If however, evolution is not only here & there a fact, but a deep & essential fact, we seem to have found out what saves the spiritual element in things up to a certain point, although it does not solve all our problems, and does not satisfy all our interests. Yet how [do we] reconcile the *significance* of evolution with the mechanical order of the world?

Discovering a deeper synthesis of the historical and mechanical—answering the Lotzean question—was the problem of the course.[86]

Royce's notes do not specify his solution, but in an 1889 essay on the same topic he claimed that belief in a "genuine historical element" in the universe implied the existence of spontaneity and of ideals that really act in the world:

> Those who have believed that the spirit of the doctrine of Evolution removed teleology from the world have failed to see that the presupposition of our historical age, ever since Rousseau and the Romantic period, has been that teleological explanations *have* their place, that history is worth studying for its own sake, and that the story of the significant ideals must form a part of every philosophical view of the world.

Having made this point, Royce was able to argue that modern psychology presupposed a paradoxical double self: "The psychical facts must be caused; the psychical facts must be significant. As significant, they are teleological; as caused, they have no significance." Royce concluded by turning to idealism, suggesting that mechanism and teleology—real causes and ideal significance—could be reconciled if seen as existing "in

---

86. Josiah Royce, "Philosophy of Nature," Lectures 4–5 (8 and 11 October 1886), Folder 4, Box 127, Royce Papers. Although Royce took a leave of absence from Harvard in February 1888, Mead would have heard the equivalent of these early classes.

and for a Universal Conscious Life, which is the world, and owns the world, and makes and solves its own infinite paradoxes."[87]

A few years after Royce began teaching at Harvard, Dewey started his own career at the University of Michigan, having been hired by Morris. Most of Dewey's courses during his early tenure at Michigan were in logic and psychology. He taught a class called Empirical Psychology each fall for four years, from 1884 to 1888, and although there was no laboratory work, his chosen textbooks highlighted the interaction between organism and environment. James Sully's *Outlines of Psychology*, which Dewey used in 1884–85, followed Spencer in viewing mental life as an adjustment of internal to external relations:

Through innumerable interactions between the nervous system and the environment the former becomes gradually modified in conformity with the latter. Thus nervous connections are built up in the brain-centres corresponding to external relations. The nervous structures are thus in a manner moulded in agreement to the external order, to the form or structure of the environment.

The next year, however, Dewey switched to John Clark Murray's *Handbook of Psychology*, which he declared "a great advance on Sully in its philosophical basis." Dewey probably preferred Murray's more active account of the organism: like James, Murray chastised other psychologists for "forgetting or underestimating the energy of intelligence in asserting itself over the force of its environment." He concluded his book by describing the debate over determinism and free will, linking the former to evolutionists like Spencer: "According to this view man's consciousness is simply the product of the forces in his environment acting on his complicated sensibility, and of that sensibility reacting on the environment." Murray endorsed a rival position, one which "tends to ally itself at the present day with that Transcendental Idealism, which refuses to accept Empirical Evolutionism as a complete solution of the problem of man's nature." Thus, although Dewey taught empirical psychology at Michigan, he still did so as an idealist.[88]

---

87. Royce, "Is There a Philosophy of Evolution?" *Unitarian Review* 32 (1889): 21, 109.
88. *Calendar of the University of Michigan for 1884–85* (Ann Arbor: University of Michigan, 1885), 50; *Calendar of the University of Michigan for 1885–86* (Ann Arbor: University of Michigan, 1886), 54; *Calendar of the University of Michigan for 1886–87* (Ann Arbor: University of Michigan, 1887), 51; *Calendar of the University of Michigan for 1887–88* (Ann Arbor: University of Michigan, 1888), 49; James Sully, *Outlines of Psychology* (New York: D. Appleton, 1884), 58; John Dewey to H. A. P. Torrey, 16 February 1886, in Dewey, *Correspondence*, no. 00434; John Clark Murray, *A Handbook of Psychology* (Montreal: Dawson Brothers, 1885), 8, 415–16.

Like Morris, Dewey distinguished between the doctrine of evolution and the philosophy based on it. For example, in an 1885 review he castigated Benjamin Franklin Tefft for conflating the two:

> The author has no inkling of any difference between a scientific theory of evolution as the theory of animal and vegetable life, the basis of morphology, and the philosophic theory of evolution which attempts to give a causal explanation of the universe. He jumbles together "natural selection," the nebular hypothesis, the variation of species, spontaneous generation, the mechanical theory of the universe, pantheism and materialism in one inextricable mass.

But despite his opposition to the philosophy of evolution, inherited from Morris, Dewey still introduced it to his Michigan students: in 1885–86 and 1887–88 he taught a class called The Philosophy of Herbert Spencer, using *First Principles* as a textbook; and in 1892–93 and 1893–94, he assigned Spencer's *Principles of Sociology* in a seminar on the history of political philosophy.[89]

Dewey also drew on Spencer's biological perspective in some of his other classes, such as the Speculative Psychology course he taught each spring from 1886 to 1888.[90] His self-imposed task in this course, as indicated by student lecture notes, was to explain the relationship between mind and world. Dewey claimed that mind must be an organic unity—that is, a fully integrated and organized whole as opposed to a mere aggregate. In support of this point, he argued that a stone "has no self at all & hence no unity," as it is "wholly dependent upon outside conditions. None of its parts have any necessary relation with one another nor with the world." Moving up the scale, we call a tree an organism because each of its parts "manifests life of whole & at same time contributes to this life." Nevertheless, even a tree is not truly an organism, according to Dewey:

---

89. John Dewey, "A Clergyman's View of Evolution," *Index* (Ann Arbor, MI), March 21, 1885; John Dewey to H. A. P. Torrey, 28 February 1886, in Dewey, *Correspondence*, no. 00435; *Calendar of the University of Michigan for 1885–86*, 55; *Calendar of the University of Michigan for 1887–88*, 50; *Calendar of the University of Michigan for 1892–93* (Ann Arbor: University of Michigan, 1893), 67; *Calendar of the University of Michigan for 1893–94* (Ann Arbor: University of Michigan, 1894), 68. For Dewey's authorship of the Tefft review, see John Dewey, *The Collected Works of John Dewey, 1882–1953*, electronic edition, InteLex Past Masters, supplementary vol. 1, 1884–1951 (Charlottesville, VA: InteLex, 2008), 308.

90. *Calendar of the University of Michigan for 1885–86*, 55; *Calendar of the University of Michigan for 1886–87* (Ann Arbor: University of Michigan, 1887), 52; *Calendar of the University of Michigan for 1887–88*, 50.

CHAPTER THREE

> Material organism not a complete Individual organism for . . . [it is] not completely related to all things in the world. Is related to certain things in its environment, those from which it draws its nourishment. But its environment is very limited. It has no direct relation to most things in existence. Higher we go in range of life wider is environment. . . . If we are to have anything which is completely organic we must have something related to all things however remote or complex. See Spencer's Psychology Vol I.

This idea that progress in the organic world involves an increase in number, range, speciality, and complexity of the adjustments of organism to environment was straight out of Spencer's *Principles of Psychology*, as Dewey's citation indicates. But Dewey gave the notion an anthropocentric twist, arguing that only in our consciousness "do we find a complete organism & hence a true unity or Individual. While there are a great many things in a world Indifferent to a material organism there is nothing which is not either actually or potentially in relation to Intelligence. Environment of mind is coextensive with Universe."[91]

Dewey argued that the basic problem of knowledge is the tension between this potentially universal character of consciousness and its inability to realize this potential in practice. We continually overcome this tension by a process of adjustment—of stimulus and response. Environment provides the stimulus: "Man's intelligence dependent for its content upon its surroundings. A mind shut off from contact with the world remains a blank." Prompted by its sensations, "mind must respond to the stimulus and construct something out of this material." Dewey then returned to Spencer's idea of organism-environment interaction, reinterpreting it in light of his idealist account of knowledge:

> Response of mind brings out & makes real for human intelligence relations which are already real for Universal intelligence. This Response includes
>
> 1—A wider & wider environment
> 2—A higher development of reacting self.
>
> i.e. range of anyone's world narrowly depends on extent to which it can react to stimuli. World of lowest Organism is simply few inches of surrounding temperature & food. Higher animals will include to certain extent environment of sights & sounds &

---

91. Edwin Charles Goddard, "Speculative Psychology, Prof. John Dewey," Lecture 6 (16 March 1887), Box 2, Edwin C. Goddard Papers (851364 Aa 2), Bentley Historical Library, University of Michigan; see Herbert Spencer, *The Principles of Psychology*, 3rd ed., 2 vols. (New York: D. Appleton, 1880), 1:294. I have substituted complete words for Dewey's abbreviations.

also certain number of remembered images. Since man has power of reacting in an indefinite number of ways, no limit can be put to his environment. i.e. merely being surrounded by a world does not constitute having a world. To have a world must be also power of selecting & responding to things in the surroundings. See Spencer's Principles of Psychology. Vol. 1 pp 291–305.[92]

Although Dewey employed Spencer's idea of organism-environment interaction, his view differed from Spencer's in two key respects: first, in the idealist notion of a universal intelligence or consciousness implied by the universal potential of our own more limited consciousness; and second, in the emphasis on the mind's active "power of selecting & responding" to the environment. Dewey, inspired by but critical of Spencer, was thus already developing his own account of organism-environment interaction in the late 1880s—an account we will examine in greater detail in chapter 4.

Like Royce and Dewey, Addams invoked biological ideas in her published work, as we will see in chapter 6. Such ideas were also included in the educational programs at Hull House, the Chicago settlement she cofounded in 1889. Biology and physiology were part of the settlement's College Extension Classes and the associated Popular Lectures: for example, in the fall of 1891 the physician Thomas Melville Hardie, who was professor of laryngology at the Post-Graduate Medical School and Hospital of Chicago, taught "Biology (with microscope)" on Wednesday evenings; and the novelist Celia Parker Woolley, whose literary characters could be found "reading Darwin and John Fiske," gave an October lecture on Charles Darwin.[93] In 1895, Margaret Warren Morley, a high school teacher and author of *A Song of Life*, delivered a lecture on "Steps in Evolution." Morley's talk would have given the Hull House audience a brief taste, but a year later they got the full meal. "The most popular lectures we ever had at Hull House," recalled Addams, "were a series of twelve upon organic evolution . . . we caught the man when he was but a university instructor, and his mind was still eager over the marvel of it all." The instructor who delivered these 1896–97 lectures was Bradley Moore Davis, an associate in the Department of Botany at the University

---

92. Edwin Charles Goddard, "Speculative Psychology, Prof. John Dewey," Lectures 10–11 (13 and 15 April 1887), Goddard Papers; see Spencer, *Principles of Psychology*, 1:291–305 (on organism-environment correspondence).

93. [Program. Sept 1891], 508* OV Scrapbook v. 1, Series XI, Hull House Collection; "Advertisements," *Medical Brief* 19 (1891): 477; Carl Stephens, ed. *The Alumni Record of the University of Illinois: Chicago Departments, Colleges of Medicine and Dentistry, School of Pharmacy* (Chicago: University of Illinois, 1921), 259; Celia Parker Woolley, *Love and Theology: A Novel* (Boston: Ticknor, 1887), 192.

of Chicago (figure 10). Davis had just finished his PhD at Harvard on the complex reproductive cycle of red algae, which implicated several of the special topics covered in his University of Chicago classes: "evolution of sex, parasitism and symbiosis, and the general relation of the lower plants to animals."[94]

Addams lamented that other scientific lecturers could not repeat Davis's performance, although Davis himself repeated it in 1899–1900. She reported that the Hull House audience was thereafter "annihilated by men who spoke with dryness of manner and with the same terminology which they used in the class room"—including the biologist Jacques Loeb, who spoke on "The Mechanism of Animal Instinct" and "The Variation of Animal Forms," and George Herbert Mead, who spoke on "The Evolution of Intelligence" and "The Present Evolution of Man." Presumably the social and political implications of evolution were more exciting: she remembered "one brilliant evening at Hull-House when Benjamin Kidd, author of the much read 'Social Evolution,' was pitted against . . . a rising man in the Socialist Party"; she also recalled the settlement being criticized in 1901 for hosting the anarchist Pyotr Alekseyevich Kropotkin, who had recently authored a series of articles on mutual aid in evolution (both Kidd and Kropotkin are discussed in chapter 5). The socialist May Wood Simons even gave a series of lectures in 1902–3 on the Evolution of Industry that explicitly linked biological and technological evolution, starting with "The Theory of Evolution" and "The Evolution of the Earth" before moving on to "Industrial Life of Primitive Man," "Evolution of Tools," and "Evolution of Textiles." Thus, evolutionary questions were frequently discussed and debated at the Hull House events coordinated by Addams.[95]

---

94. [Program & Circulars 1895], 508* OV Scrapbook v. 1, Series XI, Hull House Collection; Margaret Warner Morley, *A Song of Life* (Chicago: A. C. McClurg, 1981); Jane Addams, "A Function of the Social Settlement," *Annals of the American Academy of Political and Social Science* 13 (1899): 47; "Public Entertainments," *Hull-House Bulletin* 1, no. 7 (1896): 1; "Public Entertainments," *Hull-House Bulletin* 2, no. 1 (1897): 1; see also Jane Addams, *Twenty Years at Hull-House, with Autobiographical Notes* (New York: Macmillan, 1910), 431; Molly Laas, "A Labor of Love: Jane Addams's Evolutionary Worldview" (unpublished manuscript, 10 January 2012), PDF file; Marilyn Fischer, *Jane Addams's Evolutionary Theorizing: Constructing "Democracy and Social Ethics"* (Chicago: University of Chicago Press, 2019), 162. On Davis, see *The Harvard University Catalogue, 1895–96* (Cambridge, MA: Harvard University, 1895), 573; Bradley Moore Davis, "Development of the Procarp and Cystocarp in the Genus Ptilota," *Botanical Gazette* 22 (1896): 371–76; *Annual Register, July, 1895—July, 1896, with Announcements for 1896–7* (Chicago: University of Chicago Press, 1896), 186. On Morley, see *Historical Sketches Relating to the First Quarter Century of the State Normal and Training School at Oswego, N.Y.* (Oswego, NY: R. J. Oliphant, 1888), 194–95. Most issues of the *Hull-House Bulletin* are held in Box 43, Sub-Series A, Series X, Hull House Collection.

95. "Public Entertainments," *Hull-House Bulletin* 4, no. 1 (1900); Addams, "Function of the Social Settlement," 47; "Public Entertainments," *Hull-House Bulletin* 2, no. 2 (1897): 1; "Public Enter-

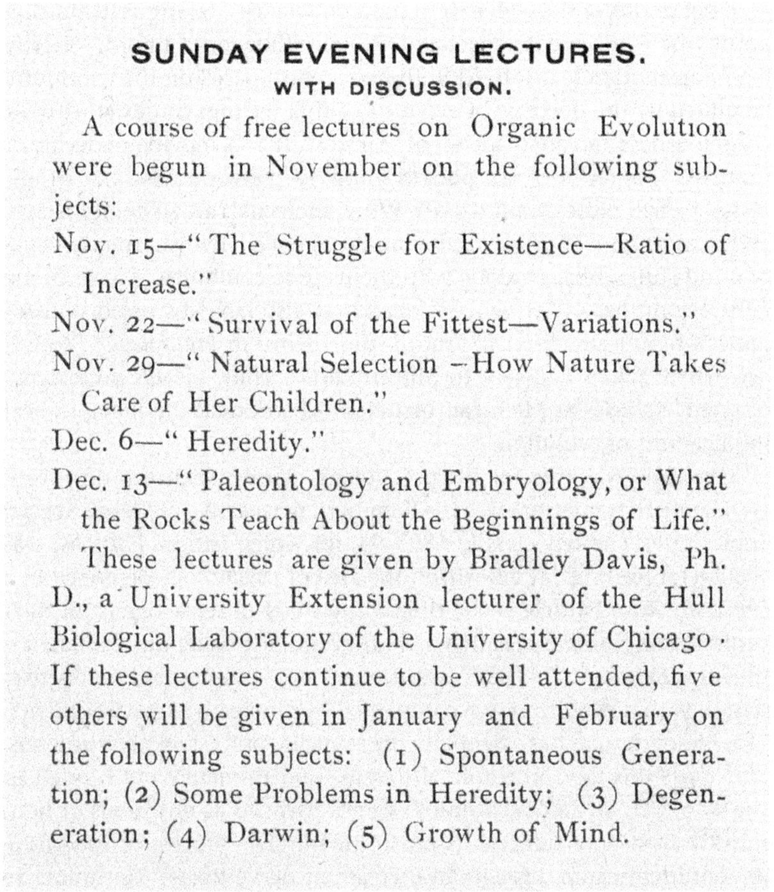

> **SUNDAY EVENING LECTURES.**
> **WITH DISCUSSION.**
>
> A course of free lectures on Organic Evolution were begun in November on the following subjects:
>
> Nov. 15—"The Struggle for Existence—Ratio of Increase."
> Nov. 22—"Survival of the Fittest—Variations."
> Nov. 29—"Natural Selection —How Nature Takes Care of Her Children."
> Dec. 6—"Heredity."
> Dec. 13—"Paleontology and Embryology, or What the Rocks Teach About the Beginnings of Life."
>
> These lectures are given by Bradley Davis, Ph. D., a University Extension lecturer of the Hull Biological Laboratory of the University of Chicago. If these lectures continue to be well attended, five others will be given in January and February on the following subjects: (1) Spontaneous Generation; (2) Some Problems in Heredity; (3) Degeneration; (4) Darwin; (5) Growth of Mind.

Figure 10 Announcement of lectures on organic evolution, delivered by Bradley Moore Davis in 1896–97 at Hull House in Chicago.
From "Public Entertainments," *Hull-House Bulletin* 1, no. 7 (1896): 1. Reproduced courtesy of Special Collections and University Archives, University of Illinois at Chicago.

Mead was hired by Dewey at Michigan in 1891 and immediately took over the class on Spencer, now renamed Philosophy of Evolution. He also taught a class for several years on Matter and Motion, which covered

---

tainments," *Hull-House Bulletin* 2, no. 3 (1897): 1; Addams, *Twenty Years at Hull-House*, 194, 402–5; "Entertainments in the Auditorium," *Hull-House Bulletin* 5, no. 2 (1902): 1; see also Beth L. Eddy, "Struggle or Mutual Aid: Jane Addams, Petr Kropotkin, and the Progressive Encounter with Social Darwinism," *Pluralist* 5 (2010). On Simons, see Mari Jo Buhle, *Women and American Socialism, 1870–1920* (Urbana: University of Illinois Press, 1981), 166–69.

"the net results of the concepts of modern science," taking as its starting point "the writings of Spencer and [William Kingdon] Clifford." Shortly before his untimely death, Clifford had argued, based on the continuity "required by the doctrine of evolution," that feelings can exist without consciousness and are made up of "Mind-stuff": "A moving molecule of inorganic matter does not possess mind, or consciousness; but it possesses a small piece of mind-stuff. When molecules are so combined together as to form the film on the under side of a jelly-fish, the elements of mind-stuff which go along with them are so combined as to form the faint beginnings of Sentience." Mead may also have discussed William James's recent attack on the mind-stuff theory in *Principles of Psychology*, which James had sent to him in Berlin. Thus, Mead's philosophy of science classes at Michigan included debates over the metaphysical implications of evolution.[96]

The same was true for his psychology classes: Spencer's organism-environment framework was the dominant perspective of Mead's Special Topics in Psychology class in 1893–94, for which James's *Principles* was "collateral reading." It fell within the area of psychology described in a *University Record* article at the time: "The study of sense organs in their earliest forms, and especially where they are just being differentiated—the line of biology."[97] Mead's evolutionary story emphasized the Spencerian idea that evolution is accompanied by "increase in environment": "For biologist man has essentially the same life process as lower animals. Man get[s] his food, the animal may be said to merely put himself in the way of it. The amoeba simply selects particles as nutritious or non-nutritious. It does not use its environment. The process of increasing environment came through an increase of nerve tissue." According to Mead, "the complication of processes of the higher animal is explained by an increase in environment," and "the environment is positively or negatively his food"—that is, things he can eat or things that can eat him. Toward the end of the class, Mead attempted to sketch the biological basis of the psychological concepts of interest and attention, the main weapons in James's attack on Spencer: "In order to assume that the fact of attention can be expressed in terms of the organism we have to

---

96. William Kingdon Clifford, "On the Nature of Things-in-Themselves," *Mind* 3 (1878): 64–65; George Herbert Mead to Henry Northrup Castle, 24 October 1890, Folder 18, Box 1, Mead Papers; William James, *The Principles of Psychology*, 2 vols. (New York: Henry Holt, 1890), chap. 6.

97. *Calendar of the University of Michigan for 1893–94* (Ann Arbor: University of Michigan, 1894), 68; "Experimental Psychology," *University Record* (Ann Arbor, MI) 2 (1893): 95. See also Daniel R. Huebner, *Becoming Mead: The Social Process of Academic Knowledge* (Chicago: University of Chicago Press, 2014), 50–51.

assume that every action is directed by the interest of the organism. . . . The ground for every act of attention is to be found in the fundamental life process. . . . Our environment is no larger than its interest for the organism."[98] According to Mead, psychological attention was based on the biological interests of organisms, which were in turn connected to an environment defined by food and predators. As we will see in chapter 7, this natural history perspective may have helped push Dewey in a similar direction in the 1890s.

Du Bois was hired in 1897 as professor of economics and history at Atlanta University, a college that (like Fisk University) enrolled mostly African American students. As soon as he arrived, he replaced traditional courses on Money and Banking and Political Economy—the latter represented by Francis Amasa Walker's decade-old textbook—with two semesters of economics and three semesters of sociology classes. Du Bois's social sciences curriculum represented a more concrete approach, designed to familiarize students "with the great economic and social problems of the world, so that they may be able to apply broad and careful knowledge to the solving of the many intricate social questions affecting their own people." The recently published textbooks that Du Bois assigned in his classes at Atlanta were full of biological ideas. Arthur Twining Hadley, in his *Economics* of 1896, took his cue from "Hegel, Comte, and Darwin," declaring in the opening chapter that evolutionary forces are at work in human society:

The modern observer sees in human history, no less than in natural history, the record of a process of elimination and survival. He sees that laws and institutions no less than genera and species are the result of natural selection instead of being ordained by Providence for all time. He sees that the explanation, not to say the justification, of national customs and feelings must be sought in the historical reasons for their survival. The modern world is coming to look at history as a record of a struggle between different ideas and different institutions, whose issue is chiefly decided by the moral qualities of the contesting races and has its chief importance in determining the moral standards of those races in the immediate future.

According to Hadley, the overall struggle for existence was between human communities and races, with the most successful forms of social

---

98. Robert Clair Campbell, "Phil. Course 9, Prof. Mead," pp. 35, 41–42, 101–2, Box 2, Campbell Family Papers (85891 Aa 2), Bentley Historical Library, University of Michigan. I have translated the shorthand symbols used in the lecture notes. On Mead's related work at Chicago, see George Herbert Mead to John and Alice Chipman Dewey, 24 March 1895, in Dewey, *Correspondence*, no. 00256.

organization regulating their internal struggle to find the right balance between competition and cooperation.[99]

Another textbook assigned by Du Bois, Richmond Mayo-Smith's *Statistics and Sociology*, was likewise inspired by biology, more specifically by Spencer's organism-environment framework. He agreed with the English philosopher that the physical environment was more important in "the earlier stages of social evolution," and so he focused on "what Herbert Spencer calls the superorganic environment" in the final chapter of the book. According to Mayo-Smith, the topics covered in the preceding chapters—namely, "race and nationality, family relationship, the institution of marriage, religious confession, illiteracy and education, social condition"—are "all phenomena of the social environment." He also regarded the "concentration of population in cities" as "one of the most important manifestations of the social environment," with urban populations differing from rural populations along a number of statistical measures. Asking whether increased emigration of Europeans to the United States would change the character of the country, Mayo-Smith answered in the negative, appealing to the overwhelming influence of the social environment: "It seems to be the super-organic influence which thus counterbalances or overcomes the influence of race. Physical environment may have some influence in developing a somewhat similar physique. But social environment has a still marked influence in bringing all into accord with the prevailing type of society." He concluded the whole book by noting that "with social self-consciousness, not only does environment modify society, but society modifies environment with a set purpose in view." It is then the job of the sociologist "to discover the purpose under which society is acting," with statistics explaining "the direction of the changes." Although it is difficult to determine the extent to which Du Bois's lectures agreed with the tendency of his assigned textbooks, his own research took a similar approach at the time, as we will see in chapter 6.[100]

---

99. *Catalogue of the Officers and Students of Atlanta University, . . . with a Statement of the Courses of Study, Expenses, Etc., 1896–97* (Atlanta: Atlanta University Press, 1897), 25, 31; *Catalogue of the Officers and Students of Atlanta University, . . . with a Statement of the Courses of Study, Expenses, Etc., 1897–98* (Atlanta: Atlanta University Press, 1898), 4, 7, 13; *Catalogue of the Officers and Students of Atlanta University, . . . with a Statement of the Courses of Study, Expenses, Etc., 1898–99* (Atlanta: Atlanta University Press, 1899), 7, 13; Arthur Twining Hadley, *Economics: An Account of the Relations between Private Property and Public Welfare* (New York: G. P. Putnam's Sons, 1896), 18–19; cf. Francis A. Walker, *Political Economy* (New York: Henry Holt, 1887).

100. *Catalogue of the Officers and Students of Atlanta University, . . . 1898–99*, 7, 13; Richmond Mayo-Smith, *Statistics and Sociology* (New York: Macmillan, 1896), 7, 358, 363, 370, 376, 382.

Finally, it is worth mentioning that one of Du Bois's acquaintances in Atlanta was the African American biologist Charles Henry Turner, who earned his MS at the University of Cincinnati in 1892 under the supervision of the comparative neurologist Clarence Luther Herrick (who was also his college mentor) and the zoologist James Playfair McMurrich. It was a small academic world at the time: McMurrich had done research with Brooks at Hopkins, overlapping with Addams's stepbrother and even teaching him in a course called Mammalian Anatomy. After working for a year at Cincinnati as an assistant to McMurrich, Turner was hired in 1893 by Clark University, another college in Atlanta that served primarily African American students. When Du Bois arrived in 1897–98, Turner was teaching nine different biology classes, including one for seniors called Philosophy of Biology that covered "such general biological problems as Evolution, Heredity, Variation, etc." For Turner, biology and philosophy were connected: he argued in an 1897 article on "Reasons for Teaching the Negro Biology" that "the modern student of ethics is a biologist, when he strives to decipher the laws that determine the development of a healthy moral nature." He believed that biology taught humility and was thus essential "to the advancement of the race": "A study of the marvelous structures and functions of animals, the demonstration that all animals are evolved branches of one common tree, and a knowledge of the laws that control the actions and relations of animals and man— . . . all these things lead one to recognize and respect the rights of others." Turner was also an appreciative early reader of Du Bois: in 1902, he cited an evocative passage from Du Bois's "The Freedman's Bureau" to introduce his own argument that "the new Southerner" and "the new Negro," if only they would realize that "a similar education and a like environment have made [them] alike in everything except color and features," would overcome the "Negro problem."[101]

---

101. *Catalogue of the Academic Department, 1890–91* (Cincinnati: University of Cincinnati, 1891), 5, 7; *Catalogue of the Academic Department, 1891–92* (Cincinnati: University of Cincinnati, 1892), 6; *Catalogue of the Academic Department, 1892–93* (Cincinnati: University of Cincinnati, 1893), 5, 48; Charles Judson Herrick, "Clarence Luther Herrick: Pioneer Naturalist, Teacher, and Psychobiologist," *Transactions of the American Philosophical Society* 45, no. 1 (1955): 37; *Twenty-Second Annual Report of the Directors of the University of Cincinnati for the Year Ending December 31, A. D. 1892* (Cincinnati: Commercial Gazette, 1893), 36–38; *Twenty-Third Annual Report of the Directors of the University of Cincinnati for the Year Ending December 31, A. D. 1893* (Cincinnati: Commercial Gazette, 1894), 41–43; "About CAU," Clark Atlanta University, accessed October 3, 2019, http://www.cau.edu/about/; Charles Henry Turner, "Morphology of the Nervous System of Cypris," *Journal of Comparative Neurology* 6 (1896): 24; "Young Colored Men and Women to the Front," *Southwestern Christian Advocate*, June 29, 1893; *Clark University Courier: Catalogue Edition, 1897–1898* (South Atlanta, GA: Clark University), 6, 25, 28–30; Charles Henry Turner, "Reasons for Teaching the Negro Biology

CHAPTER THREE

In 1898, with financial help from a subscription organized by faculty at the Gammon Theological Seminary in Atlanta, Turner was able to study at the University of Chicago for the summer quarter with the entomologist William Morton Wheeler. Turner's research at the time was focused on neurological homologies of arthropods and annelids. Teaching at Clark until 1905, he eventually returned to Chicago in the summer of 1906 and received his PhD in zoology in March 1907 for a thesis on ant behavior, supervised by Charles Otis Whitman. Turner's experiments proved that "ants are much more than mere reflex machines; they are self-acting creatures guided by memories of past individual (ontogenetic) experience." Although Dewey had left Chicago in 1904, pragmatism was still in the air: Turner's "secondary department" was Psychology, where he was supervised by Dewey's former student James Rowland Angell; he also took two comparative psychology classes with Mead and several zoology classes with Charles Manning Child, whose approach to biology was influenced by pragmatism. Turner was the second African American (after Coffin) to receive a doctorate in biology and arguably the first to receive one in psychology. This achievement earned him a 1912 profile in the *Crisis*, the magazine of the National Association for the Advancement of Colored People (NAACP) edited by Du Bois, who later recalled—after a chance encounter in 1938 with Turner's grandson Darwin—that he had known Turner in Atlanta. Du Bois used Turner's failure to secure a research position, despite his excellent research record, as a prime example of how racism hindered the careers of even the most accomplished African American scientists. Their acquaintance from 1897 to 1905 would have provided Du Bois with a window into recent developments in biology.[102]

[Part 1]," *Southwestern Christian Advocate*, April 1, 1897; Charles Henry Turner, "Will the Education of the Negro Solve the Race Problem?" in *Twentieth Century Negro Literature, or a Cyclopedia of Thought on the Vital Topics Relating to the American Negro*, ed. Daniel Wallace Culp (Toronto: J. L. Nichols, 1902), 164; citing W. E. B. Du Bois, "The Freedmen's Bureau," *Atlantic Monthly* 87 (1901): 360. For McMurrich and Haldeman at Hopkins, see "Enumeration of Classes, First Half-Year, 1884–85," *Johns Hopkins University Circulars* 4 (1884): 4; "Chesapeake Zoölogical Laboratory: Marine Station of the Johns Hopkins University, Eighth Session, Beaufort, N.C., 1885," *Johns Hopkins University Circulars* 5 (1885): 2.

102. "Personal and General," *Southwestern Christian Advocate*, July 7, 1898; *Annual Register, July, 1898—July, 1899, with Announcements for 1899–1900* (Chicago: University of Chicago Press, 1899), 447; Charles Henry Turner, "Notes on the Mushroom Bodies of the Invertebrates: A Preliminary Paper on the Comparative Study of the Arthropod and Annelid Brain," *Zoological Bulletin* 2 (1899); Charles Henry Turner, "The Mushroom Bodies of the Crayfish and Their Histological Environment," *Journal of Comparative Neurology* 11 (1901): 323–24; William Morton Wheeler, *The Social Insects: Their Origin and Evolution* (London: K. Paul, Trench, Trubner, 1928), 178; *Annual Register, July, 1906–July, 1907, with Announcements for 1907–1908* (Chicago: University of Chicago, 1907), 350; "The Associa-

The early second-cohort pragmatists, as they began their professorial careers, carried their college and graduate school experiences with evolution into the classroom. Whether at Harvard, Michigan, Atlanta, or Hull House in Chicago, they introduced a new generation to the philosophical and social implications of modern biology.

Both the first and second cohorts of pragmatists were in constant conversation with biology. However, there were important differences in how the two cohorts experienced biological debates. For the first-cohort pragmatists, whether evolution would ultimately prevail as a scientific theory was still in question. But for the second-cohort pragmatists, as this chapter has shown, evolution was an established feature of the scientific and philosophical landscape. Contributions to philosophy and the social sciences in the late nineteenth century had to come to terms in one way or another with modern biological ideas.

There were at least two other differences between the cohorts, stemming in part from this initial difference. First, many of the second-cohort pragmatists were attracted to idealism, which seemed to offer an evolutionary alternative to Spencer's philosophy. They went on to develop a dialectical account of the organism-environment relation, as we will see in chapter 4. In contrast, most members of the first cohort were either positivists (Wright, Fiske, Abbot) or strongly opposed to idealism (James). Peirce was the exception, though his interest in Kant and Hegel was not usually framed in evolutionary terms. Second, whereas the first-cohort pragmatists entered the conversation about evolution in the 1860s, when the theory itself was controversial, the second-cohort pragmatists joined

---

tion of Doctors of Philosophy," *University Record* (Chicago, IL) 11 (1907): 170; Charles Henry Turner, "The Homing of Ants: An Experimental Study of Behavior," *Journal of Comparative Neurology and Psychology* 17 (1907): 369–70, 424; Thomas C. Cadwallader, "Neglected Aspects of the Evolution of American Comparative and Animal Psychology," in *Behavioral Evolution and Integrative Levels: The T. C. Schneirla Conference Series*, ed. Gary Greenberg and Ethel Tobach (Hillsdale, NJ: Lawrence Erlbaum, 1984), 35; Harry Washington Greene, *Holders of Doctorates among American Negroes: An Educational and Social Study of Negroes Who Have Earned Doctoral Degrees in Course, 1876–1943* (Boston: Meador, 1946), 194–95; "Men of the Month," *Crisis* 3 (1912): 102; W. E. B. Du Bois, "The Negro Scientist," *American Scholar* 8 (1939): 309–10; W. E. B. Du Bois, "The Winds of Time: A Scientist," *Chicago Defender*, May 24, 1947. Information on Turner's courses and supervisors is taken from his transcript or "Record of Work" at the University of Chicago, a copy of which was provided to me by the university registrar. On Child and Dewey, see Sharon E. Kingsland, "Toward a Natural History of the Human Psyche: Charles Manning Child, Charles Judson Herrick, and the Dynamic View of the Individual at the University of Chicago," in *The Expansion of American Biology*, ed. Keith R. Benson, Jane Maienschein, and Ronald Rainger (New Brunswick, NJ: Rutgers University Press, 1991); Nadine Weidman, "Psychobiology, Progressivism, and the Anti-Progressive Tradition," *Journal of the History of Biology* 29 (1996): 271–76.

the discussion in the 1880s and '90s, by which time it was the causes, or "factors," of evolution that were up for debate. These debates over the factors of evolution are the topic of chapter 5, where Peirce is again the exception, since he was an enthusiastic participant. Both cohorts of pragmatists ultimately joined forces in the late 1890s and early 1900s to develop pragmatist ethics (chapter 6) and pragmatist logic (chapter 7), which cannot be understood apart from this evolutionary context.

FOUR

"Hegelianism Needs to Be Darwinized": Evolution and Idealism

John Dewey often implied that a growing interest in biology helped push him away from an early commitment to Hegelian idealism. In 1911, he told a young French philosopher that the early 1890s "transition in [his] thought" had been the result of three factors: the impact of "the biological conception" presented in William James's *Principles of Psychology*; his new interest in adapting "the logic of [Immanuel] Kant and [G. W. F.] Hegel to the conditions of actual scientific inquiry"; and his attempt in ethics "to develop a theory of a more organic connection between thought and action." In a later autobiographical essay, he again emphasized James's return to the "biological conception of the *psyche*, . . . a return possessed of a new force and value due to the immense progress made by biology since the time of Aristotle." Most early commentators followed Dewey's lead. According to Morton White, "a thoroughgoing Darwinism force[d] Dewey to surrender Hegel." Richard Bernstein's book on Dewey took a similar line, with an early chapter entitled "From Hegel to Darwin."[1]

---

1. John Dewey to Henri Robet, 2 May 1911, in Dewey, *Correspondence*, no. 01991; John Dewey, "From Absolutism to Experimentalism," in *Contemporary American Philosophy: Personal Statements*, ed. George Plimpton Adams and William Pepperell Montague, vol. 2 (New York: Russell & Russell, 1930), 24; Morton White, *The Origin of Dewey's Instrumentalism* (New York: Columbia University Press, 1943), 40; Richard J. Bernstein, *John Dewey* (New York: Washington Square, 1966), 9–21.

CHAPTER FOUR

It is by now well established that Dewey's philosophy remained Hegelian even after his biological turn.[2] Nevertheless, Dewey scholars who emphasize his idealism usually downplay his naturalism, while those who focus on his naturalism tend to criticize his idealism. For example, Richard Rorty saw Dewey as occupying the conceptual space "between Hegel and Darwin." But he explicitly rejected Dewey's own understanding of these thinkers as engaged with biology and evolution, instead linking them to historicism and relativism. On Rorty's reading, preoccupied with discourse and history, Dewey's philosophy was not truly biological even when it explicitly referred to biology.[3] Recent champions of Dewey's naturalism, in contrast, tend to think that his idealist ancestry only detracts from his views. Cheryl Misak, for instance, argues that "Dewey's attempt at bringing Hegelian insights to the empiricist or naturalist picture seems always less than satisfactory." For those seeking a naturalist Dewey, these Hegelian traces are an embarrassment—responsible for muddled metaphysics. As Peter Godfrey-Smith once put it, modern naturalists tend to see Dewey "as someone with good instincts but a lack of rigor and a Hegelian hangover."[4]

This contrast between science and idealism is also visible in scholarship on other second-cohort pragmatists. For example, Robert Gooding-Williams has argued that the social scientific approach in the central chapters of W. E. B. Du Bois's *The Souls of Black Folk* "successfully, but inadvertently, calls into question his Hegelianism," on display in other parts of the book.[5] In this chapter, I will argue that there is no conflict between naturalistic and idealistic pragmatism: the idea that these perspectives

---

2. John R. Shook, *Dewey's Empirical Theory of Knowledge and Reality* (Nashville, TN: Vanderbilt University Press, 2000), 20; Jim Garrison, "The 'Permanent Deposit' of Hegelian Thought in Dewey's Theory of Inquiry," *Educational Theory* 56 (2006); James A. Good, *A Search for Unity in Diversity: The "Permanent Hegelian Deposit" in the Philosophy of John Dewey* (Lanham, MD: Lexington, 2006); James A. Good and Jim Garrison, "Traces of Hegelian *Bildung* in Dewey's Philosophy," in *John Dewey and Continental Philosophy*, ed. Paul Fairfield (Carbondale: Southern Illinois University Press, 2010); James Scott Johnston, *John Dewey's Earlier Logical Theory* (Albany: State University of New York Press, 2014).

3. Richard Rorty, "Dewey between Hegel and Darwin," in *Modernist Impulses in the Human Sciences, 1870–1930*, ed. Dorothy Ross (Baltimore: Johns Hopkins University Press, 1994), 56–57. Rorty elsewhere speaks of "Dewey's naturalized Hegelianism" but does not connect this to biology or to organism-environment interaction: Richard Rorty, "Texts and Lumps," *New Literary History* 17 (1985): 14; Richard Rorty, *Philosophy and the Mirror of Nature* (Princeton, NJ: Princeton University Press, 1979), 5.

4. Cheryl Misak, *The American Pragmatists* (Oxford: Oxford University Press, 2013), 121; Peter Godfrey-Smith, "Dewey on Naturalism, Realism and Science," *Philosophy of Science* 69 (2002): S1.

5. Robert Gooding-Williams, "Philosophy of History and Social Critique in *The Souls of Black Folk*," *Social Science Information* 26 (1987): 99. See also Paul C. Taylor, "What's the Use of Calling Du Bois a Pragmatist?" *Metaphilosophy* 35 (2004): 106–7.

are necessarily in tension is an anachronism. The argument proceeds in three parts. First, I will demonstrate that some of the British idealists—key figures for Dewey and others in the 1880s and '90s—saw important connections between conceptual evolution in the work of Hegel and organic evolution in that of Herbert Spencer and Charles Darwin. This view was shared by Dewey and George Herbert Mead, and it featured in their class Movements of Thought in the Nineteenth Century, which they taught at the University of Michigan and then at the University of Chicago. Next, I will show that the dialectical notion of organism-environment interaction central to the work of Dewey and Mead stemmed from a Hegelian approach to adaptation. More specifically, I will argue that Dewey's account of the organism-environment relation derives from the work of Oxford Hegelians such as Edward Caird and Samuel Alexander, who were attempting to reconcile evolutionary ideas with a critique of Spencer's environmentalist account of human thought and action. Finally, in the last section I will describe how Josiah Royce connected evolutionism and idealism in his book *The Spirit of Modern Philosophy*, an important source for Du Bois.[6] Given the close link between evolution and idealism at the time, Du Bois's adoption of a historical or developmental perspective in his early work is not straightforward evidence of Hegelian influence. Charles Sanders Peirce summed up the late nineteenth-century attitude, arguably shared by both Dewey and Du Bois: "Hegelianism needs to be Darwinized."[7]

Hegel and Evolution

G. W. F. Hegel and Charles Darwin may seem a strange pair. Although Hegel read the work of Jean-Baptiste Lamarck and adopted the French naturalist's classification of animals, he explicitly rejected evolutionary ideas: "The formations of nature are determinate and bounded, and it is as such that they enter into existence. . . . Man has not formed himself out of the animal, nor the animal out of the plant, for each is instantly the whole of what it is."[8] Some historians of philosophy would prefer

---

6. Gooding-Williams, "Philosophy of History and Social Critique in *The Souls of Black Folk*," 103–6.
7. Charles Sanders Peirce, "Ritchie's Darwin and Hegel," *Nation*, November 23, 1893, 394.
8. Georg Wilhelm Friedrich Hegel, *Hegel's Philosophy of Nature*, trans. Michael John Petry, 3 vols. (London: George Allen & Unwin, 1970), 3:22 [§339Z]; Georg Wilhelm Friedrich Hegel, *Werke: Vollständige Ausgabe, durch einen Verein von Freunden des Verewigten*, vol. 7, Vorlesungen über die Naturphilosophie (Berlin: Duncker & Humblot, 1842), 440. When quoting from Hegel's work I will

CHAPTER FOUR

to forget about Hegel's forays into the natural sciences. Terry Pinkard comments that Hegel seems to have had a knack for betting on the wrong horse when it came to scientific debates. Nevertheless, Frederick Beiser has shown convincingly that Hegel's *Naturphilosophie* and his organicism were central to his philosophical system.[9] In this section, I will demonstrate that at least some philosophers in the late nineteenth century claimed that there was an important connection between Hegel's notion of *Entwicklung* (development or evolution) and biological evolution. Edward Caird, Samuel Alexander, and David George Ritchie in Britain and Dewey and Mead in the United States were committed to "the marriage of Hegel and Darwin."[10]

Many of the British idealists, who rose to fame in the last quarter of the nineteenth century, were trained at Balliol College in the University of Oxford.[11] Because these idealist philosophers were also deeply influenced by Hegel and Hegelian readings of Kant, I will sometimes refer to them as the Oxford Hegelians. The leader of this group of Oxford-trained philosophers was Thomas Hill Green, who taught almost all of the younger idealists at Balliol in the 1860s and '70s after attending the college himself.[12]

Green, unlike some of his followers, thought that the doctrine of organic evolution was irrelevant to philosophical concerns. He was best known in the 1870s for his criticisms of David Hume, but as Alexander Klein has shown, Green's real target was the empirical psychology of his contemporaries.[13] This is most obvious in several articles published shortly before his death, collectively entitled "Mr. Herbert Spencer and Mr. G. H. Lewes: Their Application of the Doctrine of Evolution to

---

include the section number in brackets; the "Z" indicates that the quotation is from an addition (*Zusatz*) to the section.

9. Terry Pinkard, "Speculative *Naturphilosophie* and the Development of the Empirical Sciences: Hegel's Perspective," in *Continental Philosophy of Science*, ed. Gary Gutting (Oxford: Blackwell, 2005), 19; Frederick Beiser, *Hegel* (New York: Routledge, 2005), 80–109.

10. James T. Kloppenberg, *Uncertain Victory: Social Democracy and Progressivism in European and American Thought, 1870–1920* (Oxford: Oxford University Press, 1986), 35.

11. Notable philosophers who studied at Balliol included Thomas Hill Green (1855–59), Edward Caird (1860–63), Bernard Bosanquet (1867–70), David George Ritchie (1874–78), Samuel Alexander (1878–81), and Ferdinand Canning Scott Schiller (1882–86). Some, such as Green and Caird, later taught at the college as well. Although his brother Andrew studied and taught at Balliol, the philosopher Francis Herbert Bradley went to Jesus College (1865–70) and was later a fellow at Merton. For dates and affiliations, see Joseph Foster, *Alumni Oxonienses: The Members of the University of Oxford, 1715–1886; Their Parentage, Birthplace, and Year of Birth, with a Record of Their Degrees*, 4 vols. (London: Joseph Foster, 1887).

12. Sandra M. den Otter, *British Idealism and Social Explanation* (Oxford: Clarendon Press, 1996), 10; W. J. Mander, *British Idealism: A History* (Oxford: Oxford University Press, 2011), 7.

13. Alexander Klein, "On Hume on Space: Green's Attack, James' Empirical Response," *Journal of the History of Philosophy* 47 (2009): 427–28.

Thought." In the first of these he quoted Spencer's provocative claim: "Should the idealist be right [about the subject-object relation], the doctrine of Evolution is a dream." Green responded:

To those who have humbly accepted the doctrine of evolution as a valuable formulation of our knowledge of animal life, but at the same time think of themselves as "idealists," this statement may at first cause some uneasiness. On examination, however, they will find . . . that when Mr. Spencer in such a connection speaks of the doctrine of evolution, he is thinking chiefly of its application to the explanation of knowledge—an application at least not necessarily admitted in the acceptance of it as a doctrine of animal life.[14]

Green accepted organic evolution but denied it an important role in philosophical accounts of knowledge and mind. He thus agreed with Dewey's teacher George Sylvester Morris, whom we met in chapter 3: "Conceding that the law of evolution has been established, the nature and the wording of philosophical problems have not been changed one whit." Green also agreed with Henry Sidgwick that "the principle of evolution, the process by which the human animal has come . . . to exhibit the phenomena of a moral life," does not tell us what we ought to do. In short, Green was opposed to normative approaches in evolutionary epistemology and evolutionary ethics.[15]

Many of Green's followers, however, argued that Hegel's idea of *Entwicklung*, Darwin and Spencer's theories of organic evolution, and Auguste Comte's law of development were closely connected. For example, Edward Caird was much more sympathetic to evolutionary ideas than his friend Green was.[16] In a book that later influenced Jane Addams, Caird claimed that development (a synonym of *evolution* at the time) was the central organizing principle of nineteenth-century science and philosophy:

Lessing, Kant, and Herder gave that decisive impulse under which the principle of development was carried into biology by Goethe, Schelling, and many eminent scientific men, while Hegel made it the leading idea of his philosophy. . . . After these we need

---

14. Herbert Spencer, *The Principles of Psychology*, 2nd ed., vol. 2 (London: Williams & Norgate, 1872), 311; Thomas Hill Green, "Mr. Herbert Spencer and Mr. G. H. Lewes: Their Application of the Doctrine of Evolution to Thought; Part I, Mr. Spencer on the Relation of Subject and Object," *Contemporary Review* 31 (1877): 35.

15. George Sylvester Morris, *British Thought and Thinkers: Introductory Studies, Critical, Biographical and Philosophical* (Chicago: S. C. Griggs, 1880), 346; Thomas Hill Green, *Prolegomena to Ethics* (Oxford: Clarendon Press, 1883), 9; cf. Henry Sidgwick, "The Theory of Evolution in Its Application to Practice," *Mind* 1 (1876).

16. On Green and Caird's friendship, see Sandra M. den Otter, "Caird, Edward (1835–1908)," in *Oxford Dictionary of National Biography* (Oxford: Oxford University Press, 2004).

only refer to the names of Lamarck and Comte in France, of Darwin and Spencer in England, and of Von Hartmann and Wundt in Germany, as writers who have done much to throw light on various aspects of the idea and to give it new applications. We may, indeed, say without much exaggeration that the thought of almost all the great speculative or scientific writers of this century has been governed and guided by the principle of development, if not directly devoted to its illustration.[17]

Although William Mander downplays the connection between Caird's principle of development and theories of organic evolution, Caird did discuss the Darwinian theory in several texts. Nonetheless, when Caird spoke of evolution, he was usually thinking of an abstract dialectical process, as indicated by this passage from his book on Hegel: "the unity of opposites, not as an external synthesis, but as a result of the necessary evolution of thought by means of an antagonism which thought itself produces and reconciles."[18]

Although Caird spoke often of the general idea of evolution, it was two younger Oxford philosophers—Alexander and Ritchie—who explicitly attempted a rapprochement between Hegel and Darwin. The two first met in 1878 as students at Balliol and they both held Oxford fellowships in the 1880s before moving on to professorships elsewhere in the mid-1890s. Both were influenced by Green and thus were well aware of his criticisms of "the evolution-psychology."[19] However, they also formed friendships with biologists and ended up reading Hegel with Darwin-tinted lenses.

Alexander and Ritchie each published essays during their Oxford fellowships that brought together Hegelian and evolutionary ideas. Alexander struck first with "Hegel's Conception of Nature," published in *Mind*

17. Edward Caird, *The Evolution of Religion*, 2 vols. (Glasgow: James Maclehose & Sons, 1893), 1:24; see Jane Addams, *Twenty Years at Hull-House, with Autobiographical Notes* (New York: Macmillan, 1910), 39; Marilyn Fischer, *Jane Addams's Evolutionary Theorizing: Constructing "Democracy and Social Ethics"* (Chicago: University of Chicago Press, 2019), 120–21. On *evolution* and *development*, see Robert J. Richards, *The Meaning of Evolution: The Morphological Construction and Ideological Reconstruction of Darwin's Theory* (Chicago: University of Chicago Press, 1992).

18. W. J. Mander, "Caird's Developmental Absolutism," in *Anglo-American Idealism, 1865–1927*, ed. W. J. Mander (Westport, CT: Greenwood, 2000), 52; Edward Caird, *Hegel* (Edinburgh: William Blackwood, 1883), 43. For Caird's references to Darwin's theory, see Edward Caird, "Metaphysic," in *The Encyclopaedia Britannica: A Dictionary of Arts, Sciences, and General Literature*, vol. 16 (Edinburgh: Adam & Charles Black, 1878), 92; Edward Caird, *The Critical Philosophy of Immanuel Kant*, 2 vols. (Glasgow: James Maclehose & Sons, 1889), 2:539–44.

19. Samuel Alexander, review of *Works of Thomas Hill Green*, vol. 1, edited by Richard Lewis Nettleship, *Academy*, October 10, 1885, 242. Alexander and Ritchie corresponded about a draft of this review: see David George Ritchie to Samuel Alexander, 9 September 1885, ALEX/A/1/1/236, Alexander Papers.

in 1886. He had been studying Hegel's philosophy of nature, as indicated by his notebooks and correspondence. The philosophically inclined biologist John Scott Haldane (father of the population geneticist J. B. S. Haldane) had written to him earlier that year: "I am very glad to hear that you have taken in hand the Naturphilosophie. It certainly has its fair share of unintelligibility as well as interest."[20] Alexander's essay was primarily a description of Hegel's *Naturphilosophie*, but the final section compared the views of Hegel with those of modern evolutionists:

> Between the doctrine of evolution and Hegel's theory, how great the likeness seems to be! When Hegel speaks of nature as a process in which, with ever increasing specification of external characters, there is an ever completer involution or reflexion of these parts to a centre, we seem to anticipate the law of progress from indefinite incoherent homogeneity to definite coherent heterogeneity. Hegel's philosophy is in fact an evolution, called by the name of dialectic, which is the counterpart in philosophy of what evolution is in science.[21]

The "law of progress" referred to was Spencer's law of evolution. On Alexander's reading, then, Hegel's dialectical method was itself evolutionary: Hegel's philosophy of nature foreshadowed Spencer's evolutionary philosophy.

Not all Hegelians shared Alexander's desire to bring together Hegel and modern biology. James Hutchinson Stirling, author of *The Secret of Hegel* and one of the first British thinkers to engage at length with the German philosopher, questioned Alexander's attempt in the closing pages of "Hegel's Conception of Nature" to connect Hegel and biological evolution:

> You are very gentle with these "Modern Theories" in the end. I, for my part, have no patience with what the British Association [for the Advancement of Science] glibly receives as established truth now, indisputable science now, to wit, *Natural* Selection. I never hesitate to call the Darwinian proposition a proposition of dementia.

Stirling, though admitting in a subsequent letter the possibility that he simply did not understand Darwin, sided with Hegel in rejecting biological

---

20. John Scott Haldane to Samuel Alexander, 29 January [1886], ALEX/A/1/1/110, Alexander Papers. This letter was likely written in 1886, as Haldane describes work he was doing in Dundee at that time (I am grateful to Steve Sturdy for help in dating this letter).
21. Samuel Alexander, "Hegel's Conception of Nature," *Mind* 11 (1886): 518; cf. Herbert Spencer, *First Principles*, 4th ed. (London: Williams & Norgate, 1880), 396.

evolution. His opposition to Alexander's attempt at reconcilement shows that not everyone reading Hegel in the late nineteenth century saw a harmony between Hegel's philosophy and evolutionary ideas.[22]

Ritchie, in his 1891 essay "Darwin and Hegel," made claims similar to those of Alexander five years earlier. Like Alexander, Ritchie had cultivated friendships with biologist colleagues. He sent the Oxford zoologist Edward Bagnall Poulton (of whom we will see more in chapter 5) an offprint of his Darwin-Hegel article "with the writer's kind regards"; that same year he referred to Poulton as a friend "to whom, more than to any man or book I am indebted for my biological premises."[23] Ritchie was more concerned with Hegel's general approach than with the details of the *Naturphilosophie*: "I think, however, it is worthwhile to see whether we can get any help, not from details in Hegel, but from his general method and spirit of philosophising, in making the attempt to *think* nature and human society as they present themselves to us now in the light of Darwin's theory of natural selection." Ritchie described his task as "Hegelianising natural selection," and he thus treated Darwin's factors of Heredity and Variation as forms of Hegel's categories of Identity and Difference, respectively. He related Darwin's third factor, Struggle for Existence, to Hegel's notion of negativity. Unfortunately, Ritchie did not explain these connections but merely pointed to them. Peirce's reaction, from which the title of this chapter is taken, was that "Hegelianism needs to be Darwinized much more than Darwinism needs to be Hegelianized." That aside, according to Ritchie, it is natural selection in particular, and not evolution in general, that meshes most easily with Hegel's dialectic.[24]

Dewey and Mead agreed with the Oxford Hegelians that Hegel and Darwin were part of a larger evolutionary zeitgeist, as evidenced by student lecture notes from their course Movements of Thought in the Nineteenth Century. Dewey created this course at the University of Michigan, where he taught it for three years from 1891 to 1893; Mead, Dewey, and James Hayden Tufts all taught versions of the course at the University of Chicago in the 1890s, and Mead went on to teach it most years from

---

22. James Hutchison Stirling to Samuel Alexander, 17 October 1886 and 27 October 1886, ALEX/A/1/1/277, Alexander Papers; James Hutchison Stirling, *The Secret of Hegel: Being the Hegelian System in Origin, Principle, Form, and Matter*, 2 vols. (London: Longman, Green, Longman, Roberts, & Green, 1865).

23. David George Ritchie, *Darwinism and Politics*, 2nd ed. (London: Swan Sonnenschein, 1891), iv. Poulton's personal library of books and offprints is held at the Hope Entomological Library, Oxford University Museum of Natural History.

24. David George Ritchie, "Darwin and Hegel," *Proceedings of the Aristotelian Society* 1 (1891): 63–64; Charles Sanders Peirce, "Ritchie's Darwin and Hegel," *Nation*, November 23, 1893, 394.

1898 to 1928.[25] The first version of the course, which Dewey taught at Michigan in 1891, argued that the historical approach of late eighteenth-century authors such as Johann Gottfried Herder implied "some whole which is in the process of evolution." This idea culminated, said Dewey, in Hegel's philosophy, which was based on the interaction between thought and the world:

> Hegel accounts for the apparent contradictions between intelligence and the actual conditions of life through the idea of evolution. This organization of experience is not something which is given to any man; it has to be worked out. During the process of development there will be a great deal of conflict. But this very conflict, as fast as it comes into consciousness, so fast as man becomes aware of the friction, leads to a readjustment, to the securing of a better organization.[26]

On this reading of Hegel, it is the conflict between mind and environment that leads to readjustment, development, and evolution. The following year, Dewey quoted the French thinker Ernest Renan: "The great progress of modern thought has been the substitution of the category of *evolution* for the category of the *'being.'*" Commenting on this claim, Dewey again invoked Hegel: "When we go on to consider the law of evolution . . . the transference of the Hegelian doctrine becomes even more marked. It is the same law, only considered now as the law of historic growth, not as the dialectic unfolding of the absolute."[27] Thus for Dewey, Hegel's dialectic was part of a broader nineteenth-century obsession with history, growth, and evolution.

Although Dewey spoke about Hegel's idea of evolution in abstract terms, his younger colleague made the link between Hegel and Darwin explicit. Mead framed his discussion of Hegel in Movements of Thought in the Nineteenth Century in terms of biological evolution:

> What Hegel undertook to do was to show how this opposition between subject and object could be overcome, in some sense, by means of the recognition of the nature of the process of thought itself. In biological evolution we overcome the opposition

---

25. See the *Calendar of the University of Michigan* and the *Register of the University of Chicago* for the relevant years. For more on the course, see Charles Camic, "Changing 'Movements of Thought in the Nineteenth Century': Historical Text and Historical Context," in *The Timeliness of George Herbert Mead*, ed. Hans Joas and Daniel R. Huebner (Chicago: University of Chicago Press, 2016).

26. John Dewey, "Movements of Thought in the Nineteenth Century," Lecture 12 (31 November 1891), Edwin Spencer Peck Notebooks (851997 Aa 2), Bentley Historical Library, University of Michigan.

27. Ernest Renan, *The Future of Science* (Boston: Roberts, 1891), 169; John Dewey, "Two Phases of Renan's Life: The Faith of 1850 and the Doubt of 1890," *Open Court*, December 29, 1892, 3505.

CHAPTER FOUR

Figure 11 Marginal note on Hegel and Darwin in a set of notes taken by Irene Tufts in George Herbert Mead's course Movements of Thought in the Nineteenth Century (Spring 1915) at the University of Chicago.
From p. 36, Folder 8, Box 10, Mead Papers. Reproduced courtesy of the Special Collections Research Center, University of Chicago Library.

between the identity of the life-process in all forms and the diversity of the living forms themselves by studying the process as it is taking place. . . . Now, Hegel attempted to set up a picture similar to this as it applied to the thought processes, to the process of knowing, and possibly of all sensing, perceiving, and thinking.[28]

This connection between Hegel and biological evolution was apparently obvious to at least some of Mead's students. In the margin of her notes on Hegel and evolution from the 1915 version of the Movements course, his future daughter-in-law Irene Tufts exclaimed, "Hegel + Darwin / Shock!" (figure 11). Mead made similar points in courses that dealt specifically with Hegel. Discussing Hegel's logic, he summarized, "Hegel's doctrine one of *development*, evolution—a process leading to different forms—but an identical process—the life process, the thought process, the historical process."[29] Thus Mead and Dewey in the United States, like Alexander and Ritchie in England, saw Hegel and Darwin as focusing on different aspects of the same nineteenth-century idea—development or evolution.

The Organism-Environment Dialectic

Philosophers such as Samuel Alexander and John Dewey, born in the same year as the publication of Darwin's *Origin of Species*, had no trouble

28. George Herbert Mead, *Movements of Thought in the Nineteenth Century* (Chicago: University of Chicago Press, 1936), 129. This book was assembled from notes taken during the Spring 1928 version of the course: see Series 3, Subseries 2, Mead Papers.
29. "Movements of Thought in the Nineteenth Century," p. 36 (6 May 1915), Folder 8, Box 10, Mead Papers; "Hegel's Logic," p. 48 (12 June 1923), Folder 5, Box 9, Mead Papers; see also "Hegel's Phenomenology," Folders 6–7, Box 9, Mead Papers. Irene Tufts, daughter of Chicago philosophy professor James Hayden Tufts, married Mead's son Henry in 1917 (*University of Chicago Magazine* 8 [June 1917]: 349).

finding connections between Hegel's philosophy and biological evolution. The mystery is how they and their idealist colleagues managed to reconcile the Hegel-evolution link with the opinion of their teachers Green and Morris, according to whom the fact of biological evolution was of no relevance to philosophy. In this section, I argue that the Oxford Hegelians split the difference, embracing biological evolution as relevant to philosophy but following Green in opposing Herbert Spencer's environmentalist account of knowledge and ethics. I also demonstrate that Dewey was directly influenced by these idealist philosophers: their approach united his apparently divergent interests in Hegel and biology, making him a "dialectical biologist" in the 1890s.

The approach of Dewey, Mead, and the British idealists was dialectical in at least two ways. First, they treated adaptation as a result of the reciprocal action of organism and environment: just as the environment affects the organism, the organism affects the environment. Second, they claimed that organism and environment were best seen as two aspects of one thing—life. "Reciprocal action" sounds reasonable to most scholars, but "aspects of one thing" tends to raise eyebrows. Peter Godfrey-Smith locates both ideas in the work of Richard Lewontin, coauthor of the 1985 book *The Dialectical Biologist*. Lewontin famously claimed that "the environment is a product of the organism, just as the organism is a product of the environment."[30] As Godfrey-Smith suggests, Lewontin seemed to go beyond simple causal interaction, viewing "the organism-environment pair as a single whole in which organism and environment are parts that codetermine each others' properties." Godfrey-Smith argues that the latter view tends to lump together different senses of "constructing" in the phrase "organisms constructing their environments"; he also links this view to Dewey.[31]

It is helpful to separate these two accounts of organism and environment—the *reciprocal causes* view and the *dual aspects* view—when analyzing philosophical texts, and I distinguish them in what follows. But it is important to note that the people I focus on in this chapter, from Caird and Alexander to Mead and Dewey to Levins and Lewontin, treat them

---

30. Richard Levins and Richard C. Lewontin, *The Dialectical Biologist* (Cambridge, MA: Harvard University Press, 1985), 69. Although the book is coauthored, the chapters "Adaptation" and "The Organism as Subject and Object of Evolution" (among others) had previously appeared under Lewontin's sole authorship.

31. Peter Godfrey-Smith, "Organism, Environment, and Dialectics," in *Thinking about Evolution: Historical Philosophical, and Political Perspectives*, ed. Rama S. Singh et al. (Cambridge: Cambridge University Press, 2001), 258, 261–62; see also Peter Godfrey-Smith, *Complexity and the Function of Mind in Nature* (Cambridge: Cambridge University Press, 1996), chaps. 4–5.

CHAPTER FOUR

as aspects of one general framework. I suggest that this framework represents a tradition of biological idealism or dialectical naturalism whose members (1) tend to substitute talk of organism/environment for talk of subject/object and (2) deny ontological idealism—that is, the doctrine that only the mental exists.[32] I do not attempt to evaluate or further characterize biological idealism in this chapter. For some scholars, as for Godfrey-Smith, a conflation of the dual aspects and reciprocal causes views means that the framework is incoherent or at least somewhat confused; for others, the dual aspects view is either silly or trivial: any two interacting entities can be treated as part of one process, but surely they are still independent. Nevertheless, considering the number of thinkers working at the biology-philosophy nexus who have found this type of framework appealing, it may be worthy of more serious consideration by philosophers.

Biological idealism initially emerged as an unlikely hybrid. As mentioned in the previous section, Green was strongly opposed to Spencer's empirical psychology. The basis of this psychology—and of Spencer's philosophy as a whole—was the notion of a correspondence between organism and environment. As I have shown elsewhere, Spencer popularized the word *environment* as well as the idea of a relation between two singular entities, organism and environment. Following Comte, he made organism-environment correspondence the basis of his conception of life.[33] Since Spencer viewed mind as merely an advanced form of life, he also framed intelligence in terms of adjustment to environment:

On comparing the phenomena of mental life with the most nearly allied phenomena—those of bodily life—and inquiring what is common to both groups, a generalization was disclosed which proves on examination to express the essential character of all mental actions. Regarded under every variety of aspect, intelligence is found to consist in the establishment of correspondences between relations in the organism and relations in the environment; and the entire development of intelligence may be for-

---

32. Murray Bookchin uses the term "dialectical naturalism" in a distinct but related sense: see Murray Bookchin, *The Philosophy of Social Ecology: Essays on Dialectical Naturalism* (Montreal: Black Rose, 1990).
33. Trevor Pearce, "From 'Circumstances' to 'Environment': Herbert Spencer and the Origins of the Idea of Organism-Environment Interaction," *Studies in History and Philosophy of Biological and Biomedical Sciences* 41 (2010). See also Georges Canguilhem, "Le vivant et son milieu," in *La connaissance de la vie* (Paris: Hachette, 1952); Jean-François Braunstein, "Le concept de milieu, de Lamarck à Comte et aux positivismes," in *Jean-Baptiste Lamarck, 1744–1829*, ed. Goulven Laurent (Paris: CTHS, 1997); Ferhat Taylan, *Mésopolitique: Connaître, théoriser et gouverner les milieux de vie (1750–1900)* (Paris: Éditions de la Sorbonne, 2018).

mulated as the progress of such correspondences in Space, in Time, in Speciality, in Generality, in Complexity.[34]

Thus, according to Spencer, both biological evolution and mental development involve the improvement of the correspondence between organism and environment. (This conflation of the processes that biologists now distinguish as development and evolution—that is, ontogeny and phylogeny—was common to all of the nineteenth-century thinkers discussed in this book. They treated the two processes as falling under the same general type, since both produced a correspondence between organism and environment.)[35]

Although he developed a more interactive picture in his sociology (see chapter 2), Spencer was primarily an externalist or environmentalist about life and mind: that is, he usually treated changes in an organism as the results of changes in its external environment.[36] In other words, it is the organism that changes to adapt to the environment, not vice versa. Green alluded to this one-sidedness in his critique of Spencer's account of the subject-object relation. According to Green's idealism, neither subject nor object "has any reality apart from the other. Every determination of the one implies a corresponding determination of the other." Spencer's philosophy, in contrast,

> proceeds to explain that knowledge of the world which is the developed relation between object and subject, as resulting from an action of one member of the relation upon the other. It ascribes to the object, which in truth is nothing without the subject, an independent reality, and then supposes it gradually to produce certain qualities in the subject, of which the existence is in truth necessary to the possibility of those qualities in the object which are supposed to produce them.[37]

---

34. Herbert Spencer, *The Principles of Psychology*, 2nd ed., vol. 1 (London: Williams & Norgate, 1870), 385.

35. Stephen Jay Gould, *Ontogeny and Phylogeny* (Cambridge, MA: Harvard University Press, 1977); Robert J. Richards, *The Meaning of Evolution: The Morphological Construction and Ideological Reconstruction of Darwin's Theory* (Chicago: University of Chicago Press, 1992).

36. Peter Godfrey-Smith, *Complexity and the Function of Mind in Nature* (Cambridge: Cambridge University Press, 1996), 30–99. For a contrasting viewpoint, see Snait Gissis, "Spencer's Evolutionary Entanglement: From Liminal Individuals to Implicit Collectivities," in *Biological Individuality: Integrating Scientific, Philosophical, and Historical Perspectives*, ed. Scott Lidgard and Lynn K. Nyhart (Chicago: University of Chicago Press, 2017).

37. Thomas Hill Green, "Mr. Herbert Spencer and Mr. G. H. Lewes: Their Application of the Doctrine of Evolution to Thought; Part I, Mr. Spencer on the Relation of Subject and Object," *Contemporary Review* 31 (1877): 36–37.

CHAPTER FOUR

Although Green was speaking of the subject-object relation and not the organism-environment relation, the parallel is clear: Spencer viewed the qualities of the subject as gradually produced by the object. As we saw in chapter 2, James presented a similar criticism at around the same time as Green: "[Spencer] regards the creature as absolutely passive clay, upon which 'experience' rains down."[38] Whether or not Spencer was truly an extreme externalist, he was read that way by critics like James and Green.

As Green's claim that neither subject nor object has "an independent reality" shows, however, criticisms of Spencer's organism-environment psychology went further than simply denying the one-sided externalist picture. As we will see, Green and his colleagues supported not only the reciprocal causes view but also the dual aspects view of the organism-environment relation. Both of these views can be found in the idealist tradition and in the work of Hegel himself. For example, at the end of the *Encyclopedia Logic*—a book which Dewey taught four times at Michigan from 1890 to 1893—Hegel emphasized the agency of the organism, apparently rejecting externalism:

The living being confronts an inorganic nature to which it relates as the power over it, and which it assimilates. The result of this process is not . . . a neutral product in which the independence of the two sides that confronted one another is sublated [*aufgehoben*]; instead, the living being proves itself to be what overgrasps its other, which cannot resist its power.[39]

This picture of the active organism subordinating its environment, which was later adopted by Caird and other idealists, is a form of the reciprocal causes view that inverts Spencer's externalist picture.

However, the dual aspects view, at least in the case of the subject-object relation, was also present in Hegel. It is this view that was most prominent in Green's critique of Spencer, quoted earlier: according to Green, it does not make sense to speak (as Spencer does) of the "action of one member

---

38. William James, "Brute and Human Intellect," *Journal of Speculative Philosophy* 12 (1878): 256.
39. Georg Wilhelm Friedrich Hegel, *The Encyclopedia Logic: Part I of the Encyclopaedia of Philosophical Sciences with the Zusätze*, trans. T. F. Geraets, W. A. Suchting, and H. S. Harris (Indianapolis: Hackett, 1991), 293 [§219Z]; Georg Wilhelm Friedrich Hegel, *Werke: Vollständige Ausgabe, durch einen Verein von Freunden des Verewigten*, vol. 6, *Die Logik* (Berlin: Duncker & Humblot, 1840), 394. The standard English translation in the nineteenth century was Georg Wilhelm Friedrich Hegel, *The Logic of Hegel*, trans. William Wallace (Oxford: Clarendon Press, 1874). I have quoted from a modern translation, but there are no significant differences in this passage. For Dewey's course Hegel's Logic, in which he used the book, see the *Calendar of the University of Michigan* from 1890 to 1893.

of the relation on the other," for subject and object do not have "an independent reality." This position or something like it was the foundation of idealist philosophy more generally, which viewed subject and object as aspects of the absolute. Although the subject-object relation was understood differently by different idealists, they agreed that the dualism had to be overcome somehow. In the most famous approach, that of F. W. J. Schelling and the early Hegel, "the subjective and the objective are distinct appearances, embodiments, or manifestations of the absolute."[40] The British idealists were well aware of this principle. For example, Caird presented the Schelling-Hegel view in his 1883 book *Hegel*, which Dewey read carefully:

[The philosophy of Identity] was opposed . . . to that common-sense dualism for which mind and matter, or subject and object, are two things absolutely independent of each other. . . . In like manner, it was opposed to the Kantian and the Fichtean philosophy of subjectivity, which, indeed, had expressed the idea of a unity beyond difference . . . but which had not fully developed that idea. . . . The essential principle, then, in which Hegel and Schelling meet together, is that there is a unity which is above all differences, which maintains itself through all differences, and in reference to which all differences must be explained.[41]

This principle of unity or identity, opposed to the "common-sense dualism" of subject and object, was central to the idealist critique of Spencer.

Although Green's followers endorsed this critique, they did not share his view that evolution was irrelevant to philosophy. In fact, they took up Spencer's notion of the organism-environment relation and reinterpreted it from an idealist perspective. In the collective volume *Essays in Philosophical Criticism*, published by a group of Scottish idealists the year after Green's death and dedicated to his memory, the biologist John Scott Haldane wrote the following:

[Development] cannot be expressed as a simple result of action from without. As we have seen, it is not correct to separate the surroundings in thought from the organism, and treat them as independent things, for the organism only realises itself in its

---

40. Frederick Beiser, *Hegel* (New York: Routledge, 2005), 65; see also Frederick Beiser, *German Idealism: The Struggle against Subjectivism, 1781–1801* (Cambridge, MA: Harvard University Press, 2002), 560–64.
41. Edward Caird, *Hegel* (Edinburgh: William Blackwood, 1883), 48–50. A copy of this book, with extensive underlining, is among those preserved from Dewey's personal library at the Special Collections Research Center, Morris Library, Southern Illinois University–Carbondale.

surroundings.... Development is in all cases the realisation of what was not there at the beginning of the process. Yet the resulting difference is not conceived as impressed from without, but as freely produced from within itself by that which developes [sic].[42]

Haldane, echoing Green's account of the subject-object relation, argued that the organism and its surroundings are not independent entities; he also highlighted the organism's agency (thus "freely produced"). Although Haldane was a physiologist, he was also interested in philosophy, and he corresponded with Oxford Hegelians such as Alexander. He even published a paper in *Mind* in which he again underscored the reciprocal nature of the organism-environment relationship:

In being made to react on the surroundings the organism is determined by its own influence acting through the surroundings. The surroundings in acting on the organism are therefore at the same time acted on by it. The organism is thus no more determined by the surroundings than it at the same time determines them. The two stand to one another, not in the relation of cause and effect, but in that of reciprocity.[43]

This passage, with its odd distinction between "cause and effect" and "reciprocity," suggests that Haldane subscribed to both the reciprocal causes view and the dual aspects view, apparently interpreting them as mutually reinforcing or as aspects of a more general position. Not incidentally, the reciprocal relation between organism and environment was the main topic of Haldane's scientific work, which concerned human respiration and air quality—he was even involved in the development of gas masks during the First World War.[44] Thus Haldane, in conversation with Green and other idealists, emphasized development and organism-environment interaction in both his philosophical and biological work.

Caird, who praised Green in the preface to *Essays in Philosophical Criticism*, also endorsed a dialectical relation between organism and environment. In his article "Metaphysic," which Dewey read in the 1880s, Caird referred to what he identified as the turning point of modern phil-

---

42. Richard Burdon Haldane and John Scott Haldane, "The Relation of Philosophy to Science," in *Essays in Philosophical Criticism*, ed. Andrew Seth and Richard Burdon Haldane (London: Longmans, Green, 1883), 58–59. This volume is cited in John Dewey, "Psychology as Philosophic Method," *Mind* 11 (1886): 155.

43. John Scott Haldane, "Life and Mechanism," *Mind* 9 (1884): 32–33.

44. John Scott Haldane, *Organism and Environment as Illustrated by the Physiology of Breathing* (New Haven, CT: Yale University Press, 1917); John B.C. Kershaw, "The Use of Poisonous Gases in Warfare," in *The Scientific American War Book: The Mechanism and Technique of Warfare*, ed. Albert A. Hopkins (New York: Munn, 1915), 166–68.

osophical controversy: "In what sense can we apply the idea of development to the human spirit? Are we to treat that development as merely a determination from without, or as an evolution from within, or as partly the one and partly the other?" He claimed that even though the Darwinian theory "supposes that the condition or medium in which the individual is placed determines the direction in which . . . development proceeds," this theory does not completely neglect "the *a priori* tendency of the individual to maintain itself in the struggle for existence." Conversely, he continued, no one any longer subscribes to the Leibnizian theory that "self-development is entirely conditioned by itself in such a sense that all the relations which it has to other existences are merely apparent." Caird argued that idealism transcends this opposition between individual and medium: "the history of the conscious being in his relations with [the] world is not a struggle between two independent and unrelated forces, but the evolution by antagonism of one spiritual principle. It is, on this view, the same life which within us is striving for development, and which without us conditions that development." Caird allowed that, based on Darwin's ideas, one could develop a "natural science of man" that viewed the individual human being as externally determined. But philosophy, he concluded, shows this position to be incomplete and one-sided. Caird's picture of individual and medium united in their evolution as aspects of "the same life" was essentially Green's subject-object view reinterpreted in light of contemporary biology.[45]

The vocabulary of the individual and its "medium" derives from Caird's study of Auguste Comte's philosophy. Comte, in the third volume of his *System of Positive Politics*, had connected Kant's idealism to the interaction of the organism and its *milieu*, which Caird translated into English as "medium."[46] Caird summarized this purported connection in his book *The Social Philosophy and Religion of Comte*: "Kant is supposed by [Comte] to be [the] philosopher who first extended to the mind the general biological truth of the action and reaction of organism and medium upon each other"; that is, "the mind modifies the object,

---

45. Edward Caird, "Metaphysic," in *The Encyclopaedia Britannica: A Dictionary of Arts, Sciences, and General Literature*, vol. 16 (Edinburgh: Adam & Charles Black, 1878), 92. This article is quoted in Dewey, "Psychology as Philosophic Method," *Mind* 11 (1886): 155; see also John Dewey, *Psychology* (New York: Harper & Brothers, 1887), 13.

46. Auguste Comte, *Système de politique positive, ou Traité de sociologie, instituant la religion de l'humanité*, vol. 3 (Paris: Carilian-Goeury & Dalmont, 1853), 18–22. Comte had first introduced the idea of organism-environment correspondence as the basis of life in the third volume of his *Course of Positive Philosophy*, published in 1838. This idea was subsequently picked up by Spencer.

as well as the object the mind." This action-reaction story was clearly opposed to Spencer's more one-sided account and is an example of the reciprocal causes view. Caird went further, however, endorsing the dual aspects view and accusing Comte of misunderstanding Kant's critical philosophy. For Kant, on Caird's reading, "subject and object are correlative elements in the unity of knowledge, and not two separate things, by the action and reaction of which upon each other knowledge is produced."[47] In Caird's idealist version of the relationship between organism and medium, it is not just that the causal arrow goes both ways; the unity of experience makes organism and medium inseparable.

Caird emphasized one or the other of these two positions—organism and environment as dual aspects or as reciprocal causes—depending on the context, suggesting that he did not see them as mutually exclusive. In "Metaphysic" and the Comte book, quoted previously, he highlighted the former, whereas the latter took pride of place in a two-volume work of 1889 that was the subject of one of Dewey's graduate courses at the University of Michigan: *The Critical Philosophy of Immanuel Kant*. In a section of the book concerning the problem of the external world, Caird wrote:

We think of that which develops as externally related to an environment, in which, however, it finds the means of its self-maintenance. The external relation prepares us to expect the loss of both terms in a third or resultant term; but the developing being subordinates the external environment to itself, and makes the conditions that seem to limit it a means to the maintenance and aggrandizement of its own being.

Instead of the organism subordinating itself to the environment, as in Spencer, the environment is subordinated to the organism—Caird was here echoing the passage from Hegel's *Logic* quoted earlier. He made a related point in his discussion of the relation between the organic and inorganic in Kant, suggesting that an organism's internal development is just as important as the way it is shaped by the environment:

The Darwinian theory has directed our attention almost wholly to the continuous process of adaptation to the environment by which animal and vegetable life is maintained and developed: it has laid less emphasis on the other and higher aspect of the facts, according to which the process is one of *self*-adaptation, which has self-maintenance and self-development for its end.

47. Edward Caird, *The Social Philosophy and Religion of Comte* (Glasgow: James Maclehose & Sons, 1885), 81, 84; cf. Comte, *Système de politique positive*, vol. 3, 18–22.

Caird claimed that this neglect of self-adaptation is "partially, though only partially, corrected" in Spencer's account of evolution.[48] Thus evolution, for Caird, involves not merely the environment determining the organism but also the organism's autonomous development as well as the subordination of the environment to its ends.

Published in the same year as Caird's work on Kant, Alexander's book *Moral Order and Progress* presented a dialectical account of organism-environment interaction that was even more explicitly opposed to Spencer's picture. On Alexander's reading, Spencer regarded "good conduct as an adaptation or adjustment of man to his environment." But Alexander criticized him for subscribing to an overly simplistic notion of adaptation, one that sees the environment as "something fixed and permanent, . . . the cloth according to which [man] must cut his coat."[49] Several years earlier, in the essay on Hegel discussed in the previous section, Alexander had claimed that adaptation is "as much a selection by the [organism] of the conditions under which it can develop, as the dictate of the [environment] which organisms will suffer to develop." He elaborated this position in *Moral Order and Progress*:

> The act of adaptation can only be understood as a joint action of the individual and his environment, in which both sides are adjusted to the other. What the environment is depends on the character or the qualities of the individual, for it is only in so far as it responds to him that it can affect him at all. . . . The environment, therefore, changes as the individual changes, and the act of adaptation is thus not a mere one-sided modification, but a process of selection from both sides, not the mere operation upon the individual of a foreign body which remains constant, but a contribution to a joint result. What the individual does, and what the environment is, are settled at one and the same time by the act in which they are said to be adjusted, and they both vary together.

Adaptation is a two-way street: the environment modifies the organism, but the organism also modifies the environment; organism and environment are codetermining. This dialectical account of the organism-environment relation—startlingly similar to Lewontin's view of a hundred years later—was central to Alexander's evolutionary ethics, as it was the key to moral progress: good conduct involves adaptation, but this

---

48. Edward Caird, *The Critical Philosophy of Immanuel Kant*, 2 vols. (Glasgow: James Maclehose & Sons, 1889), 1:646–47, 2:90–91. For Dewey's class, see the *Calendar of the University of Michigan for 1890–91* (Ann Arbor: University of Michigan, 1891), 57.
49. Samuel Alexander, *Moral Order and Progress: An Analysis of Ethical Conceptions* (London: Trübner, 1889), 267, 271.

adaptation "itself alters the sentiments of the agent, and creates new needs which demand a new satisfaction."[50]

These works by Caird and Alexander directly influenced Dewey, and in the early 1890s he adopted the Oxford Hegelians' dialectical account of the organism-environment relationship. In the preface to his book *Outlines of a Critical Theory of Ethics*, Dewey stated that he was "especially indebted" to Caird's books on Comte and Kant as well as to Alexander's *Moral Order and Progress*.[51] As Jennifer Welchman has shown, Dewey's ethical views were at this stage primarily an elaboration of Francis Herbert Bradley's idea that the aim of morality is self-realization—"the realization of all one's latent, potential personhood."[52] In *Outlines*, Dewey declared that the good is the realization of individuality and distinguished two aspects of individuality: capacity and environment. It was at this point that he drew from Caird and Alexander, developing his own dual aspects view:

> The moment we realize that only what one conceives as proper material for calling out and expressing some internal capacity is a part of his surroundings, we see not only that capacity depends upon environment, but that environment depends upon capacity. In other words, we see that each in itself is an abstraction, and that the real thing is the individual who is constituted by capacity and environment in their relation to one another.

Capacity and environment, according to Dewey, should be thought of as aspects rather than as independent entities; they are unified in the individual. Dewey's debt to Alexander and Caird is also obvious in a section titled "Adjustment to Environment," where he presented something more like the reciprocal causes view:

> Even a plant must do something more than adjust itself *to* a fixed environment; it must assert itself *against* its surroundings, subordinating them and transforming them into material and nutriment; and, on the surface of things, it is evident that *transformation* of existing circumstances is moral duty rather than mere reproduction of them. The environment must be plastic to the ends of the agent.

---

50. Samuel Alexander, "Hegel's Conception of Nature," *Mind* 11 (1886): 520; Alexander, *Moral Order and Progress*, 271–72, 277. Dewey referred to Alexander's 1886 essay in his 1891 course Hegel's Philosophy of Spirit at the University of Michigan: see John R. Shook and James A. Good, *John Dewey's Philosophy of Spirit, with the 1897 Lecture on Hegel* (New York: Fordham University Press, 2010), 176.

51. John Dewey, *Outlines of a Critical Theory of Ethics* (Ann Arbor, MI: Inland Press, 1891), vii.

52. Jennifer Welchman, *Dewey's Ethical Thought* (Ithaca, NY: Cornell University Press, 1995), 31, 75–83; see Francis Herbert Bradley, *Ethical Studies* (London: Henry S. King, 1876), 59–74.

That is, adjustment involves alteration of the environment and not only change in the organism. "Adjustment to environment," said Dewey, is a "phrase made familiar by evolutionists" such as Spencer. But this adjustment "is not outer conformity; it is living realization of certain relations in and through the will of the agent."[53]

This dialectical account of organism-environment interaction also formed the backdrop to Dewey's founding of the Chicago school of functional psychology. Andrew Backe has shown that Dewey's psychological views were indebted to Green's philosophy, but they were also shaped by the ideas of Caird and Alexander. Caird's idealism insisted that organism and environment—"self-determination and determination from without"—should not be seen as independent forces, for they were united as aspects of "the same life."[54] As we have just seen, Dewey adopted a similar view in *Outlines*, arguing that capacity and environment are two aspects of individuality rather than independent factors contributing to it. Continuing this line, Dewey introduced the term *function*

> to express union of the two sides of individuality. The idea of function is that of an active relation established between power of doing, on one side, and something to be done on the other. . . . A function thus includes two sides—the external and the internal—and reduces them to elements of one activity. . . . So, morally, function is capacity *in action*; environment transformed into an element in personal service.[55]

The idea that environment and organism are not separate factors but aspects of one function—one process, one coordination, one life, one experience—was central to Dewey's work beginning in the 1890s. In a course called Philosophy of Education at the University of Chicago in 1896, for example, Dewey offered yet another variation on this theme:

> Adaptation is dynamic, not static. It means control; and highest adaptation means highest control. Environment is not a fixed idea to be measured or set up by kind of life. It is different for every existing creature. There is something to which the organism and the environment are related. The function is something more than organism; it is

---

53. John Dewey, *Outlines of a Critical Theory of Ethics* (Ann Arbor, MI: Inland Press, 1891), 100, 115, 117.

54. Andrew Backe, "John Dewey and Early Chicago Functionalism," *History of Psychology* 4 (2001); Edward Caird, "Metaphysic," in *The Encyclopaedia Britannica: A Dictionary of Arts, Sciences, and General Literature*, vol. 16 (Edinburgh: Adam & Charles Black, 1878), 92.

55. John Dewey, *Outlines of a Critical Theory of Ethics* (Ann Arbor, MI: Inland Press, 1891), 100–101.

something more than environment. Organism and environment are simply the two sides of function. The organism is the method or implement of function. The environment is the supply [of] function.[56]

It is the process of life that is truly real, according to Dewey. Organism and environment are separable only as a result of analysis. The founding document of Chicago functionalism—Dewey's article "The Reflex Arc Concept in Psychology"—made an analogous claim: stimulus and response are not separate entities but rather functional phases of one coordination or adjustment.[57]

Although British idealists such as Caird, Haldane, and Alexander broke with Green in arguing for the relevance of biological evolution to philosophy, they developed an account of the relation between organism and environment that went beyond Spencer's. Evolution or development, according to the Hegelians, is not simply the environment determining the organism: first, this ignores what Caird called "self-development"; second, it neglects the fact that organisms select and modify their environments just as environments select and modify organisms. Moreover, it might even be misleading to think about organism and environment as separate, interacting entities. They are really two aspects of experience or life, two sides of the adaptive process. Dewey, as he began teaching courses on ethics and on the idealism of Kant and Hegel in the 1890s, adopted this dialectical account of the organism-environment relation. He argued that organism and environment were not independent factors but merely two aspects of one coordination or life process.

Evolutionary Strivings

As John Dewey struggled in Michigan to develop his dialectical approach to the organism-environment relationship, George Herbert Mead and W. E. B. Du Bois were learning about the connections between evolution and idealism from Josiah Royce at Harvard. In this final section, I demonstrate that Royce, like Dewey and the British idealists, saw a clear connection between evolution and idealism. I also argue that when Du Bois employed concepts that many scholars have identified as distinctively

56. "Philosophy of Education (1896)," in Dewey, *Class Lectures*, 2:95.
57. John Dewey, "The Reflex Arc Concept in Psychology," *Psychological Review* 3 (1896). For more on the reflex arc essay as involving both Hegelian and biological ideas, see Richard J. Bernstein, *John Dewey* (New York: Washington Square, 1966), 15–21.

Hegelian, he was often simply channeling the more general evolutionary spirit of late nineteenth-century thought.

In the introductory lectures of his Philosophy of Nature class, discussed briefly in chapter 3, Royce identified a tension in modern philosophy between the mechanistic outlook of the seventeenth century and the evolutionary outlook of "the present century." He argued that the earlier picture was characterized by "the absence of the human element" and dated the revival of this element to the late eighteenth and early nineteenth centuries, providing the following outline:

1. Rousseau's influence.
2. The Revival of German Poetry.
3. Kant.
4. Revival of Spinozism as against Fichte.
5. Schelling's Idealism as synthesis of Fichte & Spinoza.
6. Hegel as summarizer of this whole movement.

According to Royce, these German philosophers viewed nature "as an incorporation of spiritual truth, or as a mode of manifestation of God's spirit," and sometimes even as "a preliminary, impotent, or unconscious striving, tending *towards* the consciousness of man." In an associated essay, he declared that "the philosophical movement from Hume to Hegel . . . , coming as it did between the mechanical philosophy of an earlier time and the evolutionary doctrine of to-day, has an historical significance which no serious philosophical student can afford to overlook."[58]

Royce gave a more elaborate account of this movement in his 1892 book *The Spirit of Modern Philosophy*. The historical part of *Spirit* began with Baruch Spinoza, transitioned quickly to Kant, and culminated in "The Rise of the Doctrine of Evolution." For Royce, this doctrine was a Hegelian synthesis of seventeenth-century philosophy of nature and German idealism. Modern philosophy had begun, Royce believed, "with the external order and with dogmatic assertions about it. Growing doubtful and self-critical, it had next fallen to scrutinizing the inner life." In the late eighteenth century, Kant declared that things in themselves "aren't for me; but as for the order and unity of phenomenal nature, that is mine, and is even of my own creating." The nineteenth-century idealists went even further: "Not only the order of nature, but

---

58. Josiah Royce, "Philosophy of Nature," Lecture 3 (6 October 1886), Folder 4, Box 127, Royce Papers; Josiah Royce, "Is There a Philosophy of Evolution?" *Unitarian Review* 32 (1889): 107–8.

the very content of nature is spiritual." However, there remained a troubling gap between nature as spirit and nature as actually experienced: "From the constructions of your ideal philosophy to the empirical facts of outer nature remained a long and hopelessly tangled way. The world might be thus rational, but it was evident that the absolute spirit must be thinking of many things that you, in your finite weakness, could not well presume to construct *a priori.*" This dilemma, said Royce, led some to turn back to the empirical world with an altered outlook: "We return to the natural order, . . . but by no means as if, in returning, we left our idealism behind us." These thinkers wanted their philosophy to be effective rather than abstract: "We want our idealism to do a manly work. We want it to enter upon its true task, not of dreaming of a possible perfection, but of transforming, of enlivening, of spiritualizing, the concrete life of humanity." The "most noteworthy offspring" of this marriage of "empirical research and the truly philosophical spirit," said Royce, was "the vast industry that has gathered about what we now call the idea of evolution."[59]

Drawing on his Philosophy of Nature lectures, Royce then argued that nineteenth-century thought depended on a new "historical conception of the world," resulting from the application of "idealistic postulates" to the empirical study of human history. Earlier philosophers, said Royce, were looking for a static human nature, eventually using it to ground unalienable rights. But this abstract account, as he declared in a passage later quoted by Jane Addams, had been overturned by the modern evolutionary worldview:

Valuable, indeed, was all this unhistorical analysis of the world and of man, valuable as a preparation for the coming insight; but how unvital, how unspiritual, how crude seems to us now all that eighteenth-century conception of the mathematically permanent, the essentially unprogressive and stagnant human nature, in the empty dignity of its inborn rights, when compared with our modern conception of the growing, struggling, historically continuous humanity, whose rights are nothing until it wins them in the tragic process of civilization, whose dignity is the dignity attained as the prize of untold ages of suffering, whose institutions embody thousands of years of ardor and of hard thinking, whose treasures even of emotion are the bequest of a sacred antiquity of self-conquest! Not inalienable, but hard won and painfully kept are the true rights of man. Not a special creation, but a living organism is our nature; an organism not per-

---

59. Josiah Royce, *The Spirit of Modern Philosophy: An Essay in the Form of Lectures* (Boston: Houghton, Mifflin, 1892), 266–67, 269, 271, 273.

manent in its structure, but the outcome of labor; an organism with a long embryonic development, capable of degeneration as well as of growth, and needing therefore our constant care lest it lose all the spirituality and all the rights that it has thus far acquired.

This evolutionary worldview was not invented by biologists, for "it was coming into an historical age that made Darwin's book so great a prize, and the idea of natural selection so deeply suggestive to philosophy." Likewise, Spencer was considered "the one true prophet of the philosophy of evolution" because he had tapped into the zeitgeist: he was "a reconciler, a unifier, one who harmonizes through synthesis, and who brings to light oppositions only to enrich thought by suggesting their organic unity"—but with categories that "look so much more empirical and concrete than Hegel's." For Royce, evolutionary thought was the culmination of modern philosophy. Royce's own philosophy of evolution, like Spencer's, was an attempt to give voice to the idea "that there is a history embodied in the known world."[60]

Royce's book was first presented as a series of public evening lectures "on the History and Problems of Philosophy" at Harvard in the fall of 1890, when Du Bois was just beginning his graduate study there. It is tempting to think that Du Bois attended these lectures; in any case, Robert Gooding-Williams has convincingly shown that Du Bois was familiar with the published book, which appeared in the early spring of 1892 before he left for Berlin.[61]

Others, and in particular Shamoon Zamir, have claimed that Du Bois studied the work of Hegel in his classes at Harvard, but the evidence for this is shaky. The usual story is that Du Bois read the *Phenomenology of Spirit* with George Santayana in the spring of 1890, just before his college graduation. But although Santayana's class was titled Earlier French Philosophy, from Descartes to Leibnitz, and German Philosophy from Kant to Hegel, he and his five students only made it to Kant's *Critique of Pure Reason*, as evidenced by the President's Report for 1889–90 as well as Du Bois's own library charging records.[62] The curtailed reading may have

---

60. Royce, *Spirit of Modern Philosophy*, 273, 275, 285, 290, 295, 297; Jane Addams, *Newer Ideals of Peace* (New York: Macmillan, 1906), 32–33 (quoting p. 275 of Royce).
61. Royce, *Spirit of Modern Philosophy*, v–vi, [xvii]; *The Harvard University Catalogue, 1890–91* (Cambridge, MA: Harvard University, 1890), 75; Robert Gooding-Williams, "Philosophy of History and Social Critique in *The Souls of Black Folk*," *Social Science Information* 26 (1987): 105–6; "Books of the Week," *Nation*, March 24, 1892, 238.
62. Shamoon Zamir, *Dark Voices: W. E. B. Du Bois and American Thought, 1888–1903* (Chicago: University of Chicago Press, 1995), 113, 248–49; *Annual Reports of the President and Treasurer of Harvard College, 1889–90* (Cambridge, MA: Harvard University, 1891), 79. Du Bois checked out the

been due to Santayana's frantic preparation for the course, described in a September 1889 letter: "[Francis] Bowen has resigned his place, and his course in the Cartesians and Germans has been turned over to me of a sudden. I am expected to lecture every day, and what with reading, getting up the lectures, hunting for books in the library, and worrying over the slipshod way in which after all the work is presented to the boys, I haven't a spare moment." It may also have been due to Santayana's animosity toward Hegel. As Zamir has noted, Santayana did recall almost sixty years later that he had "liked Hegel's *Phaenomenologie*" while at Harvard. But although he took Royce's "special research" seminar on that book while teaching Du Bois's class, Santayana described the seminar as "appalling," complaining that Royce "seemed to be bent on converting me to absolute idealism nolens volens [willing or unwilling]." In the same 1890 letter, Santayana declared that if he became editor of the incipient *Philosophical Review*, "a little less Hegelian drivle [sic] might thus be administered to the American public." Santayana was thus very critical of Hegel at the time, which may explain why Royce began teaching a new course the following year—The Movement of German Thought from 1770–1830—covering Kant, Fichte, Schelling, and Hegel. Royce was staking out his territory: Santayana continued to teach "Descartes, Spinoza, and Leibnitz" at Harvard throughout the 1890s but avoided German philosophy.[63]

Du Bois did not formally study Hegel's texts in college, and the same appears true of his graduate work at Harvard (1890–92) and Berlin (1892–94). The only philosophy class he took during these four years was Wilhelm Dilthey's General History of Philosophy, a series of daily morning lectures from April to August 1893 (Mead had taken exactly the same class two years earlier). The published syllabus for Dilthey's class indicates that although it concluded with an overview of nineteenth-century philosophy, he covered individual philosophers only briefly. Like Royce, Dilthey emphasized the broader evolutionary spirit of the

---

Critique of Pure Reason from the Harvard library on both March 24 and April 24, as well as George Sylvester Morris's *Kant's Critique of Pure Reason: A Critical Exposition* (Chicago: S. C. Griggs, 1882) on May 13. The latter was the last book he checked out that semester, and he does not seem to have checked out anything by or related to Hegel. See Library Charging Lists, Students, 1889–1890, p. 658 [seq. 360]. In the charging list, the relevant call numbers are "III. 2010," "III. 1347," and "III. 2027" (for book identification, see Box 2, UAIII 50.15.47.5, Harvard University Archives).

63. George Santayana, *Persons and Places: Fragments of Autobiography*, ed. William G. Holzberger and Herman J. Saatkamp (Cambridge, MA: MIT Press, 1986), 389; Santayana to Henry Ward Abbot, 29 September 1889, and George Santayana to Charles Augustus Strong, 22 July 1890, in George Santayana, *The Letters of George Santayana*, ed. William G. Holzberger, 8 vols. (Cambridge, MA: MIT Press, 2001–8), 1:108, 111; *Annual Reports of the President and Treasurer of Harvard College, 1890–91* (Cambridge, MA: Harvard University, 1892), 57. For Santayana and Royce's courses, see the *Annual Reports of the President and Treasurer of Harvard College* from 1888–89 to 1899–1900.

period, using it to introduce the final section of the class, titled "The Theory of Evolution and the Nineteenth Century." This section of his syllabus opened with the following claim: "In the 19th century, philosophy is logically dissolved into studies of the facts of consciousness, and the theory of evolution [*Entwicklungslehre*] in particular is used to make that connection" (i.e., between philosophy and the facts of consciousness). In France, Dilthey pointed to "studies of the external world" by Buffon and Lamarck as well as to Geoffroy St. Hilaire's "theory of evolution." This more materialist approach found its "superior expression" in the positive philosophy of August Comte, whose ideas Dilthey described as "dominant in France and very common in England." Turning to the latter country, he noted that Herbert Spencer "has put the theory of evolution into effect across the whole realm of reality." Dilthey also linked evolution to German idealism, placing "the evolutionary theory of Herder and Göthe" in the background of Schelling's "beautiful unfolding of world-reason" and Hegel's "panlogism." Royce and Dilthey thus spoke in one voice: evolution was the governing principle of nineteenth-century philosophy.[64]

Even if he never formally studied Hegel, Du Bois must have been somewhat familiar with his philosophy. As Joel Williamson pointed out long ago, Du Bois's 1897 essay "Strivings of the Negro People" was in implicit dialogue with Hegel's *Philosophy of History*, which "traced the world spirit as having moved forward toward a realization of itself through the successive histories of six world historical peoples: Chinese, Indians, Persians (culminating in the Egyptians), Greeks, Romans, and Germans." After giving an almost identical list, Du Bois added "the Negro [as] a sort of seventh son, born with a veil, and gifted with second-sight in this American world."[65] Immediately following this passage, Du Bois

---

64. *Verzeichniss der Vorlesungen, welche auf der Friedrich-Wilhelms-Universität zu Berlin im Sommer-Semester vom 17. April bis 15. August 1893 gehalten werden* (n.p., 1893), 14, https://www.digi-hub.de/viewer/toc/DE-11-001799608/1/; W. E. B. Du Bois to the John Slater Fund, 10 March and 6 December 1893, in Du Bois, *Correspondence*, 1:24, 26; Wilhelm Dilthey, *Biographisch-literarischer Grundriss der allgemeinen Geschichte der Philosophie für die Vorlesungen von Professor W. Dilthey*, 3rd ed. (Trebnitz: Maretzke & Märtin, [1893]), 83–84. There were no changes to the text of this section across the first three editions of the syllabus (1885, 1890, and 1893). Thanks to Daniel Liu, David Stiver, Corey Dyck, and Falk Wunderlich for arranging scans of these rare editions and to Hans-Ulrich Lessing for help in dating them.

65. Joel Williamson, "W. E. B. Du Bois as a Hegelian," in *What Was Freedom's Price?* ed. David G. Sansing (Jackson: University Press of Mississippi, 1978), 35; W. E. B. Du Bois, "Strivings of the Negro People," *Atlantic Monthly* 80 (1897): 194; Georg Wilhelm Friedrich Hegel, *Lectures on the Philosophy of History*, trans. John Sibree (London: Henry G. Bohn, 1857), xxix–xxxix; Georg Wilhelm Friedrich Hegel, *Vorlesungen über die Philosophie der Geschichte*, 2nd ed. (Berlin: Dunker & Humblot, 1840), xxv–xxvi; see also Paul Gilroy, *The Black Atlantic: Modernity and Double Consciousness* (Cambridge, MA: Harvard University Press, 1993), 134.

introduced his famous notion of "double-consciousness," which recalls (at least to the minds of philosophers) the "unhappy consciousness" of Hegel's *Phenomenology*: just as the African American "feels his two-ness,—an American, a Negro; two souls, two thoughts, two unreconciled strivings," the unhappy consciousness lacks unity and exhibits "the doubling of self-consciousness within itself, which is essential in the concept of spirit."[66]

Nevertheless, many features of Du Bois's early work that are usually read as exclusively Hegelian can also be viewed as links to the broader scientific and evolutionary worldview described by Royce and Dilthey. Hegel contrasted natural and spiritual development in the introduction to his *Philosophy of History* lectures of 1830. He defined *Entwicklung* (development or evolution) in general as "a capacity or potentiality striving to realise itself." But whereas natural development occurs "in virtue of an unchangeable principle; a simple essence,—whose existence, *i.e.*, as a germ, is primarily simple,—but which subsequently develops a variety of parts," spiritual or cultural development "is mediated by consciousness and will" and thus involves *Geist* (mind or spirit) in "a mighty conflict with itself," striving for "the realization of its Ideal being" with "the History of the World for its theatre."[67] By the 1890s, however, by which time Darwin and Spencer were household names, it had become more difficult to view either ontogeny or phylogeny as the unfolding of a static essence. The words *development* and *evolution* would thus have conjured images of biological progress, even in a purely social or cultural context.

Scholars have emphasized the importance for Du Bois of Heinrich von Treitschke's 1892–93 lectures on "Politics." These lectures certainly had a Hegelian flavor, but as Kwame Anthony Appiah has noted, the views of Treitschke and Du Bois's other teachers at the University of

---

66. Du Bois, "Strivings of the Negro People," 194; Georg Wilhelm Friedrich Hegel, *The Phenomenology of Spirit*, trans. Terry Pinkard (Cambridge: Cambridge University Press, 2018), 123 [§206]; Georg Wilhelm Friedrich Hegel, *Werke: Vollständige Ausgabe, durch einen Verein von Freunden des Verewigten*, vol. 2, Phänomenologie des Geistes (Berlin: Duncker & Humblot, 1832), 158; this parallel is noted in Sandra Adell, *Double-Consciousness/Double-Bind: Theoretical Issues in Twentieth-Century Black Literature* (Urbana: University of Illinois Press, 1994), 18–19; Shamoon Zamir, *Dark Voices: W. E. B. Du Bois and American Thought, 1888–1903* (Chicago: University of Chicago Press, 1995), 13; Stephanie J. Shaw, *W. E. B. Du Bois and "The Souls of Black Folk"* (Chapel Hill: University of North Carolina Press, 2013), 94, 122; for criticism, see Robert Gooding-Williams, *In the Shadow of Du Bois: Afro-Modern Political Thought in America* (Cambridge, MA: Harvard University Press, 2009), 284n37.

67. Georg Wilhelm Friedrich Hegel, *Lectures on the Philosophy of History*, trans. John Sibree (London: Henry G. Bohn, 1857), 57–58; Georg Wilhelm Friedrich Hegel, *Vorlesungen über die Philosophie der Geschichte*, 2nd ed. (Berlin: Dunker & Humblot, 1840), 67–69. John Sibree's translation is often misleading, but I have left it unaltered, since Du Bois may have read it. For another translation of this passage, see Georg Wilhelm Friedrich Hegel, *Lectures on the Philosophy of World History: Introduction*, trans. Hugh Barr Nisbet (Cambridge: Cambridge University Press, 1975), 126.

Berlin also had "a Social Darwinian tilt."[68] It is hard to imagine Du Bois learning about the evolutionary struggle between nation-states without imagining a process akin to Darwin's struggle for existence—a parallel drawn by many observers at the time, including Arthur Twining Hadley, the author of a textbook Du Bois would later assign at Atlanta University (as discussed in chapter 3). Treitschke adopted Hegel's picture of social evolution: "We may say with certainty that the evolution [*Entwicklung*] of the State is, broadly speaking, nothing but the necessary outward form which the inner life of a people bestows upon itself, and that peoples attain to that form of government which their moral capacity enables them to reach." But he also—like Benjamin Kidd, of whom we will see more in chapter 5—argued that struggle and conflict drive this evolutionary process forward: "Brave peoples alone have an existence, an evolution or a future; the weak and cowardly perish, and perish justly. The grandeur of history lies in the perpetual conflict of nations, and it is simply foolish to desire the suppression of their rivalry." Thus, although Treitschke agreed with Hegel that states should not be analogized to organisms, since many of the latter lack "conscious will," his frequent invocation of *Entwicklung* took advantage of the broader scientific and popular interest in evolution at the time. When Du Bois heard Treitschke declare that "the negro . . . is employed inevitably to serve the ends of a will and intelligence higher than his own," he would thus have interpreted this as a denial of the biological and cultural evolutionary potential of African peoples, analogous to Hegel's infamous claim that Africa "has no movement or development [*Entwicklung*] to exhibit."[69]

When Williamson and others note that Du Bois's early work is "Hegelian in its language," however, they neglect to mention that much of this

---

68. Kwame Anthony Appiah, *Lines of Descent: W. E. B. Du Bois and the Emergence of Identity* (Cambridge, MA: Harvard University Press, 2014), 40; see also Jan Sapp, *Genesis: The Evolution of Biology* (Oxford: Oxford University Press, 2003), 48. On Du Bois and Treitschke, see W. E. B. Du Bois, *Dusk of Dawn: An Essay toward an Autobiography of a Race Concept* (New York: Harcourt, 1940), 98–99; Michael P. Kramer, "W. E. B. Du Bois, American Nationalism, and the Jewish Question," in *Race and the Production of American Nationalism*, ed. Reynolds J. Scott-Childress (New York: Garland, 1999), 184–88; Appiah, *Lines of Descent*, 41–43, 84–85.

69. Heinrich von Treitschke, *Politics*, trans. Blanche Dugdale and Torben de Bille, 2 vols. (New York: Macmillan, 1916), 1:12, 18, 21, 275–76; Heinrich von Treitschke, *Politik: Vorlesungen gehalten an der Universität zu Berlin*, 2 vols. (Leipzig, Ger.: S. Hirzel, 1897–98), 1:22, 28, 30, 274; Georg Wilhelm Friedrich Hegel, *Lectures on the Philosophy of History*, trans. John Sibree (London: Henry G. Bohn, 1857), 103; Georg Wilhelm Friedrich Hegel, *Vorlesungen über die Philosophie der Geschichte*, 2nd ed. (Berlin: Dunker & Humblot, 1840), 123. For Du Bois's attendance at Treitschke's 1892–93 lectures on Politics, see W. E. B. Du Bois to Daniel Coit Gilman, 28 October 1892, in W. E. B. Du Bois, *The Correspondence of W. E. B. Du Bois*, ed. Herbert Aptheker, 3 vols. (Amherst: University of Massachusetts Press, 1973–78), 1:21.

CHAPTER FOUR

language reflected a more general evolutionary outlook.[70] *Striving (Streben)* and *struggle (Kampf)* were words used prominently by Hegel, but they were also employed by biologists. Darwin, for example, famously introduced the notion of a "struggle for existence" or "struggle for life" as the engine of natural selection. He also framed this struggle in terms of striving: "As all organic beings are striving, it may be said, to seize on each place in the economy of nature, if any one species does not become modified and improved in a corresponding degree with its competitors, it will soon be exterminated."[71] This vocabulary also had biological overtones within philosophy, due to the influence of the late eighteenth-century German philosopher Johann Gottfried Herder. An important figure for Hegel as well as Du Bois, Herder linked striving to both the organic and spiritual natures of human beings: "Our complexly organized bodies, with all their senses and limbs, have been bestowed on us for use, for exercise. . . . The gods sold every thing to mortals at the price of labour; not out of envy, but from kindness; for the greatest enjoyment of existence, the sensation of active striving powers, lies in this very struggle, in this striving after the comforts of ease."[72] Herder also had a "theory of evolution"—at least according to Dilthey's syllabus, quoted earlier. Although this attribution is somewhat anachronistic, John Zammito has demonstrated that Herder did see "increasing complexity and differentiation as an immanent principle of natural development, as an intrinsically *historical* character/tendency of the entire physical world"; he also "deliberately set about erasing the borders Kant had so carefully drawn not only between life and matter but also . . . between animal and man."[73]

70. Joel Williamson, "W. E. B. Du Bois as a Hegelian," in *What Was Freedom's Price?* ed. David G. Sansing (Jackson: University Press of Mississippi, 1978), 33–34. On the notion of striving, see also Kwame Anthony Appiah, *Lines of Descent: W. E. B. Du Bois and the Emergence of Identity* (Cambridge, MA: Harvard University Press, 2014), 54–56.
71. Charles Darwin, *On the Origin of Species by Means of Natural Selection, or the Preservation of Favoured Races in the Struggle for Life* (London: John Murray, 1859), 102 (see also 66, 75, 78, 113).
72. Johann Gottfried Herder, *Outlines of a Philosophy of the History of Man*, trans. Thomas Churchill, 2 vols. (London: J. Johnson, 1803), 1:394; Johann Gottfried Herder, *Ideen zur Philosophie der Geschichte der Menschheit*, 4 vols. (Riga: J. F. Hartknoch, 1785–92), 2:233–34.
73. Wilhelm Dilthey, *Biographisch-literarischer Grundriss der allgemeinen Geschichte der Philosophie für die Vorlesungen von Professor W. Dilthey*, 3rd ed. (Trebnitz, Ger.: Maretzke & Märtin, [1893]), 74; John Zammito, *The Gestation of German Biology: Philosophy and Physiology from Stahl to Schelling* (Chicago: University of Chicago Press, 2018), 182, 185. On Herder and Hegel, see Michael N. Forster, *Hegel's Idea of a "Phenomenology of Spirit"* (Chicago: University of Chicago Press, 1998), chap. 9. On Herder and Du Bois, see Shamoon Zamir, *Dark Voices: W. E. B. Du Bois and American Thought, 1888–1903* (Chicago: University of Chicago Press, 1995), 105–6; Robert Gooding-Williams, *In the Shadow of Du Bois: Afro-Modern Political Thought in America* (Cambridge, MA: Harvard University Press, 2009), 140–41; Kwame Anthony Appiah, *Lines of Descent: W. E. B. Du Bois and the Emergence of Identity* (Cambridge, MA: Harvard University Press, 2014), 45–50.

But would these biological overtones have been salient for Du Bois? In the remainder of this section, I demonstrate that Du Bois's 1897 essay "The Conservation of Races" contains evidence that Du Bois was moved by the general evolutionary spirit of the late nineteenth century, as described by both Royce and Dilthey.

Du Bois's essay was in implicit conversation with at least two of his African American contemporaries: Anna Julia Cooper and Alexander Crummell.[74] As recently noted by Carol Wayne White, Cooper's essay "Has America a Race Problem?" had as its central image a "continual struggle" between diverse races leading to "perpetual progress" and "the evolution of civilization." Cooper borrowed this image from François Guizot's 1828 book *General History of Civilization in Europe*, which had opened with the claim that "the first idea comprised in the word *civilisation* . . . is the notion of progress, of development." For Guizot, inspired by Herder, progressive development was always governed by a struggle between diverse elements:

> It is plain enough that no single principle, no particular organisation, no simple idea, no special power has ever been permitted to obtain possession of the world, to mould it into a durable form, and to drive from it every opposing tendency, so as to reign itself supreme. Various powers, principles, and systems here intermingle, modify one another, and struggle incessantly—now subduing, now subdued—never wholly conquered, never conquering. Such is apparently the general state of the world, while diversity of forms, of ideas, of principles, their struggles and their energies, all tend towards a certain unity, a certain ideal, which, though perhaps it may never be attained, mankind is constantly approaching by dint of liberty and labour.

Cooper argued that the "diversity of social elements" in modern Europe, as outlined by Guizot, grew "out of the contact of different races." She also claimed that the United States of America, because it boasted this diversity in a single country, was evolving toward a messianic "climax of history"—namely, "the final triumph of universal reciprocity born of universal conflict with forces that cannot be exterminated." Thus, despite Guizot's later opposition to biological evolution, both his book

---

74. Wilson J. Moses, "W. E. B. Du Bois's 'The Conservation of Races' and its Context: Idealism, Conservatism and Hero Worship," *Massachusetts Review* 34 (1993); Kathryn T. Gines, "Anna Julia Cooper," in *Stanford Encyclopedia of Philosophy*, Stanford University, 1997–, article published March 31, 2015, §3, https://plato.stanford.edu/archives/sum2015/entries/anna-julia-cooper/; Chike Jeffers, "Anna Julia Cooper and the Black Gift Thesis," *History of Philosophy Quarterly* 33 (2016): 83–86. For a review of other scholarship on "The Conservation of Races," see Paul C. Taylor, "Bare Ontology and Social Death," *Philosophical Papers* 42 (2013).

CHAPTER FOUR

and Cooper's essay had an evolutionary flavor—perhaps not surprising given the Herderian background of both.[75]

Evolutionary imagery also played a role in Crummell's late work. In an 1895 address at Wilberforce College, with Du Bois in attendance, Crummell argued that "the special function of men" is to address and solve moral problems. His conclusion alluded to Darwin's notion of a struggle for existence: "Man must test, struggle with, attempt to settle [indeterminate questions], or else he will lose all mental vitality. . . . Struggle is one of the prime conditions of existence." He was more explicit ten years earlier in "The Need of New Ideas and New Aims for a New Era," telling the graduating class of Storer College that dwelling on the horrors of slavery would result in "that unique and fossilized state which is called 'arrested development.'" Instead, he urged them to focus on and adapt to "new conditions":

> These changed circumstances bring to us an immense budget of new thoughts, new ideas, new projects, new purposes, new ambitions, of which our fathers never thought. We have hardly space in our brains for the old conditions of life. . . . We have need, therefore, of new adjustments in life. The law of fitness comes up before us just now with tremendous power, and we are called upon, as a people, to change the currents of life, and to shift them into new and broader channels.

Anticipating Du Bois's approach to social ethics, which I will discuss in chapter 6, Crummell wanted the black community to guide its own evolution, adapting itself to the new possibilities and challenges of the post-Emancipation environment.[76]

Crummell was also familiar with Herbert Spencer's account of evolution, using it in another speech to combat the idea that the races in America would eventually intermingle and become one: "I might meet the theory which anticipates amalgamation by the great principle mani-

---

75. Carol Wayne White, *Black Lives and Sacred Humanity: Toward an African American Religious Naturalism* (New York: Fordham University Press, 2016), 51–52; Anna Julia Cooper, *A Voice from the South* (Xenia, OH: Aldine, 1892), 159–60, 166; François Guizot, *General History of Civilization in Europe, from the Fall of the Roman Empire to the French Revolution* (New York: D. Appleton, 1838), 39, 41–42; François Guizot, *Meditations on the Essence of Christianity* (New York: Carlton & Porter, 1865), 48–49; Günter Arnold, Kurt Koocke, and Ernest A. Menze, "Herder's Reception and Influence," in *A Companion to the Works of Johann Gottfried Herder*, ed. Hans Adler and Wulf Koepke (Rochester, NY: Camden House, 2009), 410.
76. Alexander Crummell, "The Solution of Problems: The Duty and the Destiny of Man," *A. M. E. Church Review* 14 (1898): 400, 412; Alexander Crummell, *Africa and America: Addresses and Discourses* (Springfield, MA: Willey, 1891), 19–20. "Solution of Problems" is identified as the 1895 address at Wilberforce University in Wilson J. Moses, "W. E. B. Du Bois's 'The Conservation of Races' and its Context: Idealism, Conservatism and Hero Worship," *Massachusetts Review* 34 (1993): 280.

fested in every sphere, viz: 'That nature is constantly departing from the simple to the complex; starting off in new lines from the homogeneous to the heterogeneous;' striking out in divers ways into variety." Later in the same speech, "The Race-Problem in America," Crummell appealed to a different aspect of biology to undermine the possibility of amalgamation: "When once the race type gets fixed as a new variety, it propagates itself by that divine instinct of reproduction, and . . . becomes a perpetuity, with its own distinctive form, constitution, features, and structure." Crummell was thus committed to a biological conception of race, defined as "a compact, homogeneous population of one blood, ancestry and lineage," and he argued that there were biological facts favoring the persistence of such races.[77]

Although Du Bois attacked this biological concept of race in "The Conservation of Races," and although Wilson Moses has described the essay as "Hegelian racial mysticism," Du Bois also followed Cooper and Crummell in his use of evolutionary imagery. He defined a *race* as "a vast family of human beings, generally of common blood and language, always of common history, traditions and impulses, who are both voluntarily and involuntarily striving together for the accomplishment of certain more or less vividly conceived ideals of life." This definition was primarily sociohistorical rather than biological, but Du Bois also claimed that these races had emerged through a Spencerian evolutionary process—that is, one characterized by (in Spencer's words) "continuous differentiations and integrations." "The whole process which has brought about these race differentiations," said Du Bois, "has been a growth, and the great characteristic of this growth has been the differentiation of spiritual and mental differences between great races of mankind and the integration of physical differences." Perhaps inspired by Cooper's claim that "conflict, such as is healthy, stimulating, and progressive, is produced through the co-existence of radically opposing or racially different elements," Du Bois argued that "the race idea," because it had allowed diverse life ideals to struggle and jointly contribute to "that perfection of human life for which we all long," was "the vastest and most ingenious invention of human progress." Claiming that races could still play this role, Du Bois suggested optimistically that African and European Americans could successfully coexist so long as "there is substantial agreement in laws, language and religion" and "a satisfactory adjustment of economic life." In the ideal case, as he put it a few years later, there would

---

77. Crummell, *Africa and America*, 42, 47–48; alluding to Herbert Spencer, *First Principles of a New System of Philosophy*, 4th ed. (New York: D. Appleton, 1880), 396.

be an evolutionary struggle between different race ideals that was not a mere struggle for existence:

> It is . . . the strife of all honorable men of the twentieth century to see that in the future competition of races, the survival of the fittest shall mean the triumph of the good, the beautiful and the true; that we may be able to preserve for future civilization all that is really fine and noble and strong, and not continue to put a premium on greed and impudence and cruelty.[78]

For Du Bois, as for Cooper and Crummell, the evolutionary progress of particular races—rather than their disappearance or amalgamation—was essential to human progress more generally. Thus, evolution and idealism were intertwined for Du Bois in 1897—and this biological idealism was also characteristic of his early sociological writings, as we will see in chapter 6.

In the nineteenth century, the words *development* and *evolution* were used interchangeably. The idea of an interaction between organism and environment was thus relevant to both ontogeny and phylogeny—what we would now call development and evolution. It is thus not surprising that British and American idealists, writing at the end of that century, made a connection between Hegel's *Entwicklung* (usually translated as "development") and biological evolution. Although Green and Morris thought biological evolution was irrelevant to philosophical theories of mind and morality, Caird, Alexander, and Ritchie—along with Dewey, Mead, Royce, and Du Bois on the other side of the Atlantic—ultimately argued that the conceptual evolution of Hegel and the organic evolution of Darwin and Spencer were part of a broader "movement of thought" that emphasized history, growth, and development. As we will see in chapter 6, this movement was also important for Jane Addams, who was influenced by both Caird and Royce.

After almost a century of neglect, biological idealism was reborn in 1985 with the publication of Levins and Lewontin's *The Dialectical Biolo-*

---

78. Wilson J. Moses, "W. E. B. Du Bois's 'The Conservation of Races' and its Context: Idealism, Conservatism and Hero Worship," *Massachusetts Review* 34 (1993): 286; W. E. B. Du Bois, *The Conservation of Races* (Washington, DC: American Negro Academy, 1897), 7–9, 11; Spencer, *First Principles*, 216; Anna Julia Cooper, *A Voice from the South* (Xenia, OH: Aldine, 1892), 151; W. E. B. Du Bois, "The Relation of the Negroes to the Whites in the South," *Annals of the American Academy of Political and Social Science* 18 (1901): 121. See also Adolph L. Reed Jr., *W. E. B. Du Bois and American Political Thought: Fabianism and the Color Line* (Oxford: Oxford University Press, 1997), 120–22.

gist. As Peter Godfrey-Smith notes, their views stemmed from a reading of Friedrich Engels rather than any direct engagement with pragmatism or British idealism.[79] But although there was no historical connection between dialectical biology and pragmatism, they did have a common ancestor in Hegel: both Engels and the pragmatists were attempting to unite Hegel's method with modern biological theories. Like Dewey, Engels had an interactive picture of the organism-environment relation:

> Animals, as already indicated, change external nature through their activity in the same way, even if not to the same extent, as man does; and these modifications of the environment [*Umgebung*] . . . in turn react upon and change those who made them. For in nature nothing happens in isolation. Everything affects and is affected by every other thing, and it is mostly because this manifold motion and interaction is forgotten that our natural scientists are prevented from gaining a clear insight into the simplest things.

Lewontin translated this picture of the interaction between organism and environment into mathematics:

$$\frac{dO}{dt} = f(O, E), \text{ and } \frac{dE}{dt} = g(O, E).$$

That is, changes in the environment are a function of both the organism and the environment, just as changes in the organism are a function of both the environment and the organism.[80]

Looking just at these equations and the passage from Engels, it seems as though the only thing on the table for Lewontin was the reciprocal causes view. As mentioned earlier in this chapter, however, Lewontin also endorsed the idea that organism and environment are aspects of a single whole. He and Levins declared that "a whole is a relation of heterogeneous parts that have no prior independent existence *as parts*. . . . In general, the properties of parts have no prior alienated existence but are acquired by being parts of a particular whole."[81] At least one of the

---

79. Peter Godfrey-Smith, "Organism, Environment, and Dialectics," in *Thinking about Evolution: Historical Philosophical, and Political Perspectives*, ed. Rama S. Singh et al. (Cambridge: Cambridge University Press, 2001), 255–59.

80. Friedrich Engels, "Der Antheil der Arbeit an der Menschwerdung des Affen," *Die Neue Zeit* 14 (1896): 551; translation modified from Friedrich Engels, *Dialectics of Nature* (New York: International Publishers, 1940), 289; Richard C. Lewontin, "Gene, Organism and Environment," in *Evolution from Molecules to Men*, ed. D. S. Bendall (Cambridge: Cambridge University Press, 1983), 282. See also Richard Levins and Richard C. Lewontin, *The Dialectical Biologist* (Cambridge, MA: Harvard University Press, 1985), 104–5; Engels's essay is cited at 70.

81. Levins and Lewontin, *Dialectical Biologist*, 273.

CHAPTER FOUR

pragmatists—namely, Dewey—subscribed to this dual aspects picture. In an unpublished book manuscript written in the early 1940s but only recently rediscovered, Dewey used the terms "life-functions" and "life-activities" as synonyms for *experience*. He characterized these "life-activities" as "cooperative interactivities of component factors to which the names 'environmental' and 'organic' apply." He then provided a clear statement of the dual aspects view:

> The terms organism-environment are simply *generalized* names which serve to summarize, condense, unify, a large number of particular interactivities, such as air-respiratory processes, ground-locomotor apparatus, food-stuffs-digestive-tissues etc. They do not stand for two separate and independent things which then somehow come into connection with one another and produce life functions. On the contrary, in their status and capacity of being organic and environmental, they stand for results of analysis of primary life-activities.

Dewey claimed that although it is useful to analyze experience into organism and environment, especially when engaged in scientific inquiry, "it is one of the functions of philosophy to recall us from the results of analyses, which are made for special purposes, to the larger, if coarser and in many respects cruder, events which alone have primary existence." Thus, Dewey treated organism and environment as aspects of a single whole, emerging as interacting causes only upon analysis; for him, the causes and aspects views were consistent with each other, though the latter was primary.[82]

Is this simply a metaphysical muddle, as some scholars have suggested? I cannot resolve this question here, but some parallels with German idealism may help clarify what is at stake. As discussed earlier, Hegel and Schelling both subscribed in the early 1800s to the principle of subject-object identity. However, as Beiser recounts, Hegel was dissatisfied with Schelling's version of the aspects view because it could not account for our concrete experience of a subject-object distinction:

> If philosophy is to explain the opposition between subject and object in ordinary experience, then it must show how the single universal substance, in which the subject and object are the same, divides itself and produces a distinction between subject and object. The philosopher faces an intrinsically difficult task: he must both surmount and explain the necessity of the subject-object dualism.

---

82. John Dewey, *Unmodern Philosophy and Modern Philosophy* (Carbondale: Southern Illinois University Press, 2012), 321–22, 324.

According to Beiser, Hegel's solution was to interpret the absolute as a universal organism: biological development is self-differentiation, and the subjective and objective can be seen as "different degrees of organization" in the development of the absolute. The opposition between them is necessary, but they are ultimately only *aspects* of a single whole.[83]

Dewey can be interpreted as offering an analogous solution. What is primary is the whole: experience. Only when a problem arises, or in the midst of a scientific investigation, do we resolve experience into its two most general aspects, the organic and the environmental. At this point, we arrive at the reciprocal causes view as a secondary result—without causes, after all, we would be unable to intervene, and intelligent adjustment is for Dewey "an engineering issue" involving control and "social guidance."[84] But although this analysis of a whole into its aspects is vital to the process of adjustment or reconstruction, when this process—whether ethical reflection or scientific inquiry—comes to a close, we are again left with what is primary: simply living. One key difference between Dewey's story and Hegel's is that Dewey sees this development as continual, whereas Hegel, at least on standard readings, views it as having an ultimate end point. Nevertheless, both share a commitment to a kind of developmental metaphysics. It is an open research question whether the other second-cohort pragmatists—Royce, Addams, Mead, and Du Bois—would also have endorsed this viewpoint.

This chapter has focused on idealism and evolution, treating the latter as a clear and unified theory. Chapter 5 will destroy this picture: what evolution was, how it worked, and what it implied were all the subject of intense debate in the late nineteenth century—a debate in which many of the pragmatists actively participated.

83. Frederick Beiser, *Hegel* (New York: Routledge, 2005), 65, 94, 105.
84. John Dewey, *Human Nature and Conduct: An Introduction to Social Psychology* (New York: Henry Holt, 1922), 10.

FIVE

Weismannism Comes to America: The Factors of Evolution

Thanks to Peter Bowler's book *The Eclipse of Darwinism*, it is well known that although most naturalists had accepted evolution by the mid-1870s, they usually resisted Charles Darwin's further claim that "natural selection has been the main but not exclusive means of modification." Those friendly to evolution were still often critical of natural selection, claiming that other factors were just as important—*factor* was the standard term at the time, defined by Charles Sanders Peirce as "one of several circumstances, elements, or influences which tend to the production of a given result."[1] The Harvard botanist Asa Gray, for example, in an essay that Peirce, William James, and Chauncey Wright probably read, declared that "selection, artificial or natural, no more originates [variations] than man originates the power which turns a wheel, when he dams a stream and lets the water fall upon it." This particular criticism was most

---

1. Peter J. Bowler, *The Eclipse of Darwinism: Anti-Darwinian Theories in the Decades around 1900* (Baltimore: Johns Hopkins University Press, 1983), 3; Charles Darwin, *On the Origin of Species by Means of Natural Selection, or the Preservation of Favoured Races in the Struggle for Life* (London: John Murray, 1859), 6; William Dwight Whitney, ed., *The Century Dictionary: An Encyclopedic Lexicon of the English Language*, 6 vols. (New York: Century, 1889–91), 2:2114, s.v. "factor." For a list of Peirce's entries in the *Century Dictionary*, see Kenneth Laine Ketner, ed., *A Comprehensive Bibliography of the Published Works of Charles Sanders Peirce, with a Bibliography of Secondary Studies*, 2nd ed. (Bowling Green, OH: Philosophy Documentation Center, 1986), 43–83.

influentially articulated by George Campbell, the eighth Duke of Argyll, who argued in his popular book *The Reign of Law*—reviewed by Wright for the *Nation* and read by John Dewey as a college freshman—that Darwin's preferred factor could only be part of the story: "Natural Selection can do nothing except with the materials presented to its hands. It cannot select except among the things open to selection. Natural Selection can originate nothing; it can only pick out and choose among the things which are originated by some other law."[2] Despite replies from the likes of Alfred Russel Wallace, codiscoverer of natural selection, who claimed that "when a sufficient number of individuals are produced[,] variations of any required kind can always be met with," this criticism was common. The British zoologist St. George Mivart, as discussed in chapter 1, declared that there must be "an unknown internal natural law or laws conditioning the evolution of new specific forms from preceding ones." The American paleontologist Edward Drinker Cope, as mentioned in chapter 2, insisted that Darwin's law of natural selection—what Herbert Spencer called the "survival of the fittest"—"leaves the origin of the fittest entirely untouched. . . . [natural selection] is, then, only restrictive, directive, conservative, or destructive of something already created." The question of how biological variation was produced became a major research problem. As Darwin wrote shortly before his death in 1882, "at the present time there is hardly any question in biology of more importance than this of the nature and causes of variability."[3]

As we saw in chapter 3, evolution was standard fare by the time the second cohort of pragmatists got to college. So was uncertainty about its causes. Josiah Royce's teacher Joseph LeConte wrote the following in 1878, in an essay Dewey may have read:

Evolution, in its widest and truest sense, is a grand fact, embracing every department of nature. . . . Now, in this wide sense there can be no doubt of the evolution of the organic kingdom. There may be, and in fact there is, much difference of opinion as to

---

2. Asa Gray, "Darwin and His Reviewers," *Atlantic Monthly* 6 (1860): 417; Duke of Argyll, "Opening Address [Creation by Law]," *Proceedings of the Royal Society of Edinburgh* 5 (1864): 275; also in Duke of Argyll, *The Reign of Law* (London: Alexander Strahan, 1867), 230; Chauncey Wright, "The Reign of Law," *Nation*, June 13, 1867, 470; Lewis S. Feuer, "John Dewey's Reading at College," *Journal of the History of Ideas* 19 (1958): 416.

3. Alfred Russel Wallace, "Creation by Law," *Quarterly Journal of Science* 4 (1867): 483; St. George Mivart, *On the Genesis of Species* (London: Macmillan, 1871), 51; Edward Drinker Cope, "The Method of Creation of Organic Forms," *Proceedings of the American Philosophical Society* 12 (1871): 230; Charles Darwin, prefatory notice to *Studies in the Theory of Descent*, by August Weismann, 2 vols. (London: Sampson Low, Marston, Searle, & Rivington, 1882), 1:vi. Spencer coined the phrase "survival of the fittest" in Spencer, *The Principles of Biology*, vol. 1 (London: Williams & Norgate, 1864), 444–45.

the *causes* or *factors* of evolution—there may be, and in fact there is, much difference of opinion as to the *rate* of evolution, whether always uniform or often more or less paroxysmal; but of the *fact* of progressive movement of the whole organic kingdom to higher and higher conditions . . . there is no longer any doubt.[4]

Neither their teachers nor their textbooks made any attempt to conceal these debates. For example, although its description of Darwin's theory was more detailed, Henry Alleyne Nicholson's zoology textbook—assigned to Dewey and George Herbert Mead—also mentioned Jean-Baptiste Lamarck's theory, in which "the means of modification were ascribed to the action of external physical agencies, the inter-breeding of already existing forms, and the effects of habit."[5] James Dwight Dana's geology textbooks, used by Dewey, Jane Addams, and W. E. B. Du Bois, included the following non-Darwinian conclusions among those "most likely to be sustained by further research":

The method of evolution admitted of abrupt transitions between species; as has been argued by [Alpheus] Hyatt and Cope, from the abrupt transitions that occur in the development of animals that undergo metamorphosis, and the successive stages in the growth of many others.

External agencies or conditions, while capable of producing modifications of structure, have had no more power toward determining the directions of progress in the evolution, than they now have in determining the course of progress in development from a living germ.[6]

Albert Allen Wright, in his Darwin memorial address at Oberlin College (with Mead in attendance), echoed the Duke of Argyll: "Many deny to natural selection the prominent place [Darwin] has given it, rightly arguing that it is merely a preserving force and not an originating force."[7] Nathaniel Southgate Shaler, comparing the views of Lamarck and Darwin in Du Bois's geology class at Harvard, told his students that the "weak point" of the natural selection hypothesis was that "it *does not*

---

4. Joseph LeConte, "Man's Place in Nature," *Princeton Review* 55 (1878): 786–87; Lewis S. Feuer, "John Dewey's Reading at College," *Journal of the History of Ideas* 19 (1958): 420.

5. Henry Alleyne Nicholson, *A Manual of Zoology for the Use of Students, with a General Introduction on the Principles of Zoology*, 2nd ed. (New York: D. Appleton, 1876), 39.

6. James D. Dana, *Manual of Geology: Treating of the Principles of the Science, with Special Reference to American Geological History*, 2nd ed. (New York: Ivison, Blakeman, Taylor, 1874), 604; see also James D. Dana, *New Text-Book of Geology*, 4th ed. (New York: American, 1883), 393.

7. Albert Allen Wright, "Darwin (1882)," pp. [41–42], Writings (MSS), Box 10, Albert Allen Wright Papers (RG 30/017), Oberlin College Archives; see also *Oberlin Review*, November 18, 1882, 55.

account for variation." Du Bois had already heard a similar story from William James in his philosophy class: "Difficulty in believing accidental variation in Darwin's theory is very great.... Most naturalists while they believe in evolution differ as to the method (Oscar Schmidt, [Herbert William] Conn, Wallace, [Francis] Galton, Cope, etc. are authorities). The settlement of the difficulty depends on facts not theory."[8] Du Bois, Dewey, and the rest were thus familiar with debates in the 1870s and '80s over the factors of evolution.

These debates took on a new urgency around 1890, as the scientific world reacted to a startling theory put forward by the German biologist August Weismann. Although Weismann specialized in invertebrate embryology, he was forced to take periodic breaks from his laboratory research because of recurring eye problems. During these off periods, he focused instead on theories of evolution and heredity.[9] The first of these breaks—from the mid-1860s to the mid-1870s—resulted in, among other writings, *On the Justification of the Darwinian Theory* and *Studies in the Theory of Descent*. Both of these texts focused on the problem of variation. In *Justification*, a short pamphlet, Weismann noted that species were the result not only of natural selection but also "the quality of variation of their progenitors." Darwin underlined the word *Variationsqualität* in his copy and wrote "good" in the margin; he also wrote "Causes & Law of Variation most important" on the back cover. In *Studies*, a series of case studies of variation, Weismann presented a traditionally Darwinian view of the factors of evolution: natural selection was primary, but the inherited effects of disuse had "a large share" in the suppression of "active organs."[10] But Weismann's account of evolution's factors soon changed. During his second break from laboratory work, which began in 1883, he published a series of essays presenting a new theory of heredity, based on the continuity and isolation of an inherited substance he called the "germ-plasm."[11] It was this new theory, which had as its corollary that "characters" acquired by an individual during its

---

8. W. E. B. Du Bois, "Philosophy IV Notebook," Du Bois Papers. Although most of this notebook contains notes from the second semester (February–May 1889) of James's Philosophy 4 class, the last three pages are notes from a very late session (8 May 1889) of Shaler's Natural History 8 class.

9. August Weismann, "Autobiography of Professor August Weismann," *Lamp* 26 (1903): 24–26.

10. August Weismann, *Über die Berechtigung der Darwin'schen Theorie* (Leipzig, Ger.: Wilhelm Engelmann, 1868), 29; Mario A. Di Gregorio and N. W. Gill, *Charles Darwin's Marginalia*, vol. 1 (New York: Garland, 1990), 855–57, 860; August Weismann, *Studies in the Theory of Descent*, trans. Raphael Meldola, 2 vols. (London: Sampson Low, Marston, Searle, & Rivington, 1882), 1:384; translated from August Weismann, *Studien zur Descendenz-Theorie*, 2 vols. (Leipzig, Ger.: Wilhelm Engelmann, 1875–76), 2:133.

11. August Weismann, *Ueber die Vererbung: Ein Vortrag* (Jena, Ger.: G. Fischer, 1883); August Weismann, *Die Continuität des Keimplasma's als grundlage einer Theorie der Vererbung* (Jena, Ger.: G. Fischer,

lifetime could not be passed to offspring, that brought Weismann's work to the wider attention of British and American naturalists.

It is difficult to exaggerate the importance of Weismann's theory of heredity to biological discussion at the end of the nineteenth century. Grant Allen, the opponent of William James we met in chapter 2, wrote that already by the late 1880s "nothing else was heard of in *Nature* and in the scientific societies. Weismannism became the fashionable creed of the day. . . . Young England, as a biologist, swore by the continuity of the germ-plasm, and laughed to scorn the inheritance of the acquired faculty." Just a few months later in the same journal, Benjamin Kidd— whom we will encounter again later this chapter—published what can only be described as a celebrity profile, entitled "Darwin's Successor at Home." It referred to Weismann as the "hero of the hour in biological science, upon whom Darwin's mantle seems to have descended"; spoke of the German biologist's theories as "a new gospel"; and described the details of Weismann's home life, right down to the bust of Darwin glowering on his desk. By 1892, *Popular Science Monthly* could proclaim that "the three great names in the history of biologic evolution are those of Lamarck, Darwin, and Weismann."[12] Not only was Weismann's work famous, it was also seen as opposed to that of Spencer, and the two thinkers eventually had a public debate from 1893 to 1895. Because of the prominence of Spencer's ideas, with evolution seen as relevant to psychology, sociology, and ethics, many philosophers were drawn into the factors of evolution discussions (recall that James, Royce, Dewey, and Mead all assigned Spencer's books to their classes in the 1870s and '80s, as discussed in chapters 2 and 3). In this chapter, I will describe how James, Peirce, and Dewey participated in the debates over Weismann's work in the 1890s. These debates about the causes of evolution highlighted the possible directedness of variation as well as the plasticity of individual organisms—that is, their flexible response to environmental changes.

The chapter has three parts. In the first, I will show that Weismann's ideas were salient for James and Peirce because of their opposition to Spencer. Next, I will argue that Peirce's analysis of evolution in a series of articles for the *Monist* in the early 1890s should be seen as responding and contributing to the broader conversation about the factors of evolu-

---

1885); August Weismann, *Essays upon Heredity and Kindred Biological Problems* (Oxford: Clarendon Press, 1889).

12. Grant Allen, "The New Theory of Heredity," *Review of Reviews* 1 (1890): 538; Benjamin Kidd, "Darwin's Successor at Home," *Review of Reviews* 2 (1890): 647–48, 650; "Literary Notices," *Popular Science Monthly* 40 (1892): 847.

tion. In particular, I will suggest that Peirce's views were influenced by the "speculative biology" of thinkers such as Cope.[13] In the last section of the chapter, after a brief account of the public debate between Spencer and Weismann, I will argue that Dewey incorporated the themes of this debate—plasticity, directed variation, and heredity—into his 1890s work on social ethics and social psychology.

The Reception of Weismann

August Weismann and Herbert Spencer were cast as enemies before they even encountered each other's work. In his 1887 book *The Factors of Organic Evolution*, Spencer had declared that the question of which factors were operative in evolution, given its implications for both psychology and sociology, "demands, beyond all other questions whatever, the attention of scientific men." He went on to distinguish three factors:

1. "Direct action of the medium"
2. "Natural selection" or "survival of the fittest"
3. "Modifications of structure caused by modifications of function"

Although all of these factors were constantly acting, Spencer thought that each had been dominant during a particular period: direct action was most important early in evolutionary history, with the development of cells and surfaces; natural selection was most important for plants and lower animals, after the rise of sexual reproduction and direct competition; and use and disuse leading to structural modifications, the Lamarckian factor, was most important for humans and other higher animals, since they were more active and (in the case of humans) supposedly shielded from the effects of natural selection.[14]

Unbeknownst to Spencer, Weismann's novel theory of heredity had recently appeared in German and had also been summarized in the British scientific journal *Nature*. According to Weismann, the cells that make up new offspring arise "not at all out of the body of the individual, but direct from the parent germ-cell." Thus, "the only actual carrier of the tendency of heredity is the highly organised nuclear substance," the

---

13. Herbert W. Schneider, *A History of American Philosophy* (New York: Columbia University Press, 1946), 359–63.
14. Herbert Spencer, *The Factors of Organic Evolution* (New York: D. Appleton, 1887), vi, 72–74; see also Herbert Spencer, *The Principles of Biology*, vol. 1 (London: Williams & Norgate, 1864), 464–69.

"germ-plasm," which is "something standing opposed to and separate from the entirety of cells composing the body." As Weismann noted, this isolation of the germ-plasm entails that "acquired modifications"—any environmentally induced changes that do not affect the germ-cells—are not passed to offspring.[15] British naturalists did not miss the implication: Weismann's theory made the operation of Spencer's third factor impossible. The comparative psychologist George John Romanes, reacting to *Factors of Organic Evolution*, lamented that Spencer and Weismann were unaware of each other's work, claiming that Spencer still had good arguments in favor of the Lamarckian factor. A week or so later, Weismann traveled to England to attend the annual meeting of the British Association for the Advancement of Science (BAAS), giving a paper and participating in a special discussion on September 5, 1887: "Are Acquired Characters Hereditary?" Joining the discussion were (among others) Edward Bagnall Poulton and Edwin Ray Lankester, who would soon become allies of Weismann. Although he was not in attendance, Spencer did hear about the session and used it to counter the Duke of Argyll's claim, made in response to Spencer's book, that inheritance of the effects of use and disuse was "not generally disputed" by biologists.[16]

By the summer of 1888, Poulton was already fighting with Weismann's critics. Romanes, echoing Argyll, accused Weismann and Poulton of ignoring the problem of variation: "Any proof of natural selection as an operating principle opens up the *more ultimate problem* as to the *causes* of the variations on the occurrence of which this principle depends." Weismann and his followers, said Romanes, "out-Darwin Darwin himself in their allegiance to his doctrine, attaching even more importance to natural selection than was attached to it by their master."[17] Poulton was annoyed by this claim, and he reminded readers of *Nature* that Weismann had specifically discussed the causes of variation. Romanes was annoyed in turn by Poulton's suggestion that he was unfamiliar with Weismann's work. Ultimately, Romanes declared that since Poulton and Weismann "aim at establishing for natural selection a sole and universal sovereignty which was never claimed for it by Darwin him-

---

15. Henry Nottidge Moseley, "The Continuity of the Germ-Plasma Considered as the Basis of a Theory of Heredity," *Nature* 33, no. 842 (1885): 155; summarizing August Weismann, *Die Continuität des Keimplasma's als grundlage einer Theorie der Vererbung* (Jena, Ger.: G. Fischer, 1885).

16. George John Romanes, "The Factors of Organic Evolution," *Nature* 36, no. 930 (1887): 405; *Report of the Fifty-Seventh Meeting of the British Association for the Advancement of Science, Held at Manchester in August and September 1887* (London: John Murray, 1888), 755; Duke of Argyll, "A Great Confession," *Nineteenth Century* 23 (1888): 144; Herbert Spencer, "A Counter Criticism," *Nineteenth Century* 23 (1888): 213.

17. George John Romanes, "Recent Critics of Darwinism," *Contemporary Review* 53 (1888): 841.

self," their school "may properly be called Neo-Darwinian: pure Darwinian it certainly is not." Weismann's theory of evolution, which claimed primacy for natural selection by denying the inheritance of acquired characters, now had a name: neo-Darwinism, often contrasted with the neo-Lamarckism of Cope and others.[18]

Soon after Romanes's letter appeared in *Nature*, the young American paleontologist Henry Fairfield Osborn—a professor at Princeton and an ally of Cope—attended the BAAS meeting at Bath, solidifying what had until then been a purely epistolary friendship with Poulton.[19] The two began a friendly argument over the factors of evolution that would last for years. Only a week after the meeting ended, Poulton joked with Osborn in a letter, "I do not give your neo-Lamarckian school more than three years of life." Osborn replied, "Cope rubs his hands in glee, at the prospective feast upon the Weismannians—with their non-inherited acquired characters—so give us more than 3 years of life." As we will see, Osborn ended up playing a key role in the American debates over Weismann's work.[20]

Despite these early discussions, Weismann's theories were not truly in the spotlight until 1889, when "two important works" related to the factors debates appeared: Wallace's *Darwinism*, a popular account of evolution, and Weismann's *Essays upon Heredity*, an English translation of his 1880s essays.[21] In *Darwinism*, Wallace devoted an entire chapter to "Fundamental Problems in Relation to Variation and Heredity," attacking a variety of naturalists—including Spencer and Cope—for exaggerating the importance of evolutionary factors other than natural

---

18. Edward Bagnall Poulton, "Dr. Romanes' Article in the *Contemporary Review* for June," *Nature* 38, no. 978 (1888): 295; George John Romanes, "Dr. Romanes's Article in the 'Contemporary Review,'" *Nature* 38, no. 981 (1888): 364; George John Romanes, "Lamarckism *versus* Darwinism," *Nature* 38, no. 983 (1888): 413. For early uses of the term *neo-Darwinism*, see Charles Clement Coe, "Darwinism and Neo-Darwinism," *Universal Review* 5 (1889); Lester Frank Ward, "Neo-Darwinism and Neo-Lamarckism," *Proceedings of the Biological Society of Washington* 6 (1891); Edward Drinker Cope, "Alfred Russel Wallace," in *Evolution in Science, Philosophy, and Art: Popular Lectures and Discussions before the Brooklyn Ethical Association* (New York: D. Appleton, 1891), 12; Henry Fairfield Osborn, "The Present Problem of Heredity," *Atlantic Monthly* 67 (1891): 356, 360.

19. Henry Fairfield Osborn to Lucretia Thatcher Perry Osborn, 6 September 1888, Folder 5, Box 5, Osborn Family Papers.

20. Edward Bagnall Poulton to Henry Fairfield Osborn, 16 September 1888, Folder 19, Box 17, Henry Fairfield Osborn Papers (MSS .O835), Special Collections, American Museum of Natural History; Henry Fairfield Osborn to Edward Bagnall Poulton, 22 October 1888, Folder 1, Box 19, Osborn Family Papers.

21. William Henry Flower, "[Presidential] Address," in *Report of the Fifty-Ninth Meeting of the British Association for the Advancement of Science, Held at Newcastle-upon-Tyne in September 1889* (London: John Murray, 1890), 20. The two books were published only weeks apart in the United States: see "Books of the Week," *Nation*, July 11, 1889, 40; "Books of the Week," *Nation*, July 25, 1889, 80.

CHAPTER FIVE

selection: "Whatever other causes have been at work, Natural Selection is supreme, to an extent which even Darwin himself hesitated to claim for it. The more we study it the more we are convinced of its overpowering importance."[22] Wallace's book prompted more debate in the pages of *Nature*, this time lasting almost a year. One of the earliest correspondents was Cope himself, who summarized the view we have already seen him defending: "*Selection* cannot be the cause of those conditions which are prior to selection; in other words, a selection cannot explain the *origin* of anything, although it can and does explain survival of something already originated; and evolution consists in the origin of characters, as well as their survival." Recalling Poulton's response to Romanes the previous year, Lankester accused Cope of ignoring Weismann's own account of the origin of variation: "Mr. Cope does not seem to be aware that the anti-Lamarckians attach great importance to the existence of congenital variation, . . . and that Weismann himself has developed a most ingenious theory as to the relation of fertilization and its precedent phenomena to this all-important factor in evolution." According to Poulton and Lankester, most variation resulted from the fusion of two germ-cells in sexual reproduction—a theory already outlined by Weismann and summarized in *Nature* in 1886.[23]

But the anti-Weismann group was not finished, and the polemical tone of the letters only increased. As Samuel Butler noted the next year, "we cannot take up a number of *Nature* without seeing how hot the contention is between [Lamarck's] followers and those of Weismann." The Duke of Argyll wrote in, attributing a hidden motive to Lankester and the neo-Darwinians—namely, "jealousy of any conception which tends to break down the empire of mere fortuity in the phenomena of variation." Argyll's comments highlighted the contrast between fortuitous and directed variation. Two weeks later, a letter from Osborn argued in favor of the latter, appealing to the fossil record to claim that "the new variations in the skeleton and teeth of the fossil series are observed to have a definite direction." Like Cope and others, he argued that "we are driven to the necessity of postulating some as yet unknown factor in evolution to explain these purposive or directive laws in variation." As

---

22. Alfred Russel Wallace, *Darwinism: An Exposition of the Theory of Natural Selection with Some of Its Applications* (London: Macmillan, 1889), 444.
23. Edward Drinker Cope, "Lamarck *versus* Weismann," *Nature* 41, no. 1048 (1889): 79; Edwin Ray Lankester, "Mr. Cope on the Causes of Variation," *Nature* 41, no. 1050 (1889): 129; Henry Nottidge Moseley, "Dr. August Weismann on the Importance of Sexual Reproduction for the Theory of Selection," *Nature* 34, no. 887 (1886).

we will see, Weismann's detractors often claimed that if variation were truly random, progressive evolution would be impossible.[24]

James and Peirce encountered these debates in the discussion surrounding the second important book: Weismann's *Essays upon Heredity*, edited by Poulton.[25] Du Bois's notes from Philosophy 4, quoted earlier, show that James was unaware of Weismann's work in March 1889—otherwise he would have listed him as an authority along with Cope and the rest. This changed sometime in 1890, as indicated by a late addition to the last chapter of James's *Principles of Psychology*. In this final chapter, as we saw in chapter 2, James argued that human theoretical judgments were originally "pure *idiosyncrasies*, spontaneous variations, fitted by good luck (those of them which have survived) to take cognizance of objects . . . , without being in any intelligible sense immediate derivatives from them."[26] James tried to argue the same for the "practical parts of our organic mental structure," specifically instincts. According to Spencer, as glossed by James, instincts were "'secondarily-automatic' habits," originating "out of tentative experiments made during ancestral lives" and "perfected by repetition, addition, and association through successive generations." But James sided with Darwin, for whom instincts were merely results of "the natural selection of accidentally produced tendencies to action." Over a few pages, James suggested that much of the evidence taken to support Spencer's more Lamarckian theory of instinct could also be explained by spontaneous variation and natural selection, although he admitted that his general anti-Spencerian conclusion in the chapter could not be "as confidently expressed" when it came to humans' practical minds.[27]

But in the spring of 1890, just as he was getting ready to send *Principles* to the publisher, James received a letter from a former student who was studying zoology in Freiburg "under the famous Weismann."[28] Perhaps prompted by this letter, James looked into Weismann's work and added some material to the end of his final chapter. Back when he had

---

24. Samuel Butler, "The Deadlock in Darwinism [Part 3]," *Universal Review* 7 (1890): 239; Duke of Argyll, "Acquired Characters and Congenital Variation," *Nature* 41, no. 1052 (1889): 173; Henry Fairfield Osborn, "The Palaeontological Evidence for the Transmission of Acquired Characters," *Nature* 41, no. 1054 (1890): 228.

25. Poulton was the chief editor of the volume, although he was not the primary translator of any of the chapters: see August Weismann to Edward Bagnall Poulton, 23 January 1889, in August Weismann, *Selected Letters and Documents*, 2 vols. (Freiburg, Ger.: Universitätsbibliothek Freiburg im Breisgau, 1999), 1:127–28.

26. William James, *The Principles of Psychology*, 2 vols. (New York: Henry Holt, 1890), 2:631.

27. James, *Principles of Psychology*, 2:678–86.

28. Charles Augustus Strong to William James, 13 May 1890, in James, *Correspondence*, 7:24.

finished the chapter in 1885, said James, "whether acquired ancestral habits played any part at all" in the production of instincts "was still an open question in which it would be as rash to affirm as to deny." But the situation had changed: "Already before that time . . . Professor Weismann of Freiburg had begun a very serious attack upon the Lamarckian theory, and his polemic has at last excited such a widespread interest among naturalists that the whilom almost unhesitatingly accepted theory seems almost on the point of being abandoned." According to Weismann's "captivating" new theory, said James, "the only way in which the germinal products can be influenced whilst in the body of the parent is . . . by good or bad nutrition. . . . Peculiarities of neural structure and habit in the parents *which the parents themselves were not born with*, they can never acquire unless perhaps accidentally through some coincidental variation of their own." James thus immediately embraced Weismann as an ally against Spencer.[29]

Despite this enthusiasm, James acknowledged that *Factors of Organic Evolution*—which he described as "much the solidest thing" Spencer had written—contained the strongest case available "in favor of the Lamarckian theory." James referred to one of Spencer's arguments in particular, which had already featured in the *Nature* debates and which would appear again in the public dispute between Spencer and Weismann: the argument from coadaptation of parts. Spencer began this argument by quoting Darwin: "In order that an animal should acquire some structure specially and largely developed, it is almost indispensable that several other parts should be modified and co-adapted." According to Spencer, this meant that the evolution of any specialized structure requires "numerous appropriate variations" working together. But if variations are truly fortuitous and independent of one another, and not directed in some way, argued Spencer, "there is no seeing how the required readjustments can be made":

Can we suppose that all these appropriate changes, too, would be step by step simultaneously made by fortunate spontaneous variations, occurring along with all the other fortunate spontaneous variations? Considering how immense must be the number of these required changes, added to the changes above enumerated, the chances against any adequate readjustments fortuitously arising must be infinity to one.

---

29. William James, *The Principles of Psychology*, 2 vols. (New York: Henry Holt, 1890), 2:686–87; citing August Weismann, *Essays upon Heredity and Kindred Biological Problems* (Oxford: Clarendon Press, 1889).

Although this argument had been criticized by Wallace in *Darwinism*, James saw it as forming "a great *presumption* against the all-sufficiency of the view of selection of accidental variations exclusively." He may have been following Romanes, who had described it a few years earlier as "virtually proving the truth of the Lamarckian assumption."[30]

Peirce was also exposed to Weismann's ideas in the spring of 1890, but by a more indirect route. As Christopher Hookway and others have shown, by the mid-1880s Peirce had begun to sketch the evolutionary metaphysics that culminated in his early 1890s articles for the *Monist*. In "Design and Chance," a paper read at Johns Hopkins in 1884, Peirce argued that Spencer and even Darwin had neglected the role of chance in evolution:

> It has always seemed to me singular that when we put the question to an evolutionist, Spencerian, Darwinian, or whatever school he may belong to, what are the agencies which have brought about evolution, he mentions various determinate facts and laws, but among the agencies at work he never once mentions *Chance*. Yet it appears to me that chance is the one essential agency upon which the whole process depends.

Although these views were not yet published, they were in circulation: Dewey took part in the discussion following this 1884 talk, and several philosophers living in Cambridge—including James, Royce, Francis Ellingwood Abbot, and John Fiske—heard Peirce present similar ideas in 1886.[31]

Peirce launched the first of a series of public attacks on Spencer's evolutionary philosophy in 1887. Samuel Barrows, editor of the Unitarian newspaper the *Christian Register*, had sent letters to "prominent scientific men" asking whether the discoveries of modern science had any bearing on human immortality. Cope and LeConte, in their responses, tied the answer to debates over the factors of evolution. Peirce did the same and took the opportunity to criticize Spencer's vision of a world "governed

---

30. Charles Darwin, *The Origin of Species by Means of Natural Selection, or the Preservation of Favoured Races in the Struggle for Life*, 6th ed. (London: John Murray, 1872), 179; Herbert Spencer, *The Factors of Organic Evolution* (New York: D. Appleton, 1887), 14–17; James, *Principles of Psychology*, 2:687–88; Alfred Russel Wallace, *Darwinism: An Exposition of the Theory of Natural Selection with Some of Its Applications* (London: Macmillan, 1889), 417–18; George John Romanes, "The Factors of Organic Evolution," *Nature* 36, no. 930 (1887): 405.

31. Christopher Hookway, "Design and Chance: The Evolution of Peirce's Evolutionary Cosmology," *Transactions of the Charles S. Peirce Society* 33 (1997): 18–21; "[Design and Chance]," MS 494 (1884), in Peirce, *Writings*, 4:548; "35th Meeting" (17 January 1884), Metaphysical Club Records; "Diaries," 3 February 1886, Box 16, Papers of Francis Ellingwood Abbot (HUG 1101), Harvard University Archives.

altogether by mechanism," writing, "The endless variety in the world has not been created by law. It is not of the nature of uniformity to originate variation, nor of law to beget circumstance. When we gaze upon the multifariousness of nature, we are looking straight into the face of a living spontaneity."[32] Peirce, like Argyll's caricature of the neo-Darwinian, emphasized the contrast between the spontaneity of variation and the law of natural selection, arguing that Spencer and others were neglecting the former. In his introduction to the symposium, Barrows linked the positions of Peirce and Cope, noting that "they both take strong and well-fortified ground against the mechanical and automatic philosophy."[33]

Peirce continued his attack on Spencer in 1890, convincing Charles Ransom Miller, editor of the *New York Times*, to host a write-in debate on the scientific merits of the Spencerian philosophy. It was to be initiated by Peirce—writing under the pseudonym "Outsider"—with responses printed on successive Sundays.[34] Peirce, in his opening article of March 23, asked *New York Times* readers three questions: first, whether "value has been accorded to [Spencer's philosophy], by men competent to judge it"—that is, specialists in various scientific fields; second, "whether Mr. Spencer's system is logically put together"; and third, whether "Spencer's theory . . . can point to considerable discoveries directly resulting from its predictions—not, be it understood, from the general doctrine of evolution, or from the Darwinian theory, but from the seventeen articles of the Spencerian confession." Peirce was skeptical that the answers to these questions would reflect well on Spencer. Regarding the first, Peirce slyly suggested that perhaps "each of these specialists is accustomed to think of Mr. Spencer as eminent in every branch but his own." As for the second, Peirce repeated his earlier criticisms of Spencer's mechanical worldview, claiming that a "thoroughgoing evolutionism" would attempt a real explanation of "the general laws of mechanics" rather than simply gesturing at "the Unknowable" as Spencer had done. The article ended with an unfavorable comparison between Spencer and Isaac Newton: Peirce pointed out that "the Newtonian philosophy" had directly

---

32. Charles Sanders Peirce, "[Science and Immortality]," *Christian Register*, April 7, 1887, 214; also in Samuel June Barrows, ed., *Science and Immortality: The "Christian Register" Symposium, Revised and Enlarged* (Boston: Geo. H. Ellis, 1887), 73.

33. Samuel June Barrows, "What Does Science Say?" *Christian Register*, April 7, 1887, 209; also in Barrows, *Science and Immortality*, 110–11, with the second *and* replaced by *or*.

34. Charles Ransom Miller to Charles Sanders Peirce, 17 March 1890, L289, Peirce Papers. For a brief overview of this episode, see Peirce, *Writings*, 6:lxxvi–lxxviii. One of the reasons Peirce contacted Miller is that he was desperately in need of money at this time: see Peirce, *Writings*, 8:xxv–xxvi.

resulted in many successful predictions, implicitly questioning whether the same could be said of the Spencerian. (Peirce's older brother later questioned whether this was fair, since such a test "could hardly have been satisfactorily met so soon by Newtonianism.")[35]

Most of the March 30 responses focused on the second question, with several letters claiming that asking Spencer to explain "general laws" was asking the impossible. These initial correspondents also challenged "Outsider" to give his own positive views: "If Spencer's is a 'somewhat clumsy conception of evolution' perhaps [Outsider] can tell us where to find one that is more graceful."[36] The next Sunday, however, Henry Fairfield Osborn took up the first question, promising "to examine [Spencer] solely as a biological philosopher, and especially in his treatment of the theory of organic evolution." Osborn had finished reading Weismann's *Essays upon Heredity* the previous August and had praised them in a letter to Poulton:

I am deeply impressed with your splendidly translated and edited Weismann. His arguments are very strong. He makes transmission of acquired characters appear almost a physical impossibility. . . . I am not wedded to any theory, and will not be heartbroken if it does "only survive two years more"—but Weismann's book has not shaken me an inch on the question of variation.

Peirce wanted responses from scientific experts, and he could not have asked for one more qualified: as we have seen, Osborn was a major player in the factors of evolution debates, and his response to both Wallace and Weismann in *Nature* (quoted earlier) had been reprinted in the American journal *Science* only a few months before.[37]

Weismann's ideas were central to Osborn's assessment of Spencer. He began by claiming that the English philosopher, in his *Principles of Biology*, "strikes no really original vein of thought" and "throughout . . . is what is now known as a Lamarckian." Osborn then criticized Spencer's view of the organic world for having "a strongly mechanical bias": "As the cornerstone of his system is mechanical the mechanical factors

---

35. Charles Sanders Peirce, "Herbert Spencer's Philosophy," *New York Times*, March 23, 1890, 4; citing Herbert Spencer, *First Principles* (London: Williams & Norgate, 1862), 47–67; James Mills Peirce to Charles Sanders Peirce, 13 June 1890, L339, Peirce Papers.

36. Kappa, "Flaws in 'Outsider's' Reasoning," *New York Times*, March 30, 1890, 13; R. G. E., "A Call for Specifications," *New York Times*, March 30, 1890, 13. The identities of "Kappa" and "R. G. E." are not known.

37. Henry Fairfield Osborn to Edward Bagnall Poulton, 27 August 1889, Folder 1, Box 19, Osborn Family Papers; Henry Fairfield Osborn, "The Palaeontological Evidence for the Transmission of Acquired Characters," *Science* 15, no. 367 (1890).

are regarded as the major factors in evolution. . . . The tissues of each individual are constantly adjusting themselves to the hard knocks of environment. The alterations of structure thus brought about make themselves persistent by transmission into the race." Osborn claimed that Spencer's whole evolutionary philosophy was an arch "resting on a mechanical cornerstone as the element of truth, and upon the assumption that structural alterations are inherited as its debatable element." At this point in the essay, Osborn introduced Weismann: "It has remained for another great original thinker, Weismann, to appear upon the scene and assert that one of the supports of Spencer's arch does not as yet by any means admit of demonstration." According to Osborn, Weismann had transformed Spencer's key assumption into a research question:

[Whether] acquired characters are inherited . . . is the great open problem of the day. What the issue will be is the present bone of contention between the Lamarckians, Darwinians, and Neo-Darwinians, the latter school holding that there is absolutely no transmission of acquired characters, and it is perfectly evident that if they demonstrate this proposition one great section of Spencer's system [will] fall to the ground!

Leaving no doubt as to Weismann's importance, Osborn called him "the most prominent figure of the times" and predicted that "his discoveries would mark an epoch in the history of the evolution theory." Osborn's article was not completely critical: he admitted that Spencer had anticipated some of Weismann's "views as to the origin of variation" and said that "many of [Spencer's] purely hypothetical deductions have been confirmed by, or are apparently in accordance with, the very latest discoveries." But he concluded that "speculative biology" was best done by "original thinkers in the laboratory and field" and not by armchair philosophers like Spencer.[38]

Peirce, writing in a week later, was delighted with Osborn's account of Spencerian biology—not a surprise, given his own opposition to the mechanical worldview: "Prof. Osborn fully comprehends the essence of Spencerianism. It is not that nature and man are the result of evolution; for that had been said before by biologists, and, let me add, by wide-swaying philosophers as well. What characterizes Spencerianism is the doctrine that *evolution is purely mechanical*." Peirce's response also used Osborn's final comments about "thinkers in the laboratory and field" to counter another correspondent's claim that "there are two kinds of

---

38. Henry Fairfield Osborn, "The Spencerian Biology," *New York Times*, April 6, 1890, 13.

scientists—the specialist and the generalizer, or philosopher," with Spencer in the latter category. Peirce argued that the best generalizers were primarily specialists, probably thinking of himself but also giving Cope as an example: "Who are our generalizers in this country? I have heard of Prof. Cope, whose book [*The Origin of the Fittest*] is famous, but I am assured he is one of the foremost of paleontologists, a specialist of the specialists."[39] The *New York Times* discussion was finally brought to a close on April 27 when William Jay Youmans, editor of the *Popular Science Monthly* and brother of its late founder Edward Livingston Youmans, provided a list of scientific authorities who had praised Spencer's work. To take just one example, in his *Encyclopaedia Britannica* entry on "Evolution in Biology," Thomas Henry Huxley had declared that "the profound and vigorous writings of Mr. Spencer embody the spirit of Descartes in the knowledge of our own day, and may be regarded as the 'Principes de Philosophie' of the nineteenth century." Peirce capitulated in a short letter the next week, admitting that Youmans had demonstrated "the profound respect in which Mr. Herbert Spencer is held by men of science the world over," which though not "sufficing to put his philosophy beyond doubt, does satisfactorily answer the question to which I gave special prominence."[40]

Thus, both James and Peirce were introduced to Weismann's ideas in the context of their criticism of Spencer. But whereas James embraced neo-Darwinism, Peirce was more skeptical: in "Evolutionary Love," the last of an 1891–93 series of articles in the *Monist*, Peirce endorsed a form of neo-Lamarckism, arguing that Weismann's picture of evolution was just as mechanistic as Spencer's. Peirce's *Monist* series, and its relation to debates over the factors of evolution, is the topic of the next section.

Peirce and Neo-Lamarckism

Only a few months after the *New York Times* debate, Charles Sanders Peirce was invited to write something for the Chicago weekly *Open Court* or its new sister journal the *Monist*, both of which were edited by the German émigré philosopher Paul Carus. Peirce was already familiar with *Open Court*, having cited it in one of his *Century Dictionary* entries, and

---

39. Charles Sanders Peirce, "'Outsider' Wants More Light," *New York Times*, April 13, 1890, 13; citing Kappa, "Flaws in 'Outsider's' Reasoning," *New York Times*, March 30, 1890.
40. Thomas Henry Huxley, *Science and Culture, and Other Essays* (London: Macmillan, 1881), 297–98; quoted in William Jay Youmans, "Mr. Spencer's Rank as a Philosopher," *New York Times*, April 27, 1890, 13; Charles Sanders Peirce, "'Outsider's' Thanks," *New York Times*, May 4, 1890, 2.

he told his friend Francis Calvin Russell that he would like to write a whole series of articles examining "the laws which have been found to govern the evolution of the leading ideas of mathematics and physics." Carus, who had seen Peirce's discussion of Spencer in the *New York Times*, suggested to Peirce that they stage a similar debate in *Open Court*. They settled on something closer to the former plan, and Carus received Peirce's first submission to the *Monist* in August 1890.[41]

In this article, "The Architecture of Theories," Peirce repeated his *New York Times* accusation that Spencer, because he claimed that the basis of the most general laws was unknowable, was "not a philosophical evolutionist but only a half-evolutionist." Peirce's "thoroughgoing evolutionism," in contrast, would attempt to explain the laws of nature as "results of evolution," supposing "them not to be absolute, not to be obeyed precisely." According to Peirce, the mechanistic picture of evolution failed to acknowledge that "exact law obviously never can produce heterogeneity out of homogeneity; and arbitrary heterogeneity is the feature of the universe the most manifest and characteristic." This was a direct attack on Spencer, whose very definition of evolution was "a change from an indefinite, incoherent homogeneity, to a definite, coherent heterogeneity."[42]

Although Peirce had in earlier work focused exclusively on Darwin's account of evolution, in "The Architecture of Theories"—perhaps inspired by Osborn or Cope—he considered non-Darwinian approaches as well.[43] He contrasted, for the first time in his career, the views of Darwin and Lamarck: "Darwinian evolution is evolution by the operation of chance, and the destruction of bad results, while Lamarckian evolution is evolution by the effect of habit and effort." He also added a third view, that of the geologist Clarence King, a fellow member of the Century Club in New York. According to Peirce, King had maintained that the steps of evolution are "neither haphazard on the one hand," as in Darwinian evolution, "nor yet determined by an inward striving on the other," as in Lamarckian evolution, "but on the contrary are effects of the changed environment, and have a positive general tendency to adapt the organ-

---

41. Paul Carus to Charles Sanders Peirce, 2 July 1890, 22 July 1890, and 3 August 1890, L77, Peirce Papers; Charles Sanders Peirce to Francis Calvin Russell, 3 July 1890, Folder 9, Box 91, Open Court Publishing Company Records (1/2/MSS 027), Special Collections Research Center, Southern Illinois University–Carbondale; William Dwight Whitney, ed., *The Century Dictionary: An Encyclopedic Lexicon of the English Language*, 6 vols. (New York: Century, 1889–91), 3:2943, s.v. "hylozoistic."

42. Charles Sanders Peirce, "The Architecture of Theories," *Monist* 1 (1891): 165–66; Herbert Spencer, *First Principles* (London: Williams & Norgate, 1862), 216.

43. For some of this earlier work, see "The Fixation of Belief," *Popular Science Monthly* 12 (1877): 2–3; "The Triad in Biological Development," from MS 909 (1887–88), in Peirce, *Writings*, 6:199–202.

ism to that environment, since variation will particularly affect organs at once enfeebled and stimulated." Peirce suggested that this latter mode of evolution was "called for by some of the broadest and most important facts of biology and paleontology" and had been "the chief factor in the historical evolution of institutions as in that of ideas."[44]

As Peirce indicated, King had argued—drawing on his extensive knowledge of American stratigraphy and historical geology—that dramatic changes in the inorganic environment during certain periods of evolutionary history may have led to the "survival of the plastic" as opposed to the "survival of the fittest": "When catastrophic change burst in upon the ages of uniformity, and sounded in the ear of every living thing the words 'change or die,' plasticity became the sole principle of salvation. Plasticity, then, is that quality which, in suddenly enforced physical change, is the key to survival and prosperity." Those organisms that are flexible enough to accommodate drastic changes in their environment are not necessarily those that would win in a typical "Malthusian death struggle" with other organisms. King's approach to evolution thus highlighted the importance of plasticity—a point on which he may have been influenced by Cope, who had already suggested that "highly specialized types" were more likely to go extinct because of "their less degree of plasticity and want of capacity for change under . . . changed circumstances."[45]

Peirce was almost certainly familiar with Cope's evolutionary views: as we have seen, Peirce praised Cope's book in the *New York Times*, and both of them contributed to the discussion of science and immortality in the *Christian Register*. As Wallace indicated in a review of *Origin of the Fittest*, Cope was the foremost member of the American neo-Lamarckian school, and as such he was probably one of the inspirations for Peirce's own Lamarckism.[46] Cope also developed a general theory of evolution that seems to have influenced the evolutionary metaphysics of Peirce's *Monist* series.

---

44. Charles Sanders Peirce, "The Architecture of Theories," *Monist* 1 (1891): 166–67; see also MS 972 (1890), in Peirce, *Writings*, 8:18. King was memorialized by the Century Club after his death, and Peirce frequented the club until his expulsion around 1895: see James D. Hague, ed., *Clarence King Memoirs* (New York: G. P. Putnam's Sons, 1904); Joseph Brent, *Charles Sanders Peirce: A Life*, 2nd ed. (Bloomington: Indiana University Press, 1998), 125.

45. Clarence King, "Catastrophism and Evolution," *American Naturalist* 11 (1877): 469; see also *Report of the Geological Exploration of the Fortieth Parallel*, vol. 1, *Systematic Geology* (Washington, DC: Government Printing Office, 1878); Edward Drinker Cope, "The Method of Creation of Organic Forms," *Proceedings of the American Philosophical Society* 12 (1871): 251. King's ideas anticipated to some extent those now current in paleontology: see David Jablonski, "Mass Extinctions and Macroevolution," *Paleobiology* 31 (2005).

46. Alfred Russel Wallace, "The American School of Evolutionists," *Nation*, February 10, 1887, 121.

The main tenets of Cope's theory were presented in three essays, all of which were collected in *Origin of the Fittest*: "Consciousness in Evolution," first published in 1875; "On Archaesthetism," first published in 1882; and "Catagenesis; or, Creation by Retrograde Metamorphosis of Energy," first published in 1884. The theory was also summarized in his contribution to the *Christian Register* symposium in 1887. One of Cope's boldest moves was to claim that consciousness is not a result but a cause of evolution: evolution tends toward unconsciousness and automatism, and not toward consciousness. This evolution from consciousness to automatism was made possible by "the property of protoplasm to organize machinery which shall work automatically in the absence of consciousness" and was related to what Cope called "the doctrine of the unspecialized."[47] According to this doctrine, generalized types—adaptable and without "mechanical peculiarities in their structure"—are ancestral to more specialized types. Cope applied this doctrine of the unspecialized to mental evolution as well: "The greater the proportion of unconscious automatism of habits, the less the power of adaptation. . . . The greater the degree of consciousness of stimulus, the greater will be the degree of adaptability to new relations, and to such constant rousing the unspecialized mind is always open."[48] Both Cope and King argued that plasticity underpinned evolvability, but Cope also saw consciousness as playing an essential role.

A few years later, Cope began using the term "archaesthetism" (from the Greek *archē*, beginning, and *aisthēsis*, sensation) to refer to "the hypothesis of the primitive and creative function of consciousness." He argued for archaesthetism as follows: adaptive response, and thus progressive evolution, requires a "generalized *dynamic* condition" of matter; "wherever this generalized condition exists, consciousness will be present"; consciousness thus existed at the beginning of evolution and has been its main driver. Cope supported his second premise by claiming that living beings differ from the nonliving in that "their actions have some definite reference to their well being or pleasure, or their preservation from injury or pain, and are varied with circumstances as they arise." He also invoked more familiar neo-Lamarckian points about directed variation and use/disuse: "If the law of modification of structure by use and effort be true, it is evident that consciousness or sensibility must play an important part in evolution." As Cope summarized in 1884, the hypothesis of archaesthetism "maintains that consciousness

---

47. Cope, *The Origin of the Fittest: Essays on Evolution* (New York: D. Appleton, 1887), 395–96.
48. Cope, *Origin of the Fittest*, 401.

as well as life preceded organism, and has been the *primum mobile* in the creation of organic structure."[49]

Never afraid of neologisms, Cope then introduced the term *catagenesis* (from the Greek *kata-*, downwards, and *genesis*, origin) to refer to "the process of creation by the retrograde metamorphosis of energy." Cope believed that all forms of energy had "originated in the process of running-down or specialization from the primitive energy," which he linked to evolution from consciousness to automatism. He had already made a similar point in 1875: "Consciousness constitutes then the only apparently initial point of motion with which we are acquainted. If so, we are at liberty to search for the origin of the physical forces in consciousness, as well as the vital; their present unconscious condition being possibly due, as in the case of the vital, to automatism." According to Cope, organisms evolve from generalized to specialized types, and the same may be true for energy: "If the inorganic forces are the products of a primitive condition of energy which had the essential characteristics of vital energy, it has been by a process of specialization." Chemical and physical forces, according to Cope, might be results of the specialization of vital forces—of an evolution away from adaptability and toward complete automatism.[50]

As we have seen, Peirce had been developing his account of the laws of nature as products of evolution since at least 1884: "May not the laws of physics be habits gradually acquired by systems?"[51] But as he began work on his *Monist* series in 1890, he started to combine this view with one that looked a lot like Cope's archaesthetism. It is no coincidence that Peirce actually wrote the entry for *archaesthetism* in the first volume—published in 1889—of the *Century Dictionary*, using Cope's own words: "The hypothesis of the primitive creative function of consciousness; the hypothesis that consciousness, considered as an attribute of matter, is primitive and a cause of evolution."[52] Although Peirce rejected the monism of Cope and others, which he saw as a kind of materialism, in 1890 he began defending archaesthetism without naming it as such: "The only possible way of explaining the connection of body and soul is to make matter effete mind, or mind which has become thoroughly under the dominion of habit, till consciousness and spontaneity are almost

---

49. Cope, *Origin of the Fittest*, 412, 419, 425.
50. Cope, *Origin of the Fittest*, 403–4, 434–35.
51. "[Design and Chance]," MS 494 (1884), in Peirce, *Writings*, 4:553.
52. William Dwight Whitney, ed., *The Century Dictionary: An Encyclopedic Lexicon of the English Language*, 6 vols. (New York: Century, 1889-91), 1:295–96, s.v. "archaestetism"; citing Cope, "On Archaesthetism," *American Naturalist* 16 (1882): 467, 469.

extinct." Like Cope, Peirce linked this original mind to life and feeling: "The free is living; the immediately living is *feeling*. Feeling, then, is assumed as a starting-point; but feeling uncoordinated." Although Peirce called his view "idealism," a position that Cope explicitly rejected, his description of it recalled Cope's catagenesis: "Idealism regards the psychical mode of activity as the fundamental and universal one, of which the physical mode is a specialization."[53]

In the closing passage of "The Architecture of Theories," Peirce provided a short but spectacular summary of his evolutionary cosmology. It was fundamentally archaesthetic, beginning with mind and ending with law:

> In the beginning,—infinitely remote,—there was a chaos of unpersonalized feeling, which being without connection or regularity would properly be without existence. This feeling, sporting here and there in pure arbitrariness, would have started the germ of a generalising tendency. Its other sportings would be evanescent, but this would have a growing virtue. Thus, the tendency to habit would be started; and from this with the other principles of evolution all the regularities of the universe would be evolved. At any time, however, an element of pure chance survives and will remain until the world becomes an absolutely perfect, rational, and symmetrical system, in which mind is at last crystallised in the infinitely distant future.[54]

Thus, although Peirce officially framed his metaphysics as "a Schelling-fashioned idealism which holds matter to be mere specialised and partially deadened mind," it was also inspired by Cope's scientific views, which postulated that the trajectory of evolution went from consciousness to automatism.[55]

Peirce's friends and critics commented on the striking similarities between his evolutionary metaphysics and that of Cope. In 1892, after reading the first four articles of Peirce's *Monist* series, the Scottish

---

53. "Sketch of a New Philosophy," MS 928 (1890), in Peirce, *Writings*, 8:22; "[Logic and Spiritualism]," from MS 878 (1890), in Peirce, *Writings*, 6:393; Charles Sanders Peirce, "[Notes on the First Issue of the *Monist*]," *Nation*, October 23, 1890, 326. Peirce criticized monists who "make mind a specialization of matter" in "Ribot's Psychology of Attention," *Nation*, June 19, 1890, 493. For Cope's rejection of idealism, see Edward Drinker Cope, "Evolution and Idealism," *Open Court*, January 5, 1888, 655–57.

54. Charles Sanders Peirce, "The Architecture of Theories," *Monist* 1 (1891): 175–76.

55. Charles Sanders Peirce, "The Law of Mind," *Monist* 2 (1892): 533; see also William Dwight Whitney, ed., *The Century Dictionary: An Encyclopedic Lexicon of the English Language*, 6 vols. (New York: Century, 1889–91), 3:2974, s.v. "idealism."

physiologist-philosopher Edmund Montgomery (then living in Texas) wrote the following in a letter to Peirce:

> From a materialistic standpoint, Prof. Cope has expounded a somewhat similar theory. Taking the hypothesis-indulgent Ether to be the bearer of universal and supreme "mind or consciousness," he looks upon the arising and fixation of material compounds as a process accompanied by loss of "mind or consciousness." This lapsing into unconsciousness by means of organic fixation is manifest in living forms as instinct or unconscious habit.

Referencing his 1887 debate with Cope in *Open Court*, Montgomery told Peirce that he was "too forcibly impressed with the laboriousness and ruthlessness of the process that leads to gradual and precarious mental development in our world" to think that the pinnacle of "mental or spiritual life" lay at the origin of evolution, followed by "a cruel and wanton fall from grace." Royce emphasized the same connection a few years later, also referencing Peirce's *Monist* series:

> That nature's observable Laws might even be interpreted, from an evolutionary point of view, as nature's gradually acquired Habits, originating in a primal condition of a relatively capricious irregularity, is a conception to which several recent writers, notably Mr. Cope, and, with great philosophical ingenuity, Mr. Charles Peirce, have given considerable elaboration.

Philosophers at the time thus recognized that Peirce's evolutionary views echoed those of Cope.[56]

The four articles in Peirce's *Monist* series that followed "The Architecture of Theories" discussed chance and necessity, the transmission and development of ideas, the nature of protoplasm, and the factors of evolution. From our modern-day point of view, these topics seem idiosyncratic, to say the least; at the time, however, they were seen as related. As shown in the previous discussion, chance and spontaneity were central to debates between neo-Darwinians (like Weismann) and neo-Lamarckians (like Cope) over the production of variation in evolution. Moreover, Cope, Montgomery, and other biologists of a more speculative bent thought that protoplasm was the key to understanding the relationship between mind and matter. Since I have focused in previous work on the

---

56. Edmund Montgomery to Charles Sanders Peirce, 5 October 1892, L297, Peirce Papers; Josiah Royce, "Self-Consciousness, Social Consciousness and Nature," *Philosophical Review* 4 (1895): 592.

CHAPTER FIVE

connections between consciousness, protoplasm, and evolution, I will concentrate in the remainder of this section on "Evolutionary Love," the final essay of Peirce's series.[57]

In this essay, breaking from earlier work, Peirce associated natural selection with greedy, dog-eat-dog capitalism. Already in an 1892 draft, Peirce had said that "according to [Darwin's] theory, . . . all the gain that any one has achieved is measured by the unfortunates whose unlucky chance has dragged them to their doom."[58] These notes were echoed in the opening pages of "Evolutionary Love," which suggested "that the great attention paid to economical questions during our century has induced an exaggeration of the beneficial effects of greed" and that the *Origin of Species* "merely extends politico-economical views of progress to the entire realm of animal and vegetable life." Peirce then contrasted the Christian gospel with that of Darwin and the political economists:

The gospel of Christ says that progress comes from every individual merging his individuality in sympathy with his neighbors. On the other side, the conviction of the nineteenth century is that progress takes place by virtue of every individual's striving for himself with all his might and trampling his neighbor under foot whenever he gets a chance to do so. This may accurately be called the Gospel of Greed.

Why this new obsession with greed? Probably because Peirce saw himself as one of those trampled underfoot by his neighbor. A few months before, already struggling financially, Peirce had apparently been cheated out of payment for extensive research into a new chemical bleaching process. Although his essay did not make this explicit, Peirce had cast himself as a victim of the nineteenth-century obsession with greed and individual benefit.[59]

In "Evolutionary Love," Peirce also returned to the factors of evolution debates, making a slight revision to his classification. Peirce's breakdown of the possible factors of evolution in "The Architecture of Theories," written in 1890, was as follows: evolution "by the operation of chance, and the destruction of bad results" (Darwin); evolution "by

---

57. Trevor Pearce, "'Protoplasm Feels': The Role of Physiology in Peirce's Evolutionary Metaphysics," *HOPOS* 8 (2018).

58. MS 954A, Peirce Papers. MS 954 seems to be a mixture of three sets of notes: 954A, from 1892; 954B, from 1890; and 954C, from 1893. On the latter, see André De Tienne, "'Scientific Fallibilism': Peirce's Forgotten Lecture of 1893," *Peirce Project Newsletter* 4, no. 1 (2001).

59. Charles Sanders Peirce, "Evolutionary Love," *Monist* 3 (1893): 179, 182; see also Charles Sanders Peirce, "Dmesis," *Open Court*, September 29, 1892, 3401–2. For a summary of the bleaching process episode, see Peirce, *Writings*, 8:lxxxvi–lxxxvii.

the effect of habit and effort" (Lamarck); and evolution "by external forces and the breaking up of habits" (King).[60] But by 1892, as he started making notes in preparation for "Evolutionary Love," he had altered his presentation: there was now a broader category of evolution by "mechanical causes," associated with Spencer. Here is a transcript of a page from Peirce's 1892 notes:

Three principles of Evolution
1st Darwinian. All the evolutionary steps are due to the action of chance. ~~They are insensibly small~~ At one time this seemed the chief motive of biological evolution Darwin says so. Now does not seem so
2nd Spencerian. Evolution is due to external and mechanical causes. Spencer puts the Law of Force at the bottom of everything. Which is <u>antievolutionary</u>. ~~Generally by Generally by great~~
3rd Lamarckian. By habit. And direct effort
What are the characteristic symptoms of these three[?]
All have their varieties.[61]

Peirce's shift may have been prompted by his reading of Spencer's *Essays*, which he reviewed for the *Nation* in 1891. This book contained a reprint of *Factors of Organic Evolution*, discussed earlier, which Peirce described as "the most interesting of the new essays . . . , in which the author urges almost irresistibly the indirect evidence of the transmission of acquired characters." Although Peirce's review repeated the criticisms of Spencer's evolutionism that he had been making since the late 1880s, it also attacked a new target—namely, "the assumption made by neo-Darwinians that the form of each individual is a mathematical resultant of the forms of his ancestors." Peirce would ultimately argue that the theories of King and Weismann were two varieties of the Spencerian (or mechanistic) approach to evolution.[62]

Echoing critics such as Argyll and Cope, Peirce claimed in "Evolutionary Love" that according to Darwinian evolution, "the only positive agent of change in the whole passage from moner to man is fortuitous

---

60. Charles Sanders Peirce, "The Architecture of Theories," *Monist* 1 (1891): 166–67; see also MS 954B, Peirce Papers.
61. MS 956 (1892), Peirce Papers. Although most of MS 956 consists of material related to "The Architecture of Theories," the later pages are notes and drafts related to "Evolutionary Love."
62. Charles Sanders Peirce, review of *Essays: Scientific, Political, and Speculative*, by Herbert Spencer, *Nation*, October 8, 1891, 283; citing Herbert Spencer, *Essays: Scientific, Political, and Speculative*, 3 vols. (New York: D. Appleton, 1891), 1:389–466. Peirce had already described Weismann as a "neo-Darwinian" in "The Architecture of Theories," *Monist* 1 (1891): 166n.

variation." Like Argyll, he treated natural selection as a merely negative process. Peirce also said of Darwin's hypothesis that "to a sober mind its case looks less hopeful now than it did twenty years ago," reflecting what would have appeared to him as a consensus.[63] Spencer, LeConte, Osborn, Cope, and Montgomery, despite their other differences, all agreed that Darwin's preferred factor was inadequate, and they were also skeptical of Weismann's new theory: "It is," Montgomery wrote, "quite inconceivable that the fortuitous arising of variations wholly irrelated to the propensities and needs of life, should by the negative process of successive weeding, ever be competent to construct out of shapeless material those most specific organs with which we find ourselves endowed."[64] Peirce had even read criticism of Weismann in the *Monist*—for example, in an article by the German biologist Ernst Krause (writing under the anagrammatic pseudonym "Carus Sterne") that appeared in the same issue as Peirce's "Man's Glassy Essence." Krause described Weismann's view as "a theory of perfect mechanical variability" and concluded that "the adherents of Neo-Darwinism will . . . have to furnish many additional facts if they wish to invest their theory with any degree of probability."[65]

Peirce then contrasted Darwin's appeal to fortuitous variation with a more mechanical picture, associated in his notes with Spencer but now linked to a group of naturalists who had made "mechanical necessity [the] chief factor of evolution." Weismann, for instance, "though he calls himself a Darwinian, holds that nothing is due to chance, but that all forms are simple mechanical resultants of the heredity from two parents." As in his review of Spencer, Peirce was here alluding to the view that most biological variation is the result of—in Weismann's words—"the coalescence of two distinct germ-cells" in sexual reproduction. Others in this mechanistic camp pointed instead to external causes: geol-

---

63. Charles Sanders Peirce, "Evolutionary Love," *Monist* 3 (1893): 183, 185. See also Andrew Reynolds, *Peirce's Scientific Metaphysics: The Philosophy of Chance, Law, and Evolution* (Nashville, TN: Vanderbilt University Press, 2002), 101.

64. Edmund Montgomery, "Cope's Theology of Evolution," *Open Court*, June 23, 1887, 274. For skepticism about Weismann, see Joseph LeConte, "The Factors of Evolution: Their Grades and the Order of Their Introduction," *Monist* 1 (1891): 331–33; Edward Drinker Cope, "Alfred Russel Wallace," in *Evolution in Science, Philosophy, and Art: Popular Lectures and Discussions before the Brooklyn Ethical Association* (New York: D. Appleton, 1891), 12–13; Henry Fairfield Osborn, "The Present Problem of Heredity," *Atlantic Monthly* 67 (1891): 362–63; Lester Frank Ward, "Neo-Darwinism and Neo-Lamarckism," *Proceedings of the Biological Society of Washington* 6 (1891): 45–50; Ernst Haeckel, *The History of Creation, or the Development of the Earth and Its Inhabitants by the Action of Natural Causes: A Popular Exposition of the Doctrine of Evolution in General, and of that of Darwin, Goethe, and Lamarck in Particular*, trans. Edwin Ray Lankester, 4th ed., 2 vols. (New York: D. Appleton, 1892), 1:220–21.

65. Carus Sterne, "Recent Evolutionary Studies in Germany," *Monist* 3 (1892): 106, 110; see also Charles Sanders Peirce, "Evolutionary Love," *Monist* 3 (1893): 185n; Peirce, *Writings*, 8:414.

ogists such as King, for example, "think that the variation of species is due to cataclysmic alterations of climate." Thus, Peirce contrasted "evolution by sporting and evolution by mechanical necessity," which he called "conceptions warring against each other."[66]

According to Peirce, Lamarck's was a "third method" of evolution that, in good Hegelian fashion, "superseded" the opposition of these other two while retaining their important aspects.[67] Central to this method was "habit-taking," the action of which "is essentially dissimiliar to that of a physical force; and that is the secret of the repugnance of such necessitarians as Weismann to admitting its existence." Earlier in the essay, Peirce had quoted Henry James (William's father), who claimed that truly "creative Love" reserves its tenderness "for what intrinsically is most bitterly hostile and negative to itself." Inspired by this view, Peirce argued that "the movement of love is circular, at one and the same impulse projecting creations into independency and drawing them into harmony." He could now connect this creative love to Lamarckian evolution, since in the latter, habit "serves to establish the new features, and also bring them into harmony with the general morphology and function of the animals and plants to which they belong."[68] Peirce went further than most neo-Lamarckians, however, claiming that "evolution by creative love," what he called "*agapastic* evolution," operates by the disposition of an organism "to catch the general idea of those about it and thus to subserve the greater purpose." According to Peirce, "in genuine agapasm, . . . advance takes place by virtue of a positive sympathy among the created." As his terminology indicates, Peirce was attempting to link his preferred form of Lamarckism to Christian love and charity—to the *agapae* ("feasts of brotherly love") of the early Christians.[69]

Peirce, who had recently been preparing a series of lectures on the history of science, chose to illustrate his three modes of evolution with case studies from "the historical development of human thought" rather

---

66. Peirce, "Evolutionary Love," 185–86; August Weismann, *Essays upon Heredity and Kindred Biological Problems* (Oxford: Clarendon Press, 1889), 272; see also "Carnegie Institution Correspondence," L75 (1902), in Peirce, *New Elements*, 4:66.

67. Peirce, "Evolutionary Love," 186, 188–89. "Supersede" was the usual translation of Hegel's term *aufheben*: see William Wallace, prolegomena to *The Logic of Hegel*, by Georg Wilhelm Friedrich Hegel (Oxford: Clarendon Press, 1874), clxxviii.

68. Peirce, "Evolutionary Love," 177, 186–87; Henry James, *Substance and Shadow: Or Morality and Religion in Their Relation to Life: An Essay upon the Physics of Creation* (Boston: Ticknor & Fields, 1863), 442.

69. Peirce, "Evolutionary Love," 188–89; Augustus Neander, *General History of the Christian Religion and Church*, trans. Joseph Torrey, vol. 1 (Boston: Crocker & Brewster, 1847), 325.

than with examples from the biological world.⁷⁰ The evolution of thought in the Darwinian mode consisted "in slight departures from habitual ideas in different directions indifferently, . . . these new departures being followed by unforeseen results which tend to fix some of them as habits more than others." Peirce argued that this mode had been unimportant historically and was seen primarily "in backwards and barbarizing movements." His two examples were the evolution of Christianity from the "pristine integrity" of Mark to its later obsession with bitterness and vengeance, and the development of the French Revolution from the Declaration of the Rights of Man to the Reign of Terror.⁷¹

When thought evolves in a more mechanistic mode, said Peirce, new ideas will be "adopted without foreseeing whither they tend, but have a character determined by causes either [a] external to the mind, such as changed circumstances of life, or [b] internal to the mind as logical developments of ideas already accepted, such as generalisations." Peirce detected the external form of mechanistic evolution in the impact of Aristotle on medieval European thought and in the influence of observed facts on scientific theories. He suggested that this form almost always works in combination with other modes of evolution. For example, Peirce thought that William Whewell's writings on the history of science had shown that ideas—and not just facts—had played an essential role in the progress of science: the history of science is not just a history of adjustments to new observations. Peirce also suggested that "the recent Japanese reception of western ideas" was the "purest instance" of external mechanistic evolution. He provided fewer examples of the internal form of mechanistic evolution, which he described as "logical groping, which advances upon a predestined line without being able to foresee whither it is to be carried nor to steer its course," but he did suggest that it had governed the history of philosophy.⁷²

Finally, the Lamarckian mode in the evolution of thought—the "evolutionary love" of Peirce's title—involved "the adoption of certain men-

---

70. Peirce, "Evolutionary Love," 190. Peirce delivered a set of Lowell Lectures on the history of science from November 1892 to January 1893: see Peirce, *Writings*, 8:lxi, and Peirce, *Historical Perspectives*, 1:143–295.

71. Peirce, "Evolutionary Love," 192–94; see also MS 1286 (1893), in Peirce, *Historical Perspectives*, 1:287, and in Peirce, *Collected Papers*, 7:269. Peirce referred to the Darwinian mode of evolution as "tychastic" (from the Greek *tychē*, chance).

72. Peirce, "Evolutionary Love," 190–91, 195–96; see also "Lecture IX [Post-Hellenic to the Fifteenth Century]," MS 1280 (1892), in Peirce, *Historical Perspectives*, 1:240–41. Peirce referred to the mechanistic mode of evolution as "anancastic" (from the Greek *anagkē*, necessity). For Peirce's interpretation of Whewell, see MS 1274 (1892), in Peirce, *Historical Perspectives*, 1:143–44; see also William Whewell, *The Philosophy of the Inductive Sciences, Founded upon Their History*, 2 vols. (London: John W. Parker, 1840), 1:xvii–xviii.

tal tendencies ... by an immediate attraction for the idea itself, whose nature is divined before the mind possesses it, by the power of sympathy." Peirce distinguished three forms of this mode of evolution according to who is affected and how. The idea may affect

1. "a whole people or community in its collective personality"
2. "a private person directly, ... by virtue of his sympathy with his neighbors"
3. "an individual, by virtue of an attraction it exercises on his mind"

Peirce gave the conversion of Saint Paul as an example of (b) and described (c) as "the phenomenon which has been well called the *divination* of genius."[73] According to Peirce, there was abundant evidence for this mode of evolution in the history of thought. Although he admitted that one could not directly demonstrate the existence of "such an entity as the 'spirit of an age' or of a people," he argued that many key ideas had occurred "simultaneously and independently to a number of individuals of no extraordinary general powers." He gave the example of Gothic architecture: "at the time the style was living, there was quite an abundance of men capable of producing works of this kind of sublimity and power"; "cathedral chapters, in the selection of architects, treated high artistic genius as a secondary consideration, as if there were no lack of persons able to supply that." Peirce also pointed out that "great discoveries" were often "made independently and almost simultaneously," giving as nineteenth-century examples the existence of the planet Neptune, the principle of the conservation of energy, the mechanical theory of heat, the kinetic theory of gases, the doctrine of natural selection, the method of spectrum analysis, and the periodic law of the chemical elements. The same was true for inventions. For instance, according to Peirce, although (diethyl) ether had been available for decades, "three different New England physicians" independently put it into use as an anesthetic for the first time in the 1840s—perhaps because "philanthropy was [then] undoubtedly in an unusually active condition."[74]

Peirce believed that all these modes of evolution had been operative in the history of thought. Nevertheless, he suggested that the Darwinian and mechanistic modes were almost always mixed with the Lamarckian, since variations are usually directed rather than fortuitous and both variation and survival usually happen in the context of the purposeful development of an idea, rather than as part of a purposeless

---

73. Peirce, "Evolutionary Love," 191.
74. Peirce, "Evolutionary Love," 197–200.

mechanism.⁷⁵ Thus, although technically a pluralist about the factors of evolution, Peirce thought—at least in 1892–93—that Lamarckian processes were most important. A few years later, however, in an unpublished manuscript on the lessons of the history of science, Peirce seems to have reverted to his position in "The Architecture of Theories," wherein he had described King's "evolution by external forces and the breaking up of habits" as "the chief factor in the historical evolution of institutions as in that of ideas."⁷⁶ In the later manuscript, probably written around 1896, Peirce claimed that the evolution of science was sometimes Lamarckian, involving the perpetual modification of "our opinion in the effort to make that opinion represent the known facts as more and more observations [come] to be collected" (his example was progress in the classification of chemical elements). But he also argued that science usually advances not gradually but "by leaps," where "the impulse for each leap is either some new observational resource, or some novel way of reasoning about the observations." In such cases, said Peirce, science is evolving in a Kingian mode. As an example, he cited the impact of Louis Pasteur, an outsider to the medical establishment, whose germ theory of disease provided both new methods and new ideas.⁷⁷ Despite appearances, however, there may ultimately be no contradiction between Peirce's two positions, since insofar as the rise of the germ theory involved the purposeful development of ideas, it should also count as agapastic evolution.⁷⁸ This conclusion is also supported by the last of Peirce's 1892–93 Lowell lectures on the history of science, where although he declared that "great and startling advances in scientific thought" are the result of "the violent breaking up of certain habits"—that is, Kingian evolution—he also claimed that Lamarckian evolution was "the method of the ordinary successful prosecution of scientific inquiry" and that "we should see more of it if we were to trace out the history of science into its later era" (Peirce's survey ended with Johannes Kepler).⁷⁹

Because Peirce focused on examples from the history of thought, readers today might be tempted—given the eventual triumph of neo-Darwinism—to argue that his Lamarckism was meant to apply to in-

---

75. Peirce, "Evolutionary Love," 183, 194, 197.
76. Charles Sanders Peirce, "The Architecture of Theories," *Monist* 1 (1891): 167.
77. MS 1288 (1896), in Peirce, *Collected Papers*, 1:108–9.
78. Charles Sanders Peirce, "Evolutionary Love," *Monist* 3 (1893): 185; cf. Charles Sanders Peirce, "The Architecture of Theories," *Monist* 1 (1891): 167.
79. MS 1286 (1893), in Peirce, *Historical Perspectives*, 1:287–89, and in Peirce, *Collected Papers*, 7:270–74.

tellectual rather than biological history. However, as demonstrated in this section, discussions in biology at the time would have confirmed Peirce's commitment not only to the role of consciousness in evolution (Cope) but to the central importance of directed variation, possibly stimulated by environmental changes (Osborn, King). Weismann and the neo-Darwinians claimed that evolution required only fortuitous variation and natural selection, but they were challenged from all sides. Only a few months after "Evolutionary Love" was published, Spencer himself launched a public attack on behalf of neo-Lamarckism against Weismann—an attack that highlighted the centrality of directed variation and plasticity to debates over the factors of evolution. The Spencer-Weismann debate and Dewey's response to it is the topic of this chapter's final section.

Dewey and the Spencer-Weismann Debate

Late in 1892, a few months after Charles Sanders Peirce had finished writing "Evolutionary Love," Herbert Spencer sent a letter to Percy William Bunting, the editor of the English journal *Contemporary Review*. "I have in contemplation an article," wrote Spencer, "the object of which will be to raise for more definite consideration certain aspects of the doctrine of natural selection: the purpose being to show that natural selection *taken alone* is utterly inadequate to account for the facts of organic evolution."[80] Spencer asserted that his "chief aim" was "forcing a discussion," and thus he sought permission from Bunting to circulate his article as a pamphlet a month after its publication. Spencer clearly wanted the dispute to be as public as possible, for he wrote Bunting the following February to say that he had sent the proofs of the article and presented a plan to make an event of its publication: "It occurs to me that not improbably you will have proposals for replies. Should such be the case may I suggest that it might not be a bad plan to have a *symposium*, in which the thing should be discussed by various men?" He proposed Lankester, Romanes, Wallace, and several others as good candidates for such a discussion.[81] Spencer was seeking not only to attack Weismann's views in print but also to ensure that the attack reached

---

80. Herbert Spencer to Percy William Bunting, 21 November 1892, Folder 17, Box 5, Bunting Papers.
81. Herbert Spencer to Percy William Bunting, 19 February 1893, Folder 17, Box 5, Bunting Papers.

the widest possible audience. He got his wish, as the ensuing discussion reached all the way to Russia and Argentina.[82]

Spencer's article, titled "The Inadequacy of 'Natural Selection,'" was split between the February and March 1893 issues of the *Contemporary Review* and was quickly reprinted in the United States by several popular periodicals. In addition to calling into question the isolation of the germ-plasm, Spencer returned to his argument from the coadaptation of parts, which had been praised by both Romanes and James. Spencer examined a case in which a quadruped finds itself in a new environment, one in which the ability to leap would be a great benefit. Myriad coordinated variations in the muscles and joints of the fore and hind limbs are necessary to improve the animal's leaping ability, said Spencer, and it is impossible to account for such changes if acquired variations cannot be inherited. The fortuitous variation of the neo-Darwinians is not enough:

> See, then, the total requirements. We must suppose that by natural selection of miscellaneous variations, the parts of the hind limbs shall be co-adapted to one another, in sizes, shapes, and ratios; that those of the fore limbs shall undergo co-adaptations similar in their complexity, but dissimilar in their kinds; and that the two sets of co-adaptations shall be effected *pari passu* [in equal step].

The chances of such coordinated variations appearing spontaneously, according to Spencer, are "billions to one." He concluded that "either there has been inheritance of acquired characters, or there has been no evolution."[83]

Weismann's answer, "The All-Sufficiency of Natural Selection," was split between the September and October 1893 issues of the *Contemporary Review*. Like Spencer, Weismann expressed his wish to send copies of his paper "to many scientific men in England, America etc.," and then to "publish it afterwards also in German."[84] His reply focused on Spencer's coadaptation argument, admitting that "often these co-operative changes are so numerous that it is difficult to understand how all, at one time and independently, should possibly arise through spontaneous variations and natural selection." However, despite this difficulty, Weismann insisted that natural selection is the only possible explanation of

---

82. Alexander Vucinich, *Darwin in Russian Thought* (Berkeley: University of California Press, 1989), 151–69; Alex Levine and Adriana Novoa, ¡*Darwinistas! The Construction of Evolutionary Thought in Nineteenth Century Argentina* (Leiden, Neth.: Brill, 2012), 35–44.

83. Herbert Spencer, "The Inadequacy of 'Natural Selection,'" *Contemporary Review* 63 (1893): 445–46.

84. August Weismann to Percy William Bunting, 22 June 1893, Folder 15, Box 6, Bunting Papers.

the coadaptation of parts in one particular case, that of "the neuters of the state-forming insects, especially the ants and termites." Following Darwin, Weismann argued that the special adaptations of neuter insects, since they leave no offspring of their own, must have arisen indirectly via natural selection at the level of the colony. In such cases, Weismann declared, *"it is just because no other explanation is conceivable, that it is necessary for us to accept the principle of natural selection."* Weismann admitted that it is almost impossible to imagine the process of natural selection or to estimate the "selection-value" of any particular variation; he even claimed that "we shall never be able to establish by observation the progress of natural selection." Nevertheless, he continued, no other explanation can account for the coadaptation of parts in neuter insects, and thus we should also appeal to natural selection in other sorts of cases, no matter how complex they seem.[85]

In his reply, "A Rejoinder to Professor Weismann," Spencer called the problem of the "co-adaptation of co-operative parts" the "crucial case" for Weismann's argument. The key issue, according to Spencer, was the source and nature of variation:

As [Weismann] admits, these parts [of the neuter insects] must have varied simultaneously in due proportion to one another. What must have been the proximate causes of their variations? They must have been variations in what he calls the "determinants." . . . Consequently to produce simultaneously these many variations of parts, adjusted in their sizes and shapes, there must have simultaneously arisen a set of corresponding variations in the "determinants" composing the germ-plasm. What made them simultaneously vary in the requisite ways? . . . Nothing but *a fortuitous concourse of variations*; reminding us of the old "fortuitous concourse of atoms."[86]

This, then, was Spencer's challenge to Weismann: if you cannot explain the directed origin of germ-plasm variation, you have not explained the apparent directedness of variation at the level of the organism. But such directedness, according to Spencer, is easily explainable if the environment is an important cause of heritable variation.

---

85. August Weismann, "The All-Sufficiency of Natural Selection," *Contemporary Review* 64 (1893): 311, 313–14, 319, 327; see also Charles Darwin, *On the Origin of Species by Means of Natural Selection, or the Preservation of Favoured Races in the Struggle for Life* (London: John Murray, 1859), 235–42.

86. Herbert Spencer, "A Rejoinder to Professor Weismann," *Contemporary Review* 64 (1893): 905–6. The phrase "fortuitous concourse of atoms" referred to the atomistic materialism of Epicurus: see Jonathan Swift, *The Works of Jonathan Swift*, ed. John Nichols, 19 vols. (London: J. Johnson, 1808), 2:455–56.

CHAPTER FIVE

Weismann's task was thus to explain the coadaptation of parts using only the principle of natural selection. To do this, he invoked—in his Romanes Lecture at Oxford, delivered in May 1894—a struggle between parts of the organism, analogous to the struggle between organisms. Armed with this idea, termed "intra-selection," he was able to tackle Spencer's problem of coadaptation using the example of deer antlers:

> It is by no means necessary that all the parts concerned—skull, muscles and ligaments of the neck, cervical vertebrae, bones of the fore-limbs, &c.—should simultaneously adapt themselves *by variation of the germ* to the increase in size of the antlers; for in each separate individual the necessary adaptation will be temporarily accomplished by intra-selection—by the struggle of parts—under the trophic influence of functional stimulus.

It is the capacity for ontogenetic variation in the face of varying environmental stimuli, stimuli which include changes in other parts of the organism, that makes possible "the harmonious co-adaptation of parts in the course of the phyletic metamorphosis of a species." However, this capacity—what King and Cope called "plasticity"—is according to Weismann the product of "ordinary selection of individuals," and thus organism-level selection is still the more fundamental process.[87]

Weismann then extended this "struggle of parts" to the germ-plasm itself, claiming that some ontogenetic coadaptation is required after fertilization to ensure well-functioning offspring: "The primary constituents of [parents'] germ-substance could not be united together to produce a young organism, exhibiting harmony in its various parts, if they did not all have a certain scope for variation, so as to render them capable of adaptation to one another."[88] In Spencer's view, however, Weismann's new tactic had just moved the problem down a level: "Professor Weismann tells us merely that we must suppose that the germ-plasm acquires a certain sensitiveness such as gives it a proclivity to development in the requisite way. How is such proclivity obtainable? Only by having a multitude of its 'determinants' simultaneously changed in fit modes."[89] Surprisingly, Weismann at this point effectively capitulated

---

87. August Weismann, *The Effect of External Influences upon Development* (Oxford: Clarendon Press, 1894), 18–19; see also Robert J. Richards, *Darwin and the Emergence of Evolutionary Theories of Mind and Behavior* (Chicago: University of Chicago Press, 1987), 400–401; Wilhelm Roux, *Der Kampf der Theile im Organismus: Ein Beitrag zur Vervollständigung der mechanischen Zweckmässigkeitslehre* (Leipzig, Ger.: W. Engelmann, 1881).
88. Weismann, *Effect of External Influences upon Development*, 21.
89. Herbert Spencer, "Weismannism Once More," *Contemporary Review* 66 (1894): 598.

to Spencer: "The main difficulty as to whence the necessary variations come still remains." Spencer's criticisms had thus forced Weismann to admit that "we yet know but little" about "the obscure problem of the origin of variation."[90]

Weismann, however, had one last trick up his sleeve: an elaboration of the process of intraselection that he would later call his theory of "germinal selection."[91] This elaboration depended on the idea that variations in a certain direction are more likely to appear than others because of a kind of cooperation between germ-level selection and organism-level selection. According to Weismann, the constituents of the germ that determine particular parts of the organism (i.e., the determinants) are constantly competing for nourishment. If individual selection favors the increase of a set of parts, "the determinants cannot vary in the *minus* direction, and at the same time the average capacity for assimilation of these groups of determinants is increased. The supply of nutritive fluid is therefore increased, and the *plus* variations are again favoured more than *minus* variations." That is, in cases where individual selection favors organism-level variation in one direction, intraselection produces a greater number of variations in the same direction at the germ level. Weismann concluded that "the influence of selection on the elements of the germ in directing variation plays an important part in the whole process of natural selection." Spencer had thus prompted Weismann to elaborate an entirely new process stemming from the interplay between two levels of selection. This process allowed Weismann to claim that variation could "become directed" without committing himself to the inheritance of acquired characters.[92]

Social Weismannism

It is likely that John Dewey became aware of the Spencer-Weismann dispute while it was still going on: Henry Fairfield Osborn published a response to Spencer's and Weismann's initial essays in the May 1894 issue of the *Psychological Review*, an important venue for Dewey's own work.[93] Osborn emphasized the debate's broad implications: "While inconclusive it is most stimulating and has attracted wide attention, because the

---

90. August Weismann, "Heredity Once More," *Contemporary Review* 68 (1895): 425, 430.
91. August Weismann, "Germinal Selection," *Monist* 6 (1896); August Weismann, *On Germinal Selection as a Source of Definite Variation* (Chicago: Open Court, 1896).
92. August Weismann, "Heredity Once More," *Contemporary Review* 68 (1895): 431–32.
93. Dewey had an article in the very first issue and was also a member of the journal's editorial board: see *Psychological Review* 1 (1894): [i], 63–66.

question bears with equal force upon problems of ethics and psychology as upon all lines of biological thought." According to Osborn, neither Spencer nor Weismann had "brought forward inductive evidence"—thus the verdict of "inconclusive." Nevertheless, Osborn was more critical of neo-Darwinism: "In the absence of fact, [Weismann] presents the group of speculations which have grown up in the Neo-Darwinian school . . . —all processes spun out of the human mind, without an iota of direct evidence in their favor."[94]

Osborn's claim that the Spencer-Weismann dispute had important social and political implications was apparently vindicated by the arguments of Benjamin Kidd's book *Social Evolution*. In the very next issue of the *Psychological Review*, Dewey wrote a review of this book along with several others, including *The Psychic Factors of Civilization* by Lester Frank Ward, a paleobotanist with the United States Geological Survey.[95] Ward had been a staunch critic of the laissez-faire approach to society for many years. Dewey, while in graduate school at Johns Hopkins a decade earlier, had heard him attack Spencer's contention that "nature's method" of "utterly soulless competition" should be applied in society.[96] More recently, Ward had highlighted what he took to be the political upshot of Weismann's theories:

> If nothing that the individual gains by the most heroic or the most assiduous effort can by any possibility be handed on to posterity, the incentive to effort is in great part removed. If all the labor bestowed upon the youth of the race to secure a perfect physical and intellectual development dies with the individual to whom it is imparted why this labor? . . . In fact the whole burden of the Neo-Darwinian song is: Cease to educate, it is mere temporizing with the deeper and unchangeable forces of nature.[97]

LeConte had made similar points in a *Monist* article that Dewey probably read: "If Weismann and Wallace are right, if natural selection be indeed the only factor used by nature in organic evolution and there-

---

94. Henry Fairfield Osborn, "The Discussion between Spencer and Weismann," *Psychological Review* 1 (1894): 312, 314.

95. John Dewey, "Social Psychology," *Psychological Review* 1 (1894); Benjamin Kidd, *Social Evolution* (New York: Macmillan, 1894); Lester Frank Ward, *The Psychic Factors of Civilization* (Boston: Ginn, 1893).

96. Lester Frank Ward, "Mind as a Social Factor," *Mind* 9 (1884): 565. Ward's presentation at Hopkins had the same title: see "38th Meeting" (22 April 1884), Metaphysical Club Records; "Proceedings of Societies, Etc.," *Johns Hopkins University Circulars* 3 (1884): 138.

97. Lester Frank Ward, "Neo-Darwinism and Neo-Lamarckism," *Proceedings of the Biological Society of Washington* 6 (1891): 65; also quoted in Maurizio Meloni, *Political Biology: Science and Social Values in Human Heredity from Eugenics to Epigenetics* (London: Palgrave Macmillan, 2016), 56.

fore available for use by Reason in human evolution, then alas for all our hopes of race-improvement, whether physical, mental, or moral!"[98] Both Spencer and his opponents agreed that Weismann's views had implications for social policy.

Dewey's review described Kidd's book as having an "extreme Weismannism of premise," since the main argument of *Social Evolution* depended on Weismann's analysis of panmixia. According to Weismann, "the suspension of the preserving influence of natural selection may be termed *Panmixia*, for all individuals can reproduce themselves and thus stamp their characters upon the species, and not only those which are in all respects, or in respect to some single organ, the fittest."[99] Weismann had actually opened his first response to Spencer with a discussion of panmixia, since the retrogression or degeneration of organs was usually attributed to the Lamarckian factor—that is, the inheritance of the effects of disuse. In contrast, Weismann argued that because unused organs no longer provide a fitness benefit, organisms with any version of them will make the same contribution (all else being equal) to the next generation, resulting in the gradual decline of the organ: "Superfluous parts are no longer controlled by selection, are not preserved at the height of their development, but slowly sink through Panmixia."[100] As Weismann had declared a few years earlier in *Open Court*, "if the fitness [*Zweckmässigkeit*] of living things in all their parts rests upon the principle of natural selection, then *this fitness must be maintained by the same process that created it*, and it must disappear so soon as this process of natural selection ceases."[101]

This idea that the relaxation of selection results in degeneration seemed to rule out certain views of social progress. In one of his forays into politics, Thomas Henry Huxley had argued that "the mitigation or abolition" of the struggle for existence is "the chief end of social organization." He claimed that "of all the successive shapes which society has taken, that most nearly approaches perfection in which the war of individual against individual is most strictly limited." Citing Huxley and

---

98. Joseph LeConte, "The Factors of Evolution: Their Grades and the Order of Their Introduction," *Monist* 1 (1891): 334. An article by Dewey appeared in the very next issue.

99. John Dewey, "Social Psychology," *Psychological Review* 1 (1894): 410; August Weismann, *Essays upon Heredity and Kindred Biological Problems* (Oxford: Clarendon Press, 1889), 90; translated from August Weismann, *Ueber die Vererbung: Ein Vortrag* (Jena, Ger.: G. Fischer, 1883), 35.

100. August Weismann, "The All-Sufficiency of Natural Selection," *Contemporary Review* 64 (1893): 311.

101. August Weismann, "Retrogression in Animal and Vegetable Life," *Open Court*, September 12, 1889, 1830; translated from August Weismann, "Ueber den Rückschritt in der Natur," *Berichte der naturforschenden Gesellschaft zu Freiburg* 2 (1887): 15.

CHAPTER FIVE

various socialists, Kidd worried that society might begin to decline just to the extent that the struggle for existence was restricted in this way.[102] According to Kidd, although writers such as Karl Marx, Henry George, and Edward Bellamy were right that "the lower classes of our population have no sanction from their reason for maintaining existing conditions," they were wrong to think that the elimination of competition was a progressive goal. On Kidd's Weismannian view, "if the continual selection which is always going on amongst the higher forms of life were to be suspended, these forms would not only possess no tendency to make progress forwards, but must actually go backwards."[103] Part of Kidd's strategy was to show that although many socialists claimed the sanction of biology—from Marx's standpoint, "the evolution of the economic formation of society is viewed as a process of natural history"—they were relying on science that had been superseded by neo-Darwinism.[104] As Weismann wrote in his foreword to the German translation of *Social Evolution*,

> [Kidd] sides with the newest standpoint of biology, in which selection is taken not only in the Darwinian sense as the principle that produces progress, but also as that which alone maintains the achieved height of evolution. As the eye of animals that live in darkness atrophies (in our view) because it has no more value for this species and thus will no longer be preserved at its height by selection, so also would human society decline from its attained height if constant competition were suspended. The conclusion is not favorable to the aspirations of socialists.[105]

According to Kidd and Weismann, neo-Darwinism had undermined socialism.

Dewey would have been familiar with the socialist writers criticized by Kidd, many of whom were inspired by the evolutionary ethics of Herbert Spencer—an interesting irony, given Spencer's famous support of laissez-faire. Spencer's ethics depended on his claim, discussed in chapter 2, that "the life of the organism will be perfect only when the

---

102. Thomas Henry Huxley, "The Struggle for Existence: A Programme," *Nineteenth Century* 23 (1888): 166; also in Thomas Henry Huxley, *Social Diseases and Worse Remedies* (London: Macmillan, 1891), 22–23; cited in Benjamin Kidd, *Social Evolution* (New York: Macmillan, 1894), 3n1.

103. Kidd, *Social Evolution* (New York: Macmillan, 1894), 36–37, 68.

104. Karl Marx, *Capital: A Critical Analysis of Capitalist Production*, trans. Samuel Moore and Edward Aveling, vol. 1 (London: Swan Sonnenschein, 1887), xix.

105. August Weismann, foreword to *Soziale Evolution*, by Benjamin Kidd (Jena, Ger.: Gustav Fischer, 1895), iv; see also August Weismann to Gustav Fischer, 30 May 1894, in August Weismann, *Selected Letters and Documents*, 2 vols. (Freiburg: Universitätsbibliothek Freiburg im Breisgau, 1999), 1:219.

correspondence [between organism and environment] is perfect."[106] At the beginning of *The Data of Ethics*, published in 1879, Spencer applied this lesson to conduct, or "the adjustment of acts to ends." On Spencer's view, conduct evolves as purposive acts lead to an improved correspondence with the environment. Spencer claimed that the highest form of conduct is not strictly individualistic, since the life of the species matters to evolution as well. Echoing the libertarian principle of his 1851 book *Social Statics*, he praised "adjustments such that each creature may make them without preventing them from being made by other creatures." But even this was not the limit, in that "a still higher phase" in the evolution of conduct is "mutual help in the achievement of ends . . . either indirectly by industrial co-operation, or directly by volunteered aid." This "mutual aid," said Spencer, "increases the totality of the adjustments made, and serves to render the lives of all more complete." Thus, the highest species—human beings chief among them—have complicated cooperative societies.[107]

Of course, Spencer was aware that people often behave in selfish, uncooperative ways. But such behavior would be a thing of the past for "the completely adapted man in the completely evolved society." This "ideal social being" is one whose "spontaneous activities are congruous with the conditions imposed by the social environment formed by other such beings." Spencer believed that evolution was pushing human beings toward this end point. As Dewey summarized the process in 1891, "the being which survives must be the being which has properly adapted himself to his environment, which is largely social, and there is assurance that the conduct will be adapted to the environment just in the degree in which pleasure is taken in acts which concern the welfare of others." Spencer declared that right and wrong were to be judged from the viewpoint of this "ideal man as existing in the ideal social state."[108]

With Spencer speaking in worshipful tones of "mutual aid" in the "ideal social state," it is not surprising that some socialists and anarchists adopted him as a reluctant ally. Henry George, the American proponent of land socialization mentioned by Kidd, had argued in 1879—citing

---

106. Herbert Spencer, *The Principles of Psychology* (London: Longman, Brown, Green, & Longmans, 1855), 376; also in Herbert Spencer, *The Principles of Biology*, vol. 1 (London: Williams & Norgate, 1864), 82.

107. Herbert Spencer, *The Data of Ethics* (London: Williams & Norgate, 1879), 5, 18–20; cf. Herbert Spencer, *Social Statics: Or, the Conditions Essential to Human Happiness Specified, and the First of them Developed* (London: John Chapman, 1851), 103.

108. Spencer, *Data of Ethics*, 275, 280; John Dewey, *Outlines of a Critical Theory of Ethics* (Ann Arbor, MI: Inland Press, 1891), 70.

earlier work by Spencer—that the key to human progress was "association in equality": "Men tend to progress just as they come closer together, and by cooperation with each other increase the mental power that may be devoted to improvement, but just as conflict is provoked, or association developes [sic] inequality of condition and power, this tendency to progression is lessened, checked, and finally reversed."[109] Then in the late 1880s, the Russian anarchist Pyotr Alekseyevich Kropotkin argued in "The Scientific Bases of Anarchy" that the philosophy of evolution had inadvertently provided a justification of anarchism. In Kropotkin's view, Spencer and other evolutionists—by highlighting "the plasticity of organisation" and the adaptation of "each of the constituent parts of the aggregate to the needs of free co-operation"; by showing that the struggle for existence has the "wider sense of adaptation of all individuals of the species to the best conditions for survival of the species, as well as the greatest possible sum of life and happiness for each and all"; and by enforcing "the opinion of social reformers as to the necessity of modifying the conditions of life for improving man, instead of trying to improve human nature by moral teachings while life moves in the opposite direction"—had demonstrated that progress depended on the "socialisation of wealth and integrated labour, combined with the fullest possible freedom of the individual." In short, Kropotkin argued that Spencer's *Data of Ethics* "derives from the study of nature the very same conclusions which the forerunners of anarchy, [Charles] Fourier, and Robert Owen, derived from a study of human character," although he cheerfully admitted in a footnote that "Spencer does not fully endorse all the conclusions which ought to be drawn from his system of philosophy."[110]

Dewey was sympathetic to anarchism and actually cited Kropotkin's essay in his political philosophy class at Michigan in 1892–93.[111] In the

---

109. Henry George, *Progress and Poverty: An Inquiry into the Cause of Industrial Depressions and of Increase of Want with Increase of Wealth* (New York: Sterling, 1879), 457. Dewey was lavish in his praise of George in several later writings: John Dewey, "An Appreciation of Henry George," in *Significant Paragraphs from Henry George's "Progress and Poverty,"* ed. Harry Gunnison Brown (New York: Robert Schalkenbach Foundation, 1928); John Dewey, foreword to *The Philosophy of Henry George*, by George Raymond Geiger (New York: Macmillan, 1933).

110. Pyotr Kropotkin, "The Scientific Bases of Anarchy," *Nineteenth Century* 21 (1887): 243–44; citing Herbert Spencer, *The Data of Ethics*, 3rd ed. (London: Williams & Norgate, 1881), appendix. Kidd probably deliberately chose not to cite the radical Kropotkin, who had himself criticized Weismann's theories (citing Osborn and others) in "Recent Science," *Nineteenth Century* 32 (1892): 1007–13.

111. "Political Philosophy (1892–1893)," in Dewey, *Class Lectures*, 1:118. Dewey also cited Herbert Spencer, "The Coming Slavery," *Contemporary Review* 45 (1884); and Pyotr Kropotkin, "The Coming Anarchy," *Nineteenth Century* 22 (1887).

same set of lectures, Dewey gave an extended criticism of Spencer's account of the social organism, providing instead his own dialectical interpretation that attempted to synthesize individual freedom and the social whole:

> The individual is free so far as his environment is in consciousness; so far as he is ignorant that far is he a slave. It is the intelligence which the whole [social] organism puts at the disposal of the individual which measures his *freedom*; while it is the intelligence which he contributed to the whole which *measures* his *efficiency*, his *service*. It can only bring the individual to consciousness; it can't do it for him; the last would subordinate him, it would be slavery. The environment does not mean a physical surrounding,—*it means the conditions of action.*

Dewey's critique of Spencer, with its protopragmatist focus on action, thus combined the dialectical model of the organism-environment relationship that we encountered in chapter 4 with an anarchist picture of the social organism. Like Thomas Hill Green, Dewey thought that true freedom of the individual was only possible given certain social limitations; like Kropotkin, Dewey believed (as he wrote in 1891) that "the real criterion of evolutionary ethics" in Spencer's system was "one of social relationships."[112]

But Kidd argued in *Social Evolution* that the "Utopian dreams" of Spencer and the socialists had been undermined by Weismann's theory of heredity. Echoing Ward and LeConte, Kidd admitted that if Weismann's views were incorrect, the "future society" imagined by socialism might be possible:

> If we tend to inherit in our own persons the result of the mental and moral culture of past generations, then we may venture to anticipate a future society which will not deteriorate, but which may continue to make progress, even though the struggle for existence be suspended, the population regulated exactly to the means of subsistence, and the antagonism between the individual and the social organism extinguished, even as Mr. Herbert Spencer has anticipated.

However, if Weismann is right about heredity, Kidd continued, any relaxation of the struggle for existence will result in social decline. Kidd

---

112. John Dewey, *Outlines of a Critical Theory of Ethics* (Ann Arbor, MI: Inland Press, 1891), 72–74. For Green's analysis of individual freedom and state regulation, see Thomas Hill Green, "Lecture on 'Liberal Legislation and Freedom of Contract,'" in *Works of Thomas Hill Green*, ed. Richard Lewis Nettleship, vol. 3 (London: Longmans, Green, 1888), 370–72.

was not an advocate of laissez-faire; in fact, he insisted that the struggle for existence is most effective as a driver of social progress when legislation secures "to *all* the members of the community the right to be admitted to the rivalry of life, as far as possible, on a footing of equality of opportunity." Nevertheless, he claimed that neo-Darwinism had provided scientific evidence that conflict and struggle were necessary conditions of social progress, contradicting the cooperation-focused ideals of Huxley, Spencer, George, Kropotkin, and others.[113]

In his 1894 review, Dewey expressed sympathy with Ward's criticisms of Spencer and laissez-faire. According to Ward, the human capacity for intelligence and invention—the "psychic factors" of his title—had changed the evolutionary equation, with "true legislation . . . simply the application in the sphere of social forces of the principle of invention." Dewey also praised Ward's "general theory of the evolution of intelligence," which echoed James's critique of Spencer's psychology (discussed in chapter 2) by showing that "the 'raining in' of an external environment upon the organism until its main features are reproduced in the organization of the latter offers more difficulties than it solves." For Ward, on Dewey's reading, the development of intelligence through organism-environment interaction was instead an "experimental" process, controlled by the organism itself—a process we will examine in greater detail in chapter 7.[114]

Turning next to Kidd's book, Dewey described its key biological premise:

Progress is always effected through competition and struggle. There is infinite narrow variation, some variations tending slightly below, others slightly above, the existing average standard. Progress comes only through selection of favorable differentiations, and there is no selection save where there is rivalry and struggle. This biological law (with regard to which Dr. Kidd follows Weismannism in its extreme form) holds of human as of animal history. Its scene of operation is simply transferred to the rivalry of nations and of industrial life.[115]

---

113. Benjamin Kidd, *Social Evolution* (New York: Macmillan, 1894), 141, 191–92; citing Herbert Spencer, *The Data of Ethics* (London: Williams & Norgate, 1879), chap. 14, on the conciliation of egoism and altruism.

114. John Dewey, "Social Psychology," *Psychological Review* 1 (1894): 406–7. Dewey was echoing James's recent description of Spencer's organism as "absolutely passive clay, upon which 'experience' rains down": see William James, *The Principles of Psychology*, 2 vols. (New York: Henry Holt, 1890), 1:403.

115. Dewey, "Social Psychology," 408.

This complicated premise was neo-Darwinism in a nutshell: evolution as a result of the natural selection of small, fortuitous, undirected variations. At the time of writing, neither Kidd nor Dewey could have known about Weismann's endorsement of a form of directed variation, discussed earlier. Dewey, however, had read Osborn's treatment of the Spencer-Weismann dispute, in which Osborn had declared:

> The fundamental postulate of the selectionists that adaptive structures arise out of the fortuitous play of the adaptive and non-adaptive variations is negatived by direct evidence to the contrary. Palaeontology shows conclusively that there is an adaptive trend in variation under the operation of some law; whether this is the Lamarckian law or some unknown law remains to be determined.[116]

Taking this cue, Dewey explicitly considered the possibility of directed variation at the end of his review of Kidd:

> If we suppose that consciously acquired activity, and habits formed under the direction of intelligence, are conserved, the case against [Kidd's] point is much strengthened. . . . There is even no need to suppose that the conservation of rationalized activity is direct or through the organism; if the environment is so changed as to set up conditions which stimulate and facilitate the formation of like habits on the part of each individual, the same end is reached.[117]

On Dewey's view, variation in human activity is plausibly directed by intelligence and involves the deliberate cultivation of certain habits. Moreover, the inheritance of such acquired habits need not involve germ-plasm changes, as it could be outsourced to the environment, with environmental modifications scaffolding the reliable production of habits. Dewey's proposed model of evolution thus included what we would now call nongenetic inheritance.[118]

Why did Dewey favor this particular account of social evolution? Probably because it was consistent with contemporary work in developmental and comparative psychology. Psychologists such as William

---

116. Henry Fairfield Osborn, "The Discussion between Spencer and Weismann," *Psychological Review* 1 (1894): 315.

117. John Dewey, "Social Psychology," *Psychological Review* 1 (1894): 410.

118. On scaffolding and nongenetic inheritance, see John Odling-Smee, "Niche Inheritance: A Possible Basis for Classifying Multiple Inheritance Systems in Evolution," *Biological Theory* 2 (2007); William C. Wimsatt and James R. Griesemer, "Reproducing Entrenchments to Scaffold Culture: The Central Role of Development in Cultural Evolution," in *Integrating Evolution and Development: From Theory to Practice*, ed. Roger Sansom and Robert N. Brandon (Cambridge, MA: MIT Press, 2007).

CHAPTER FIVE

James and James Mark Baldwin had recently emphasized the importance of human responsiveness to the social environment in the formation of habits, made possible by plasticity along with imitation as the selection of preferred stimuli. "The phenomena of habit in living beings," wrote James, "are due to the plasticity of the organic materials of which their bodies are composed"; imitation, said Baldwin, is "the method by which our living *milieu* in all its aspects gets carried over and reproduced within us."[119] Dewey was probably also familiar with the work of comparative psychologist Conwy Lloyd Morgan, who had pointed out that repeated environmental shaping of organisms each generation could mimic the inheritance of acquired characters: "If each plastic embryo is moulded in turn by a similar influence, how can we conclusivly [sic] prove hereditary summation?"[120] In a criticism published in the *Monist*, Morgan had also highlighted Weismann's admission that the mingling of germ-cells in sexual reproduction could not on its own produce "effective variation"—that is, variation that exceeds current limits and makes evolutionary progress possible. In his book *The Germ-Plasm*, published just before his debate with Spencer, Weismann had already granted that adaptations cannot be due "to rare, fortuitous variations, occurring only once" but must involve variations "exhibited over and over again by many individuals," perhaps due to common nutritional changes.[121] Dewey had probably also seen Morgan's report on experiments with young chicks and ducks, published in *Open Court*. After presenting a detailed series of observations, Morgan concluded that "inherited co-ordinations [of activities] are perfected and rendered more effective by intelligent guidance" and that "imitation is an important factor in the early stages of mental development."[122] Dewey's idea of the environment facilitating the formation and reliable inheritance of certain habits was thus "in the air" for those working at the intersection

---

119. William James, *The Principles of Psychology*, 2 vols. (New York: Henry Holt, 1890), 1:105, italics removed; James Mark Baldwin, "Imitation: A Chapter in the Natural History of Consciousness," *Mind*, n.s., 3 (1894): 37. See also Charles Sanders Peirce, "Man's Glassy Essence," *Monist* 3 (1892): 16–18.
120. Conwy Lloyd Morgan, *Animal Life and Intelligence* (Boston: Ginn, 1891), 167; see also Conwy Lloyd Morgan, "The Law of Psychogenesis," *Mind*, n.s., 1 (1892): 92.
121. Conwy Lloyd Morgan, "Dr. Weismann on Heredity and Progress," *Monist* 4 (1893): 26–27; quoting August Weismann, *The Germ-Plasm: A Theory of Heredity*, trans. William Newton Parker and Harriet Rönnfeldt (New York: Charles Scribner's Sons, 1893), 431–32.
122. Conwy Lloyd Morgan, "Instinct and Intelligence in Chicks and Ducklings: A Contribution to Elementary Psychology," *Open Court*, April 26, 1894, 4060. As Jo Ann Boydston has noted, Dewey was at this time a regular contributor to the *Monist* and its sister journal *Open Court*: see John Dewey, *The Early Works, 1882–1898*, ed. Jo Ann Boydston, 5 vols. (Carbondale: Southern Illinois University Press, 1969–72), 3:lxv–lxvi.

of biology and psychology. Only a year later, Baldwin declared that "heredity does not stop with birth" and began to emphasize the importance of what he called "social heredity."[123]

Social Ethics and Social Psychology

The opposition between Herbert Spencer and August Weismann was also connected with two important topics in John Dewey's 1890s work: (a) the role of conflict in moral progress and (b) functional unity in psychology. The most obvious example of moral conflict in 1890s America was the battle between capital and labor, spurred by economic depression. In July 1894, just as his review of Ward and Kidd appeared, Dewey took up a new position at the University of Chicago in the midst of labor unrest. As he told his wife Alice in a letter, "the Pullman car builders are on a strike, & in order to help them the Railroad Union has boycotted all the [rail]roads using Pullman cars & won't handle them."[124] George Mortimer Pullman, manufacturer of railway sleeping cars, had recently reduced wages for residents in his planned workers' town near Chicago, but without lowering rents. This action led to a strike by Pullman workers on May 11, 1894, and to the boycott by railway workers (mentioned by Dewey) the following month. Although the strike continued, the boycott ended in July after federal troops were sent to Chicago, followed by the arrest of railway union leaders and a decision by the American Federation of Labor to forgo a sympathy strike.[125] In mid-August, with the strike of the Pullman workers still ongoing, Dewey praised Jane Addams's economic analysis of the situation, as reported in the labor-leaning *Chicago Times*. According to Addams, "the [Pullman] company had applied competition to wages and monopoly to rents. Wages were cut down because the tendency of wages was downward and the number of men seeking work increased by the hard times. But rents remained the same because the men were obliged to live in Pullman to work there." She had visited Pullman as part of a committee attempting to convince both the company and the workers to submit the matter to arbitration. But

---

123. James Mark Baldwin, *Mental Development in the Child and the Race* (New York: Macmillan, 1895), 361, 364. Baldwin, coeditor of the *Psychological Review*, was also following the Spencer-Weismann debate: see Baldwin, *Mental Development*, 31.

124. John Dewey to Alice Chipman Dewey, 30 June 1894, in Dewey, *Correspondence*, no. 00145.

125. Richard Schneirov, Shelton Stromquist, and Nick Salvatore, eds., *The Pullman Strike and the Crisis of the 1890s: Essays on Labor and Politics* (Urbana: University of Illinois Press, 1999), 8–9; see also Andrew Feffer, *The Chicago Pragmatists and American Progressivism* (Ithaca, NY: Cornell University Press, 1993), 91–92.

although the workers were "not only willing, but anxious, to arbitrate," the company held fast, "insisting that there was nothing to arbitrate."[126]

Could anything positive be said about this apparently intractable opposition between capital and labor? In his July review, Dewey had criticized Ward for underplaying the importance of conflict: "The positive evolutionary significance of conflict seems hardly to be recognized by Mr. Ward.... To me it appears as sure a psychological as biological principle that men go on thinking only because of practical friction or strain somewhere, that thinking is essentially the solution of tension."[127] But his view was challenged by Addams herself only a few months later. On October 9, after telling Dewey how her request for support of Hull House's relief efforts had just been spurned by a wealthy donor because of her criticisms of Pullman in the *Chicago Times* and elsewhere, the two of them got into a debate over the importance of conflict to historical progress.[128] According to Addams, as Dewey reported in another letter to Alice, "antagonism was not only useless and harmful, but entirely unnecessary; ... it lay never in the objective differences, which would always grow into unity if left alone, but from a person's mixing in his own personal reactions.... *historically* also, only evil had come from antagonisms." Dewey, a bit skeptical, told Alice that he had

> asked [Addams] if she didn't think that besides the personal antagonisms, there was that of ideas & institutions, as Christianity & Judaism, & Labor & Capital, the Church & Democracy now & that a realization of that antagonism was necessary to an appreciation of the truth, & to a consciousness of growth, & she said no. The antagonism of institutions was always unreal; it was simply due to the injection of the personal attitude & reaction; & then instead of adding to the recognition of meaning, it delayed & distorted it.

Dewey was impressed but not converted. He did not see how "all this conflict & warring of history" could be "perfectly meaningless," without

---

126. "Pullman Ground Fine," *Chicago Times*, August 19, 1894, 3; partially quoted in John Dewey to Alice Chipman Dewey, 18–19 August 1894, in Dewey, *Correspondence*, no. 00175. For more on Addams, Dewey, and the Pullman Strike, see Cheryl Hudson, "The 'Un-American' Experiment: Jane Addams's Lessons from Pullman," *Journal of American Studies* 47 (2013).

127. John Dewey, "Social Psychology," *Psychological Review* 1 (1894): 408.

128. This debate between Dewey and Addams has been extensively discussed by scholars: see Robert B. Westbrook, *John Dewey and American Democracy* (Ithaca, NY: Cornell University Press, 1991), 80–81; Louise W. Knight, *Citizen: Jane Addams and the Struggle for Democracy* (Chicago: University of Chicago Press, 2005), 322–25; Donald J. Morse, *Faith in Life: John Dewey's Early Philosophy* (New York: Fordham University Press, 2011), 21–23; Erin McKenna and Scott L. Pratt, *American Philosophy: From Wounded Knee to the Present* (London: Bloomsbury, 2015), 49–50; Beth L. Eddy, *Evolutionary Pragmatism and Ethics* (Lanham, MD: Lexington, 2016), 87–89.

any "functional value" at all. But a few days later, Dewey wrote to Addams and capitulated: "not only is actual antagonizing bad," he said, "but the assumption that there is or may be antagonism is bad—in fact, the first antagonism always come[s] back to the assumption. I'm glad I found this out before I began to talk on social psychology as otherwise I fear I should have made a mess of it." As he wrote to Alice in a later addendum to this letter, "I can see that I have always been interpreting the ~~Hegelian~~ dialectic wrong end up—the unity as the reconciliation of opposites, instead of the opposites as the unity in its growth."[129]

Despite this purported shift in his thinking, however, throughout the 1890s Dewey continued to appeal to struggle as an essential driver of progress—bringing him closer to Weismannians like Kidd than to Spencerians like Kropotkin. As Beth Eddy has shown, Addams's viewpoint had much in common with that of Kropotkin, who had recently taken Spencer's notion of "mutual aid" and argued that it characterized evolutionary history more broadly (and not just its later stages).[130] Dewey, in contrast, attacked Spencer's teleological account of moral progress in the *Monist* in 1898, criticizing the idea that "the goal of evolution is a complete state of final adaptation in which all is peace and bliss and in which the pains of effort and of reconstruction are known no more." The article was primarily a critique of Huxley's claim, mentioned earlier, that ethics and civilized society oppose and even eliminate the struggle for existence. Dewey followed Kidd in arguing that the struggle had not ended but only changed its form: "The environment is now distinctly a social one, and the content of the term 'fit' has to be made with reference to social adaptation." Even in civilized society, said Dewey, tension and struggle are at the heart of moral progress: "An act which was once adapted to given conditions must now be adapted to other conditions. The effort, the struggle, is a name for the necessity of this re-adaptation. . . . The tension is between an organ adjusted to a past state and the functioning required by present conditions. And this tension demands reconstruction."[131] Thus, although Dewey and Addams both linked evolution to moral progress, they disagreed about the role

---

129. John Dewey to Alice Chipman Dewey, 10 October 1894, in Dewey, *Correspondence*, no. 00206; John Dewey to Jane Addams, 12 October 1894, in Dewey, *Correspondence*, no. 00619.
130. Beth L. Eddy, *Evolutionary Pragmatism and Ethics* (Lanham, MD: Lexington, 2016), chap. 4.
131. John Dewey, "Evolution and Ethics," *Monist* 8 (1898): 333–34. For more on Dewey's criticism of Huxley, see John Teehan, "Evolution and Ethics: The Huxley/Dewey Exchange," *Journal of Speculative Philosophy* 16 (2002); Eddy, *Evolutionary Pragmatism and Ethics*, chaps. 2–3; Trevor Pearce, "American Pragmatism, Evolution, and Ethics," in *The Cambridge Handbook of Evolutionary Ethics*, ed. Michael Ruse and Robert J. Richards (Cambridge: Cambridge University Press, 2017), 47–51.

CHAPTER FIVE

of competition and rivalry: for Dewey, it was a necessary factor in social evolution, whereas for Addams it was useless and harmful. In 1896, Dewey had praised a new analysis by Addams of the 1894 labor unrest, in which she attributed it partly to Pullman's personal flaws and famously compared him to King Lear; but in 1898, Dewey was still insisting—like the Weismannians—that struggle and tension were essential to social evolutionary progress.[132]

Starting in the mid-1890s, Dewey also frequently connected the Spencer-Weismann dispute to his developing functional psychology. As Jim Garrison has shown, Dewey's early functionalism—as presented in his articles "The Theory of Emotion" and "The Reflex Arc Concept in Psychology"—was a hybrid of the Hegelian approach described in chapter 4 and the biological-functional approach of James and Darwin.[133] At the center of Dewey's functional psychology was his famous claim in "Reflex Arc" that stimulus and response should "be viewed, not as separate and complete entities in themselves, but as . . . functioning factors" within a single "mode of behavior" or "organized coördination of activities." In "Theory of Emotion," Dewey had claimed that habitual coordinations could "become so organically registered—*pace* Weissman [*sic*]—as to become hereditary," and in "Reflex Arc" he applied his framework to the Spencer-Weismann dispute. According to Dewey, because neither disputant considered the functional activity of a particular organ when discussing its evolution, each of them had emphasized one contributing factor: Spencer saw only "an external pressure of 'environment'" and Weismann saw only "an unaccountable spontaneous variation from within the 'soul' or the 'organism.'"[134]

Dewey elaborated on this point in his lectures at the University of Chicago. In Political Ethics, the third course of a three-semester ethics sequence, he argued—following the dialectical approach described in chapter 4—that there is only "a relative distinction between organism

---

132. John Dewey to Jane Addams, 18 August 1896, in Dewey, *Correspondence*, no. 00547; Jane Addams, "A Modern Lear," *Survey* 29 (1912); see also Eddy, *Evolutionary Pragmatism and Ethics*, 87–89.

133. Jim Garrison, "Dewey's Theory of Emotions: The Unity of Thought and Emotion in Naturalistic Functional 'Co-ordination' of Behavior," *Transactions of the Charles S. Peirce Society* 39 (2003); see also Richard J. Bernstein, *John Dewey* (New York: Washington Square, 1966), 15–21; Andrew Backe, "Dewey and the Reflex Arc: The Limits of James's Influence," *Transactions of the Charles S. Peirce Society* 35 (1999).

134. John Dewey, "The Theory of Emotion. (II.) The Significance of Emotions," *Psychological Review* 2 (1895): 19–20, 32; John Dewey, "The Reflex Arc Concept in Psychology," *Psychological Review* 3 (1896): 358, 360. The phrases "mode of behavior" and "organized coördination of activities" are from "Theory of Emotion," which goes into more detail about the notion of coordination.

and environment": "Situation and functioning represent the whole process. Organism as organism does not represent the whole; neither does environment as environment. In any case of readaptation or readjustment, the old environment as well as the old organism have to adapt themselves to the new environment." In Dewey's view, any new variation is always a mediation of an old habit, with evolution as "a conflict within function . . . a conflict between the constant and the variable, between habit and [the] changed circumstances under which that habit must be exercised." Without this focus on mediation, said Dewey, you end up with "two schools of evolution, Spencer and Weismann; a spontaneous breaking loose, or complete control by environment."[135] The terms *organism* and *environment*, for Dewey, have no meaning "taken simply at large, without any statement of the function that is under consideration, and the historical development of that function." Spencer assumed that more-advanced organisms—and more-advanced people—were simply better adapted to a fixed environment. But for Dewey, as we saw in chapter 4, organism and environment "vary together," in Samuel Alexander's phrase. What matters is the overall process: Spencer's mistake was to "conceive the environment, which is really the outcome of the process of development, which has gone on developing along with the organism, as if it was something which had been there from the start." Instead, "what has taken place has been the development of the environment, the creation of the environment, the evolution of the environment."[136] As Dewey wrote in his attack on Huxley, "so far as the progressive varieties are concerned, it is not in the least true that they simply adapt themselves to current conditions; evolution is a continued development of new conditions which are better suited to the needs of organisms than the old. The unwritten chapter in natural selection is that of the evolution of environments." And in human society, said Dewey, "selection along the line of variations which enlarge and intensify the environment is active as never before."[137]

In the Huxley paper, Dewey also returned to the idea—first presented in his 1894 review—that the social environment can reliably form certain habits, effectively sidestepping the Spencer-Weismann debate:

---

135. "Political Ethics (1896)," in Dewey, *Class Lectures*, 1:1538; see also "Political Ethics (1898)," in Dewey, *Class Lectures*, 1:1651–52. For descriptions of the courses making up the ethics sequence, see *Annual Register, July, 1894—July, 1895, with Announcements for 1895–6* (Chicago: University of Chicago Press, 1895), 44–45.

136. "Political Ethics (1898)," in Dewey, *Class Lectures*, 1:1651, 1655.

137. John Dewey, "Evolution and Ethics," *Monist* 8 (1898): 339–40.

We do not need to go here into the vexed question of the inheritance of acquired characters. We know that through what we call public opinion and education certain forms of action are constantly stimulated and encouraged, while other types are as constantly objected to, repressed, and punished. What difference in principle exists between this mediation of the acts of the individual by society and what is ordinarily called natural selection, I am unable to see. In each case there is the reaction of the conditions of life back into the agents in such a way as to modify the function of living.[138]

Dewey focused on selection "within the life of one and the same individual," but his discussion of the "evolution of environments" implied that there would also be a kind of environmental inheritance, akin to what Baldwin called "social heredity." As Dewey wrote in a review of Baldwin's 1897 book *Social and Ethical Interpretations in Mental Development*, "physical heredity must, on the negative side, not be of a sort to throw the individual into antagonism beyond a certain point with the interests of the community; positively, it must lend itself, must have an active trend, towards just the sort and variety of relationships which the social tradition imposes."[139] Dewey was skeptical of certain "directed variation" views: for example, he criticized Cope's idea that the organism has "a type latent within . . . so that variations come along determinate directions." Nevertheless, the idea of an "active trend" in heredity, corresponding to the social environment, was consonant with the evolutionary framework of Osborn and other anti-Weismannians.[140]

The Spencer-Weismann dispute was about the origin and nature of evolutionary variation. Was it directed or undirected? What role did the environment play? Weismann ended up admitting that variation at the level of the organism must be directed in some way, consistent with the views of biologists such as Osborn. From 1894 to 1898, Dewey intervened in the factors of evolution debates, criticizing the neo-Darwinian assumptions of Kidd's *Social Evolution* and placing plasticity, variation, and heredity at the center of his own theories of psychological and social evolution. Dewey followed Kidd—and rejected the views of George, Kropotkin, and Addams—in arguing that conflict was the main driver of social and ethical progress. But he refused to accept Kidd's Weismannism: not only did he think that acquired psychological characters were often inherited, he also

---

138. Dewey, "Evolution and Ethics," 337.
139. John Dewey, "Social and Ethical Interpretations in Mental Development," *New World* 7 (1898): 511; see James Mark Baldwin, *Social and Ethical Interpretations in Mental Development: A Study in Social Psychology* (New York: Macmillan, 1897), 57–64. Dewey would later stress the importance of social heredity, for example in "Principles of Education (1902)," Dewey, *Class Lectures*, 2:592.
140. "Political Ethics (1898)," in Dewey, *Class Lectures*, 1:1652.

suggested that this inheritance could be secured by the continuity and modification of the social environment rather than through changes to the organism itself. For Dewey, as we saw in chapter 4, organism and environment coevolve: we make a mistake when we try to attribute variation and evolution to either the environment or the organism, taken alone.

Members of both the first and second cohorts of pragmatists were deeply involved in the 1890s debates over the factors of evolution that emerged in reaction to Weismann's new theory of heredity. Peirce and James had always been very critical of Spencer's philosophy. But whereas James embraced neo-Darwinism even before the Spencer-Weismann debate, Peirce ended up supporting neo-Lamarckism in 1893 and remained skeptical of Weismann's views even in 1902.[141] Dewey, although he characteristically refused to take sides, was critical of Weismann's theory of heredity in the 1890s while at the same time drawing from the work of Weismannians such as Kidd.

By the time Spencer died in 1903, Weismann's views had triumphed. In a 1904 speech, W. E. B. Du Bois—after outlining the supposed importance of the transmission of "acquired ability" for both education and race betterment—declared that the Lamarckian conception of heredity had "been very seriously questioned" and "practically overthrown" by the work of Weismann. Although he noted the pessimistic possibilities of Weismannism as outlined by Ward and LeConte, Du Bois ended up arguing in favor of Baldwin's social heredity:

The human child receives its body and the physical bases of life from its parents, but it receives its thoughts, the larger part of its habits, its tricks of doing, its religion, its whole conception of what it is and what the whole world about it is from the society in which it is placed; and this heredity which is not physical at all has been aptly called social heredity.

According to Du Bois, the fact of social heredity shows that Weismannism does not undermine education: "The public school of today is the largest and most efficient single organ for transmitting the social heritage of men." Even beyond the schools, said Du Bois, "the larger part of the training of human beings must come from the social surroundings in which they live."[142]

---

141. "Carnegie Institution Correspondence," L75 (1902), in Peirce, *New Elements*, 4:66.
142. W. E. B. Du Bois, *Heredity and the Public Schools: A Lecture Delivered under the Auspices of the Principals' Association of the Colored Schools of Washington, D.C.* (Washington, DC: R. L. Pendleton,

CHAPTER FIVE

When Dewey and Du Bois gave their speeches to the National Negro Conference in 1909, as mentioned in the introduction, they were fully in Weismann's camp. "The whole tendency of biological science at the present time," said Dewey, "is to make it reasonably certain that the characteristics which the individual acquired are not transmissible, or if they are transmissible, then in such a small degree as to be comparatively and relatively negligible." Dewey admitted, echoing Ward and LeConte, that this may seem "a disappointing and discouraging doctrine," since "what one individual attains by his own effort and training, does not modify the level from which the next generation then starts." But channeling Kidd, although rejecting his Eurocentrism and racism, Dewey gave social Weismannism a positive spin:

This doctrine that acquired characteristics are not transmitted becomes a very encouraging doctrine because it means, so far as individuals are concerned, that they have a full, fair and free social opportunity. . . . In other words, there is no "inferior race," and the members of a race so-called should each have the same opportunities of social environment and personality as those of a more favored race.

Given the realities of racism, there was of course no true equality of opportunity. However, Dewey thought it was society's responsibility—again following Kidd's harmonizing of progressive capitalism and Weismannism—"to see to it that the environment is provided which will utilize all of the individual capital that is being born into it." Like Du Bois, Dewey explicitly embraced "social heredity": heredity and evolution are about intelligently evolving environments rather than spontaneously varying germ-cells.[143]

Also mentioning Weismann, Du Bois argued in his speech that social evolution depends on "self-development" and "social self-realization." For Du Bois, this freedom—"equality of opportunity for unbounded future attainment"—was "the central assertion of the evolutionary theory."[144]

1904), 6–7, 9, 11. Adolph Reed has claimed that despite this later endorsement of Weismann, Du Bois was drawing on Lamarckian ideas in his late 1890s work: see Adolph L. Reed Jr., *W. E. B. Du Bois and American Political Thought: Fabianism and the Color Line* (Oxford: Oxford University Press, 1997), 119–24.

143. John Dewey, "Address of John Dewey," in *Proceedings of the National Negro Conference, 1909: New York, May 31 and June 1* (n.p., [1909]), 71–72, https://hdl.handle.net/2027/nyp.33433081797809. For Kidd's claim that "the coloured races" are inferior "as regards the qualities contributing to social efficiency," see Benjamin Kidd, *Social Evolution* (New York: Macmillan, 1894), 315.

144. W. E. B. Du Bois, "Evolution of the Race Problem," in *Proceedings of the National Negro Conference*, 149, 152, 156; also in W. E. B. Du Bois, *John Brown* (Philadelphia: George W. Jacobs, 1909), 375, 379, 383.

Dewey was surely sympathetic to this view: in the 1890s, he had frequently stressed the importance of guiding intelligence in social evolution. As he wrote in the *Monist*, "that which was unconscious adaptation and survival in the animal, taking place by the 'cut and try' method until it worked itself out, is with man conscious deliberation and experimentation."[145] In the next chapter, we will explore how pragmatist treatments of morality featured both evolution and experimentation, bringing ethics together with social science and social reform.

---

145. John Dewey, "Evolution and Ethics," *Monist* 8 (1898): 340.

SIX

Pragmatist Ethics: Evolution, Experiment, and Social Progress

Experimental philosophy was first distinguished from speculative philosophy in the seventeenth century. Robert Boyle, for example, argued that philosophers should be tentative collectors of experiments rather than overconfident builders of systems. At the end of the nineteenth century, some philosophers still embraced the label: Chauncey Wright chastised St. George Mivart in 1871 for forgetting that they were living in "the age of 'experimental philosophy,'" and Charles Sanders Peirce was happy to adopt "experimental philosophy" in 1889 as the name of his own approach, defining it as "that philosophy which accepts nothing as absolutely certain, but holds that opinions will gradually approximate to the truth in scientific researches into nature."[1] Since positivism had failed to acquire institutional credibility, the experimental method only truly entered American philosophy departments with the rise of experimental psychology in the last quarter of the nineteenth century.

---

1. Robert Boyle, "A Proemial Essay, Wherein, with Some Considerations Touching Experimental Essays in General, Is Interwoven Such an Introduction to All Those Written by the Author, as Is Necessary to Be Perus'd for the Better Understanding of Them," in *The Works of Robert Boyle*, ed. Michael Hunter and Edward B. Davis, vol. 2 (London: Pickering & Chatto, 1999); Chauncey Wright, "The Genesis of Species," *North American Review* 113 (1871): 68; William Dwight Whitney, ed., *The Century Dictionary: An Encyclopedic Lexicon of the English Language*, 6 vols. (New York: Century, 1889–91), 2:2079, s.v. "experimental."

As W. E. B. Du Bois recalled, "psychology was reducing metaphysics to experiment," a development that had a direct impact on the pragmatists. John Dewey praised "the new psychology" in 1884 for introducing "a new *method*,—that of experiment," and Peirce, who had himself carried out psychological experiments, predicted in 1890 that "psychology is destined to be the most important experimental research of the twentieth century; fifty years hence its wonders may be expected to occupy popular imagination as wonders of electricity do now."[2]

The experimental approach was not only a method but also an attitude, available to anyone who possessed the "Character of the Experimentalist"—the title of one of the chapters of William Stanley Jevons's *The Principles of Science*, an 1874 book that was well known to William James, Peirce, and probably Du Bois. Jevons had already emphasized this point in his logic textbook, which was used by Dewey in his classes at the University of Michigan, by James in his Logic and Psychology class at Harvard University (which Du Bois attended), and by George Herbert Mead when he started teaching logic at the University of Chicago. "The great experimentalist," wrote Jevons, "is he who ever has a theory or even a crowd of theories or ideas upon his mind, but is always putting them to the test of experience and dismissing those which are false."[3] Du Bois was particularly interested in the character of the experimentalist, and he argued in an assignment for his Harvard composition class with Josiah Royce that Leonardo da Vinci should be considered "the founder of modern experimental science." Hermann Grothe, the main source for Du Bois's essay, sounded much like Jevons when he praised the mind-set of the Italian polymath: Leonardo did not

---

2. W. E. B. Du Bois, *Dusk of Dawn: An Essay toward an Autobiography of a Race Concept* (New York: Harcourt, 1940), 26; John Dewey, "The New Psychology," *Andover Review* 2 (1884): 282; "[Logic and Spiritualism]," MS 878 (1890), in Peirce, *Writings*, 6:394. For one of Peirce's experiments, see Charles Sanders Peirce and Joseph Jastrow, "On Small Differences of Sensation," *Memoirs of the National Academy of Sciences* 3 (1884).
3. William Stanley Jevons, *The Principles of Science: A Treatise on Logic and Scientific Method*, 2 vols. (London: Macmillan, 1874), chap. 26; William Stanley Jevons, *Elementary Lessons in Logic: Deductive and Inductive* (London: Macmillan, 1870), 237. For references to *Principles of Science*, see William James, review of *The Principles of Science*, by William Stanley Jevons, *Atlantic Monthly* 35 (1875); William James, "Great Men, Great Thoughts, and the Environment," *Atlantic Monthly* 46 (1880): 456n; "[Beginnings of a Logic Book]," MS 749 (1883), in Peirce, *Writings*, 4:401; W. E. B. Du Bois, "Lecture Notebook," p. 3 [5], Series 10, Du Bois Papers. For references to *Elementary Lessons in Logic*, see the *Calendar of the University of Michigan for 1884–85* (Ann Arbor: University of Michigan, 1885), 50; *The Harvard University Catalogue, 1889–90* (Cambridge, MA: Harvard University, 1889), 118; *Annual Register, July, 1893—July, 1894, with Announcements for 1894-5* (Chicago: University of Chicago Press, 1894), 42; John Dewey to Alice Chipman Dewey, 14 December 1894, in Dewey, *Correspondence*, no. 00246.

CHAPTER SIX

"cling to his views as unalterable truths" but modified them in response to experiment.[4]

Evolution was also sometimes portrayed as an experimental process. As Charles Darwin wrote in the opening pages of *The Variation of Animals and Plants under Domestication*, a book twice reviewed by James, "man selects varying individuals, sows their seeds, and again selects their varying offspring. . . . [He], therefore, may be said to have been trying an experiment on a gigantic scale; and it is an experiment which nature during the long lapse of time has incessantly tried."[5] That is, since the artificial selection of certain promising varieties in domestication is properly described as an experimental process, evolution can be seen by analogy as a ramifying series of natural experiments. Those who followed the neo-Lamarckians in emphasizing the role of consciousness and environmental modification in evolution—for instance, Dewey and the psychologist James Mark Baldwin, as discussed in chapter 5—were even more likely to think about evolution in such terms. By 1898, Dewey was describing variation in evolution as a "process of experimentation."[6]

Progress was framed as both evolutionary and experimental by social scientists in the reform tradition that Dewey, Mead, Du Bois, and Jane Addams joined in the 1890s. There was already a long history of talking about experiments in politics. In his 1889 entry for *experiment* in the *Century Dictionary*, Peirce cited John Adams, writing in 1787:

The systems of legislators are experiments made on human life and manners, society and government. Zoroaster, Confucius, Mithras, Odin, Thor, Mahomet, Lycurgus, Solon, Romulus, and a thousand others, may be compared to philosophers making

---

4. Hermann Grothe, *Leonardo da Vinci als Ingenieur und Philosoph: Ein Beitrag zur Geschichte der Technik und der induktiven Wissenschaften* (Berlin: Nicolaische Verlags-Buchhandlung, 1874), 90; frequently cited in W. E. B. Du Bois, "Does the Scientific Work of Leonardo da Vinci Entitle Him to Be Called the Founder of the Modern Scientific Method[?]" [Thesis: English C, 1888–89], Series 10, Du Bois Papers. Du Bois's essay is misdated in the archival record: the supposed date is his Harvard graduation year. Du Bois's transcript shows that he took English C in 1888–89, and his library records indicate that he checked out Grothe's book, along with a biography of Leonardo by Jean Paul Richter, in the early months of 1889: see Library Charging Lists, 1888–1889, p. 838 [seq. 445]. In the charging list, "V. 3207" is the designation of Grothe's book, checked out on February 16 and March 2, 1889 (for book identification, see Box 2, UAIII 50.15.47.5, Harvard University Archives). For Royce's teaching of English C in 1888–89, see *The Harvard University Catalogue, 1888–89* (Cambridge, MA: Harvard University, 1888), 105.

5. Charles Darwin, *The Variation of Animals and Plants under Domestication*, 2 vols. (London: John Murray, 1868), 1:3.

6. "Political Ethics (1898)," in Dewey, *Class Lectures*, 1:1661.

experiments on the elements. Unhappily a political experiment cannot be made in a laboratory, nor determined in a few hours.[7]

Dewey told his Michigan students in 1892–93 that he favored "experimenting in legislation": we should "experiment differently in separate states and note the results."[8] Du Bois heard a similar story from one of his economics professors in Berlin. Gustav Schmoller, after defining *experiment* in an 1893–94 seminar as "the power to alter at will the factors in a problem and so to measure them," claimed that "experiment tho[ugh] harder in G[eistige] [i.e., Social or Mental] Sc[iences], is still not impossible—viz.: in history and government."[9] Jevons endorsed an even stronger view in his book *Methods of Social Reform* (also cited in Peirce's dictionary entry), claiming that experimentation was the only route to social progress:

I maintain that, in large classes of legislative affairs, there is really nothing to prevent our making direct experiments upon the living social organism. Not only is social experimentation a possible thing, but it is . . . the universal mode of social progress. It would hardly be too much to say that social progress is social experimentation, and social experimentation is social progress.[10]

For scientists like Jevons, social reform was essentially experimental.

---

7. John Adams, *A Defence of the Constitutions of Government of the United States of America*, 2 vols. (London: C. Dilly and John Stockdale, 1787), xxiv; partially quoted in William Dwight Whitney, ed., *The Century Dictionary: An Encyclopedic Lexicon of the English Language*, 6 vols. (New York: Century, 1889–91), 2:2079, s.v. "experiment."

8. "Political Philosophy (1892–1893)," in Dewey, *Class Lectures*, 1:182–83.

9. W. E. B. Du Bois, "On Method, Schmoller in Seminar, Winter Semester 93–94," Lecture 3, p. 15 [31], in "Lecture Notebook," Series 10, Du Bois Papers. This part of Du Bois's notebook tracks the central sections of a handbook entry by Schmoller on the theory and method of political economy: Gustav Schmoller, "Volkswirtschaft, Volkswirtschaftslehre und -methode," in *Handwörterbuch der Staatswissenschaften*, ed. Johannes Conrad et al., vol. 6 (Jena, Ger.: Gustav Fischer, 1894), 532–49, 563. Du Bois's notebook is misdated in the archival record: it is titled "[Natio]nal Oekonomie, Schmoller u. Wagner," and was begun in the summer semester of 1893, when Du Bois took Schmoller's Allgemeine oder theoretische Nationalökonomie (General or Theoretical Political Economy) and Adolph Wagner's Nationalökonomische und finanzwissenschaftliche Übungen im staatswissenschaftlich-statistischen Seminar (Tutorial on Political Economy and Financial Science in the Seminar on Political Science and Statistics). See W. E. B. Du Bois to the John F. Slater Fund, 10 March 1893, in Du Bois, *Correspondence*, 1:24; *Verzeichniss der Vorlesungen, welche auf der Friedrich-Wilhelms-Universität zu Berlin im Sommer-Semester vom 17. April bis 15. August 1893 gehalten werden* (n.p., 1893), 21–22, https://www.digi-hub.de/viewer/toc/DE-11-001799608/1/.

10. William Stanley Jevons, *Methods of Social Reform, and Other Papers* (London: Macmillan, 1883), 256.

CHAPTER SIX

Others explicitly linked social experimentation and social evolution. Take for example Richard Theodore Ely, a reform-oriented American economist. Ely was connected to the pragmatists in various ways: like Du Bois fifteen years later, he was trained in the historically and statistically minded approach of German economics by members of the Verein für Sozialpolitik (Social Policy Association), initially described by its founders as the "Social Reform Association"; *Hull-House Maps and Papers*, as well as Addams's books *Democracy and Social Ethics* and *Newer Ideals of Peace*, were published in book series edited by Ely; and when Dewey was in graduate school at Johns Hopkins, Ely was teaching there and participated in Herbert Baxter Adams's "seminary" on American economic and institutional history, which Dewey regularly attended on Friday evenings.[11]

On one of those evenings during his first semester at Hopkins in 1882, Dewey probably heard Ely give a talk on "The Past and the Present of Political Economy." Ely praised the new "Historical School" of German economics, highlighting its members' opposition to laissez-faire as well as their claim that "the whole life of the world had necessarily been a series of grand economic experiments, which, having been described with more or less accuracy and completeness, it was possible to examine." He also embraced the idea, famously associated with Herbert Spencer but also endorsed by German social scientists such as Karl Knies and Johann Caspar Bluntschli, that "the nation in its economic life is an organism . . . composed of interdependent parts, which perform functions essential to the life of the whole."[12] Thus, not only did Ely favor Jevons's idea that "social progress is social experimentation," he also thought about society in biological terms: his own *Introduction to Politi-*

---

11. *Hull-House Maps and Papers: A Presentation of Nationalities and Wages in a Congested District of Chicago, with Comments and Essays on Problems Growing Out of Social Conditions* (New York: Thomas Y. Crowell, 1895), [i]; Jane Addams, *Democracy and Social Ethics* (New York: Macmillan, 1902), [iii]; Jane Addams, *Newer Ideals of Peace* (New York: Macmillan, 1906), [iii]; "Enumeration of Classes, Second Half-Year, 1882-3," *Johns Hopkins University Circulars* 2 (1883): 93; Richard Theodore Ely, *Ground under Our Feet: An Autobiography* (New York: Macmillan, 1938), 39–51; Benjamin G. Rader, *The Academic Mind and Reform: The Influence of Richard T. Ely in American Life* (Lexington: University of Kentucky Press, 1966), 11–14; W. E. B. Du Bois to the John F. Slater Fund, 10 March 1893, in Du Bois, *Correspondence*, 1:23; Gustav Schmoller, *Zur Social- und Gewerbepolitik der Gegenwart* (Leipzig, Ger.: Duncker & Humblot, 1890), 1–2; Erik Grimmer-Solem, *The Rise of Historical Economics and Social Reform in Germany, 1864–1894* (Oxford: Oxford University Press, 2003), 174–76. For the moniker "Social-Reform-Verein," see *Verhandlungen der Eisenacher Versammlung zur Besprechung der socialen Frage* (Leipzig, Ger.: Duncker & Humblot, 1873), 161.

12. Richard Theodore Ely, "The Past and the Present of Political Economy," in *Johns Hopkins University Studies in Historical and Political Science*, ed. Herbert Baxter Adams, vol. 2, no. 3 (Baltimore: John Murphy, 1884), 45, 49. Ely had given a presentation with a similar title on October 20, 1882: see "Proceedings of Societies," *Johns Hopkins University Circulars* 2 (1882): 38.

cal Economy, published a few years later, took an explicitly evolutionary view of economic life.[13] Although the experimentalism of reformers like Jevons and Ely was opposed to Herbert Spencer's laissez-faire approach, their broader perspective still paralleled that of Spencer's *Principles of Sociology*, which Dewey assigned in some of his courses at Michigan. Spencer described social progress as a "super-organic evolution" in which technology, knowledge, and legislation "are ever modifying individuals and modifying society, while being modified by both." In short, reformers like Ely juxtaposed experiment and evolution in their theories of social progress.[14]

Finally, moral progress was often linked to evolution and experiment. Spencer, as discussed in chapter 5, claimed that ethics was evolving toward greater "mutual aid" and ultimately the "ideal social state," in which, as Royce joked, "there are no moral conflicts, nothing but a tedious cooing of bliss from everybody."[15] James, in his 1891 essay "The Moral Philosopher and the Moral Life"—admired by Dewey and probably also known to Mead, Addams, and Du Bois—criticized the Spencerian idea that "our moral judgments have gradually resulted from the teaching of the environment" but gave an experimental account of moral progress. According to James, ethics stems from the wants and needs of real people rather than from some "abstract moral order": there is an ethical obligation whenever there is "a claim actually made by some concrete person." The development of ethics, said James, has been the history of attempts to satisfy jointly as many demands as we can: "*Invent some manner* of realizing your own ideals which will also satisfy the alien demands—that and that only is the path of peace!" For James, ethical progress is experimental: radicals and conservatives alike "are simply deciding through actual experiment by what sort of conduct the maximum amount of good can be gained and kept in this world. These experiments are to be judged, not *a priori*, but by actually finding, after the fact of their making, how much more outcry or how much appeasement comes about." That is, ethics is a series of human experiments in attempting to satisfy our diverse and often conflicting desires. Dewey

---

13. Richard Theodore Ely, *An Introduction to Political Economy* (New York: Chautauqua, 1889), chaps. 5–6.

14. Herbert Spencer, *The Principles of Sociology*, vol. 1 (London: Williams & Norgate, 1877), 15. For Dewey's courses, see *Calendar of the University of Michigan for 1892–93* (Ann Arbor: University of Michigan, 1893), 67; *Calendar of the University of Michigan for 1893–94* (Ann Arbor: University of Michigan, 1894), 68.

15. Josiah Royce, *The Religious Aspect of Philosophy* (Boston: Houghton, Mifflin, 1885), 74; cited in John Dewey, *Outlines of a Critical Theory of Ethics* (Ann Arbor, MI: Inland Press, 1891), 78; see also William James, "Herbert Spencer's Data of Ethics," *Nation*, September 11, 1879, 179.

CHAPTER SIX

was very appreciative of this picture: in June 1891 he told James that he had already read the essay several times and recommended it to his students at the University of Michigan.[16]

For Spencer, ethics was evolutionary; for James, it was experimental. I will argue in this chapter that for the second cohort of pragmatists it was both: in the years around 1900, Dewey, Mead, Addams, and Du Bois developed a view of moral and social progress as experimental evolution. Although they rejected the teleological approach of Spencer, who saw ethics as proceeding to a specified evolutionary end point, they still employed a modified version of his organism-environment framework, as described in chapters 2 and 4.[17] Their application of this framework to ethics led them to a distinctive picture in which moral philosophy was inextricable from social science and social reform. Thus, I agree with James Kloppenberg and Axel Schäfer that pragmatist ethics was in part a response to developments in politics and the social sciences, and I also follow Charlotte Seigfried and Colin Koopman in viewing pragmatism more broadly as essentially experimental. Although it is perhaps too strong to call the second cohort of pragmatists "Reform Spencerians," they were certainly "Reform Evolutionists."[18]

The chapter has four sections: in the first, I will describe how Addams, Dewey, Mead, and Du Bois constructed experimental field sites for moral and social inquiry; in the second, I will argue that they all deployed Spencer's organism-environment framework in their thinking

---

16. William James, "The Moral Philosopher and the Moral Life," *International Journal of Ethics* 1 (1891): 330–31, 338, 346, 348; John Dewey to William James, 3 June 1891, in Dewey, *Correspondence*, no. 00460. Both Mead and Du Bois forged personal connections with James in the late 1880s, shortly before this essay appeared: see Gary A. Cook, *George Herbert Mead: The Making of a Social Pragmatist* (Urbana: University of Illinois Press, 1993), 15–19; David Levering Lewis, *W. E. B. Du Bois: Biography of a Race, 1868–1919* (New York: Henry Holt, 1993), 91–92. The essay was also reprinted in James's widely read book *The Will to Believe, and Other Essays in Popular Philosophy* (New York: Longmans, Green, 1897).

17. On pragmatism and the notion of progress, see David W. Marcell, *Progress and Pragmatism: James, Dewey, Beard, and the American Ideal of Progress* (Westport, CT: Greenwood, 1974); Beth L. Eddy, *Evolutionary Pragmatism and Ethics* (Lanham, MD: Lexington, 2016).

18. James T. Kloppenberg, *Uncertain Victory: Social Democracy and Progressivism in European and American Thought, 1870–1920* (Oxford: Oxford University Press, 1986), chaps. 4–7; Axel R. Schäfer, *American Progressives and German Social Reform, 1875–1920: Social Ethics, Moral Control, and the Regulatory State in a Transatlantic Context* (Stuttgart, Ger.: Franz Steiner, 2000), chap. 1; Charlene Haddock Seigfried, *Pragmatism and Feminism: Reweaving the Social Fabric* (Chicago: University of Chicago Press, 1996), chaps. 4–5; Colin Koopman, "Pragmatist Resources for Experimental Philosophy: Inquiry in Place of Intuition," *Journal of Speculative Philosophy* 26 (2012). On "Reform Spencerism" and "Reform Darwinism," see Mark Francis, "The Reforming Spencerians: William James, Josiah Royce, and John Dewey," in *Global Spencerism: The Communication and Appropriation of a British Evolutionist*, ed. Bernard Lightman (Leiden, Neth.: Brill, 2016); Robert C. Bannister, *Social Darwinism: Science and Myth in Anglo-American Social Thought* (Philadelphia: Temple University Press, 1979), 11.

on ethics and social reform; in the third, I will demonstrate that each of them had a vision of moral progress as evolution guided by social scientific research; and in the fourth, I will show that their experimental-evolutionary approach, although explicitly opposed to racism, was also committed to the discourse of eugenics and civilization.

Fieldwork in Ethics

Jane Addams and Ellen Gates Starr founded Hull House, their Chicago settlement house at the corner of Halsted and Polk, in 1889. The year before, they had toured Toynbee Hall in London, the world's first University Settlement, where—in the words of its founders Henrietta and Samuel Barnett—"men who have knowledge may become friends of the poor and share that knowledge and its fruits as, day by day, they meet in their common rooms for talk or for instruction, for music or for play." The Barnetts were committed to social reform, which they defined as "the removal of certain conditions in and around society which stand in the way of man's progress towards perfection." In their 1888 book *Practicable Socialism*, they lamented that scientists knew more about the habits of social insects than about social conditions and their effects:

> The study of the condition of the people receives hardly as much attention as that which Sir J[ohn] Lubbock gives to the ants and the wasps. Bold good men discuss the poor, and cheques are given by irresponsible benefactors; but there are few students who reverently and patiently make observations on social conditions, accumulate facts, and watch cause and effect. Scientific method has won the great victories of the day, and scientific method is supreme everywhere except in those human affairs which most concern humanity.

According to the Barnetts, reform needed to be placed on a scientific basis.[19]

Social reform had also become a topic of interest in philosophy departments. In 1886–87, Francis Greenwood Peabody—who had been involved with the Metaphysical Club in the early 1870s and was now professor of Christian morals at Harvard—began teaching Philosophy 11:

---

19. Samuel Augustus Barnett and Henrietta Rowland Barnett, *Practicable Socialism: Essays on Social Reform* (London: Longmans, Green, 1888), 23, 107, 158; referring to Sir John Lubbock, *Ants, Bees, and Wasps: A Record of Observations on the Habits of the Social Hymenoptera* (London: Kegan Paul, Trench, 1882). On Addams and Toynbee Hall, see Louise W. Knight, *Citizen: Jane Addams and the Struggle for Democracy* (Chicago: University of Chicago Press, 2005), 166–75.

CHAPTER SIX

The Ethics of Social Reform. A few months earlier, he had argued in "Social Reforms as Objects of University Study" that moral philosophy needed "a new method of approach. It ought to have its part in the inductive method which is controlling research elsewhere. The moral life has its living specimens, waiting for our scientific observation, subjects, as it were, for the field-work of the student."[20] Peabody then published a longer article in the *Andover Review*, advocating for a new connection "between economic science and ethics." He claimed that social changes "proceed through the mechanism of economic laws" but are motivated by Christian sentiment: "Sentiment without science is like steam unapplied to its proper work. It seethes and boils and threatens with its tumultuous vitality until it is compressed in its proper engine. Science without sentiment is mechanism without steam, ingenious and complete, but without the dynamic which gives it motion and power."[21] Peabody's "inductive ethics" required fieldwork, in which pressing social problems were addressed by those knowledgeable about economics but also motivated by Christian values. This new approach, with its emphasis on "the scientific habit of mind," helped shape the second cohort of pragmatists: Du Bois actually took Peabody's Ethics of Social Reform class in 1889–90, later recalling "Peabody's social reform with a religious tinge," and Dewey probably read Peabody's *Andover Review* article, as the journal was an important venue for his own work at the time.[22]

Social reform was also linked to new developments in social scientific methodology. In May 1890—probably in connection either with Peabody's class or with Political Economy 1, which included "lectures on social questions"—Du Bois checked out two books from the Harvard library by the statistician Carroll Davidson Wright, who was running the new United States Bureau of Labor. Wright's work illustrated the importance of statistics in social reform arguments. For example, in *Uniform*

---

20. Francis Greenwood Peabody, "Social Reforms as Subjects of University Study," *Independent* (New York), January 14, 1886, 5. For Peabody's class, see *The Harvard University Catalogue, 1886–87* (Cambridge, MA: Harvard University, 1886), 107. For his participation in the Metaphysical Club, see Peirce, *Writings*, 3:xxx–xxxi.

21. Francis Greenwood Peabody, "The Philosophy of the Social Questions," *Andover Review* 8 (1887): 563–64.

22. Peabody, "Philosophy of the Social Questions," 567; Du Bois Transcript; *The Harvard University Catalogue, 1889–90* (Cambridge, MA: Harvard University, 1889), 119; W. E. B. Du Bois, "A Negro Student at Harvard at the End of the 19th Century," *Massachusetts Review* 1 (1960): 354. Peabody also wrote Du Bois a letter of recommendation to the John F. Slater Fund: see Francis Greenwood Peabody to Rutherford Birchard Hayes, 20 April 1891, in Du Bois, *Correspondence*, 1:12. Dewey published many articles and book reviews in the *Andover Review* between 1884 and 1891, including one that appeared just a few months before Peabody's: John Dewey, "Ethics and Physical Science," *Andover Review* 7 (1887).

Hours of Labor—also an early influence on Hull House resident Florence Kelley—he demonstrated that the textile mills of Massachusetts, which had a mandated ten-hour workday, were just as productive as those of other states. Moreover, Wright reported that some mills in those other states, having experimentally shortened their workday to ten hours, had seen a production increase that was partly attributable to "the improved physical conditions which so great a reduction of the hours of labor afforded."[23] The next year, probably for Political Economy 2, Du Bois read George Gunton's *Wealth and Progress*, which argued that an eight-hour workday would promote social evolution by allowing workers to have "more frequent contact with an increasingly differentiated social environment."[24] Du Bois was apparently convinced that improvements in working conditions were beneficial to both owners and workers, and he drew on these ideas in a speech given at Harvard commencement in June 1891 (when he received his master of arts degree):

The capitalist of Massachusetts is . . . coming to see that it is not to his own economic advantage, in the long run, to neglect his employees. Thus through that process of education by legislative experiment, conference, and mutual concession, an industrial evolution has been going on here *within* the law, which, in the absence of the ballot from the hand of the laborer, must have brought revolution.

Southern capitalists, in contrast, by preventing black workers from "exercising their full political rights," were committing an "economic fallacy" and inadvertently inciting revolution.[25] As this passage indicates, Du Bois was already thinking about social progress in terms of both experiment and evolution at this early stage of his career.

---

23. Carroll Davidson Wright, *Uniform Hours of Labor* (Boston: Wright & Potter), 137, 142. Du Bois checked out this book, as well as Wright's *Comparative Wages, Prices, and Cost of Living*, on May 10, 1890: see Library Charging Lists, Students, 1889–1890, p. 658 [seq. 360]. In the charging list, "VI. 5438" and "VI. 5439" are the call numbers for Wright's two books (for book identification, see Box 2, UAIII 50.15.47.5, Harvard University Archives). For the description of Political Economy 1, taught by Silas Marcus MacVane and Frank William Taussig, see *Annual Reports of the President and Treasurer of Harvard College, 1889–90* (Cambridge, MA: Harvard University, 1891), 80. For Wright's early influence on Kelley, see Florence Kelley, "My Novitate," *Survey* 58 (1927): 32.

24. George Gunton, *Wealth and Progress: A Critical Examination of the Labor Problem* (New York: D. Appleton, 1887), 232. Du Bois checked this book out from the Harvard library on January 7, 1891: see Library Charging Lists, 1890–1891, p. 250 [seq. 151]. In the charging list, "VI. 4561" is the call number for Gunton's book: see "Accessions to the University Library," *Harvard University Bulletin* 5 (1888): 96. Political Economy 2 was the only economics class that Du Bois took in 1890–91: see W. E. B. Du Bois to Faculty of Arts and Sciences [Harvard], 23 March 1891, Du Bois Folder.

25. W. E. B. Du Bois, "Harvard and the South," pp. 6–8, Series 10, Du Bois Papers.

CHAPTER SIX

Many philosophers and social scientists at the time were, like Peabody, attempting to unite ethics and economics in the service of social reform. In 1892, Carroll Wright and Jane Addams—along with Frank William Taussig, one of Du Bois's professors at Harvard—lectured for the Department of Economics of the recently founded School of Applied Ethics, which was an organized series of summer courses in Plymouth, Massachusetts, targeting "clergymen, teachers, journalists, philanthropists, and others who are now seeking careful information upon the great themes of Ethical Sociology."[26] This particular department of the School of Applied Ethics was directed by Henry Carter Adams, a colleague of Dewey's at the University of Michigan. Adams argued that economics was essential to modern ethics: "Inasmuch as the most significant changes of the nineteenth century are industrial in character, the most pressing of the practical questions of right and wrong find their root in industrial relationships."[27] Addams, in her School of Applied Ethics lecture on "The Subjective Necessity for Social Settlements," described Hull House as both a site of ethical fieldwork and a response to a new urban environment: "The Settlement, then, is an experimental effort to aid in the solution of the social and industrial problems which are engendered by the modern conditions of life in a great city." She also stressed "its flexibility, its power of quick adaptation, [and] its readiness to change its methods as its environment may demand."[28] Thus not only did they view the settlement as a kind of experimental field site, Addams and the other residents also tried to cultivate an appropriately experimental attitude toward their dynamic urban environment.

Ethicists like Peabody and economists like Adams agreed that both of these disciplines were needed to address what were termed "social questions." This phrase, as evidenced by the course description for Du Bois's Political Economy 1 class at Harvard, referred to topics such as

---

26. "Program of School of Applied Ethics," *International Journal of Ethics* 1 (1891): 483. For the 1892 program, see S. Burns Weston, "School of Applied Ethics," *International Journal of Ethics* 2 (1892): 408. Taussig was one of the professors for Political Economy 1, which Du Bois took in 1889–90, and he also taught Political Economy 2, which Du Bois took in 1890–91: see Du Bois Transcript; W. E. B. Du Bois to Faculty of Arts and Sciences [Harvard], 23 March 1891, Du Bois Folder; *The Harvard University Catalogue, 1889–90* (Cambridge, MA: Harvard University, 1889), 119; *The Harvard University Catalogue, 1890–91* (Cambridge, MA: Harvard University, 1890), 77.

27. Henry Carter Adams, introduction to *Philanthropy and Social Progress: Seven Essays* (New York: Thomas Y. Crowell, 1893), vi; see also Henry Carter Adams, "An Interpretation of the Social Movements of Our Time," *International Journal of Ethics* 2 (1891): 48. On Dewey and Adams, see George Dykhuizen, "John Dewey and the University of Michigan," *Journal of the History of Ideas* 23 (1962): 523–24.

28. Jane Addams, "The Subjective Necessity for Social Settlements," in *Philanthropy and Social Progress*, 22–23; also in Jane Addams, "A New Impulse to an Old Gospel," *Forum* 14 (1892): 356.

"Coöperation, Profit-Sharing, Trades-Unions, [and] Socialism." At the time, many economists expressed sympathy for socialist ideas even as they carefully distanced themselves from the more radical views of Karl Marx and others. (This distancing mattered: even a partial defense of the Knights of Labor got Adams fired from an earlier job at Cornell University.)[29] Although Du Bois recalled late in life that the tendency of his Harvard economics classes "was toward English free trade," he probably also had opportunities to engage with more reform-oriented ideas—especially in classes taught by newer instructors, who were somewhat more sympathetic to the cause of labor, as one of his other recollections suggests.[30]

For instance, Du Bois's first economics class (which ran from February to May 1889) was Political Economy 9: Management and Ownership of Railways. It was taught by John Henry Gray, a recent graduate of Harvard College who, thirteen years later, still recalled teaching Du Bois about "the mysteries of railroading."[31] The class almost certainly included discussion of the pros and cons of government management, which was then a major topic in economics. Arthur Twining Hadley's *Railroad Transportation*, for example, contrasted the views of Adolph Wagner, who was cautiously in favor of state management of railways, and Jevons, who had argued against it in *Methods of Social Reform*. One of Du Bois's essays for the class included a great deal of discussion of why state management had been effective in Germany, citing Hadley's book among others.[32] Although Gray's own sympathies at the time are unknown, he would soon travel Germany (a few years before Du Bois), studying with various members of the Verein für Sozialpolitik and reporting on their activities for American social scientists: "The Association was founded . . . as a direct protest against Manchesterism [i.e.,

---

29. Henry Carter Adams, "The 'Labor Problem,'" *Scientific American Supplement* 22 (1886): 8862–63; S. Lawrence Bigelow, I. Leo Sharfman, and R. M. Wenley, "Henry Carter Adams," *Journal of Political Economy* 30 (1922): 205. For the description of Du Bois's class, see *Annual Reports of the President and Treasurer of Harvard College, 1889–90* (Cambridge, MA: Harvard University, 1891), 80.

30. W. E. B. Du Bois, *Dusk of Dawn: An Essay toward an Autobiography of a Race Concept* (New York: Harcourt, 1940), 40; W. E. B. Du Bois, "Apologia," in *The Suppression of the African Slave-Trade to the United States of America, 1638–1870* (New York: Social Science Press, 1954), 328.

31. John Henry Gray to W. E. B. Du Bois, 14 May 1902, Series 1A, Du Bois Papers. For Du Bois's attendance in Gray's class, see Du Bois Transcript.

32. Arthur Twining Hadley, *Railroad Transportation: Its History and Its Laws* (New York: G. P. Putnam's Sons, 1886), 253–55; W. E. B. Du Bois, "Origin and Methods of the German Railway System," in Series 10, Du Bois Papers. Du Bois's essay is discussed by Kenneth Barkin but incorrectly linked to a class with Taussig rather than Gray: see Kenneth D. Barkin, "W. E. B. Du Bois and the Kaiserreich," *Central European History* 31 (1998): 160. The archival record mistakenly links the essay to History 9 rather than to Political Economy 9. For a description of the latter, see *Annual Reports of the President and Treasurer of Harvard College, 1888–89* (Cambridge, MA: Harvard University, 1890), 72.

laissez-faire].... The younger economists of the historical school, who organized the new association, were the same who, because of their advocacy of greater activity on the part of the State, came to be known as 'Socialists of the Chair.'" Gray went on to argue in his doctoral thesis for the public regulation of gas lighting, endorsing the general perspective of Wagner and other *Kathedersozialisten* (literally, "lectern socialists"): "We live in a social, political, and industrial age that the eighteenth-century English philosophers of *laissez faire* would never have dreamed of . . . the enormous tasks of modern city life require a new treatment, and not merely the application of the simple philosophy of previous centuries."[33] Du Bois was thus exposed to debates over the merits of state intervention in his very first economics class at Harvard.

His graduate classes also featured these debates: the second half of Political Economy 2: History of Economic Theory, which Du Bois took in 1890–91, featured critical lectures on socialism by John Graham Brooks, and that same year he probably read a speech by the economist Francis Amasa Walker that cheered "the abandonment of *laissez-faire*, as a principle of universal application."[34] Then in 1891–92, Du Bois took Political Economy 3: The Principles of Sociology with Edward Cummings, who had recently written about trade unions and cooperative production in the *Quarterly Journal of Economics*, even "welcoming profit-sharing as the next great phase of industry." Cummings's background was both academic and practical: although recently returned from study at the University of Berlin (where he preceded Gray by one year), he had also been a resident of Toynbee Hall in 1888–89.[35] In the spring of 1892, while Du

---

33. John Henry Gray, "The German Economic Association," *Annals of the American Academy of Political and Social Science* 1 (1891): 515; John Henry Gray, *Die Stellung der privaten Beleuchtungsgesellschaften zu Stadt und Staat: Die Erfahrungen in Wien, Paris und Massachusetts* (Jena, Ger.: Gustav Fischer, 1893), 132.

34. Francis Amasa Walker, "Recent Progress of Political Economy in the United States," in *Publications of the American Economic Association*, ed. Richard T. Ely, vol. 4, no. 4, *Report of the Proceedings of the American Economic Association, Third Annual Meeting, Philadelphia, December 26–29, 1888* (Baltimore: Guggenheimer, Weil, 1889), 27. Du Bois checked this number of the *Publications* out of the Harvard library on April 13, 1891: see Library Charging Lists, 1890–1891, p. 250 [seq. 151]. In the charging list, "VI. 2654.4" is the call number of the volume, and "IV" indicates the number: see "Accessions to the University Library," *Harvard University Bulletin* 6 (1890): 32. For Political Economy 2, see W. E. B. Du Bois to Faculty of Arts and Sciences [Harvard], 23 March 1891, Du Bois Folder; *Annual Reports of the President and Treasurer of Harvard College, 1890–91* (Cambridge, MA: Harvard University, 1892), 58; Irwin Collier, "Harvard: History of Economic Theory; Final Exam Questions, Taussig, 1891–94," *Economics in the Rear-View Mirror* (blog), [October] 2018, http://www.irwincollier.com/harvard-history-of-economic-theory-final-exam-questions-taussig-1891-94/. For Brooks's criticisms of socialism, see John Graham Brooks, review of *Zur Social- und Gerwerbepolitik der Gegenwart*, by Gustav Schmoller, *Political Science Quarterly* 6 (1891).

35. Edward Cummings, "The English Trades-Unions," *Quarterly Journal of Economics* 3 (1889); Edward Cummings, "Co-operative Production in France and England," *Quarterly Journal of Economics* 4

Bois was enrolled in his class, Cummings advocated for a more scientific approach to social settlements and praised Charles Booth's book *Labour and Life of the People*, which had deployed maps and statistics in its sociological analysis of East London. Like Addams, Cummings described Toynbee Hall as a promising "experiment," but he also lamented its neglect of "systematic study." Booth's work had shown, said Cummings, "that it is in cities, in East Ends, North Ends, South Coves, that the sociological arena and laboratory are both to be found. It is a happy omen that scientific investigation and popular interest have both felt the need, the opportunity, and the duty at the same moment." Although Booth avoided explicit policy prescriptions, he did see his research as directly relevant to reform: "If the facts thus stated are of use in helping social reformers to find remedies for the evils which exist, or do anything to prevent the adoption of false remedies, my purpose is answered." Thus, Political Economy 3 probably exposed Du Bois not only to the idea of the city as laboratory, but also to what Cummings called "the best sociological methods," exemplified by Booth's research.[36]

Early in 1892, while still at Harvard, Du Bois read John Rae's *Contemporary Socialism*, which contained a whole chapter analyzing and criticizing lectern socialism. By the fall of that year, he was following in the footsteps of Ely, Adams, Gray, and Cummings, studying political economy at the University of Berlin with the lectern socialists themselves.[37] As the famous early 1870s speeches by Wagner and Schmoller on "the social question" had made clear, these economists were trying to find "a third way" between laissez-faire and revolutionary socialism. We are opposed to the "*radical* projects" of socialists, said Wagner, taking instead "the way of *reform*: i.e., appropriate *continued development* and, if necessary,

---

(1890): 386; "Prof. Cummings Called," *Cambridge Tribune*, August 25, 1900, 2. On Political Economy 3, see W. E. B. Du Bois to Faculty of Arts and Sciences [Harvard], n.d. [Spring 1892], Du Bois Folder; *The Harvard University Catalogue, 1891–92* (Cambridge, MA: Harvard University, 1891), 81.

36. Edward Cummings, "University Settlements," *Quarterly Journal of Economics* 6 (1892): 258, 266, 277; see also Edward Cummings, "Labour and Life of the People of London," *Annals of the American Academy of Political and Social Science* 2 (1892); Charles Booth, ed., *Labour and Life of the People*, vol. 1, *East London* (London: Williams & Norgate, 1889), 6.

37. John Rae, *Contemporary Socialism*, 2nd ed. (New York: Charles Scribner's Sons, 1891), chap. 6. Du Bois checked Rae's book out of the Harvard library on January 18, 1892: see Library Charging Lists, Students, 1891–1892, p. 389 [seq. 221]. In the charging list, "VI. 6372" is the call number for Rae's book: see "Accessions to the University Library," *Harvard University Bulletin* 6 (1892): 376. Ely and Adams both studied at Berlin in 1878–79, Cummings studied there in 1890–91, and Gray studied there in 1891–92: see Richard Theodore Ely, *Ground under Our Feet: An Autobiography* (New York: Macmillan, 1938), 51; S. Lawrence Bigelow, I. Leo Sharfman, and R. M. Wenley, "Henry Carter Adams," *Journal of Political Economy* 30 (1922): 204; *Harvard College, Class of 1883, Secretary's Report, No. III, July, 1890*, 29, https://hdl.handle.net/2027/hvd.32044107296931; *Harvard College, Class of 1887, Secretary's Report, No. 3, 1893* (Burlington, VT: Free Press Association, 1893), 48.

modification of existing conditions." Although we are "dissatisfied with existing social relations [and] filled with the necessity of reform," said Schmoller, "we protest against all socialist experiments."[38] Wagner and Schmoller were critical of how evolutionary ideas were employed by the laissez-faire school and the radical socialists, but they were still committed to the possibility of progress. One theme in Schmoller's work was that industrial progress had gone forward more quickly than progress in other areas of life, creating a kind of mismatch: "We have advanced more quickly in technology than in our ethical views and social institutions." Like Peabody, Schmoller insisted that "social, political, and economic progress rests not only on the increase of knowledge, but first and foremost on the accumulating victories of moral ideas." Schmoller also claimed that "the great epochs of economic progress are above all tied to the reform of social institutions," although this reform had to be tailored to particular national and legislative circumstances. As mentioned in chapter 3, Wagner's political economy textbook (used by Du Bois) described the fundamental view of "historical political economy" as "that of 'relativity,' the avoidance of the 'absolutism of solutions' in practical, political-economic questions." But progress had to be guided by science: as Schmoller noted, "The higher and ideal purpose of all strict scholarly work . . . is to search for truth and knowledge"—not just for its own sake, but also "to shed light on practical life, illuminating, smoothing, and showing the way."[39]

---

38. Adolph Wagner, *Rede über die sociale Frage: Gehalten auf der freien kirchlichen Versammlung evangelischer Männer in der K. Garnisonkirche zu Berlin am 12. October 1871* (Berlin: Wiegandt & Grieben, 1872), 17; Gustav Schmoller, "Eröffnungsrede," in *Verhandlungen der Eisenacher Versammlung zur Besprechung der socialen Frage* (Leipzig, Ger.: Duncker & Humblot, 1873), 5.

39. Gustav Schmoller, *Zur Social- und Gewerbepolitik der Gegenwart* (Leipzig, Ger.: Duncker & Humblot, 1890), 34, 184, 191, 242; Adolph Wagner, *Grundlegung der politischen Oekonomie: Erster Theil, Grundlegung der Volkswirthschaft (Erster Halbband)*, 3rd ed. (Leipzig, Ger.: C. F. Winter, 1892), 12. For criticism of the invocation of "the Darwinian theory of the struggle for existence" by social scientists, see Schmoller, *Zur Social- und Gewerbepolitik*, 204. For criticism of the use of evolutionary ideas by "radical socialists," see Wagner, *Grundlegung der politischen Oekonomie*, 10. Du Bois took Wagner's class Allgemeine oder theoretische Nationalökonomie (General or Theoretical Political Economy) in 1892–93, just after the publication of the first half-volume of the *Grundlegung*, and part of that half-volume (§134, on value) was listed in the bibliography that opened Du Bois's 1893 economics notebook: see W. E. B. Du Bois to the John F. Slater Fund, 10 March 1893, in Du Bois, *Correspondence*, 1:24; Verzeichniss der Vorlesungen, welche auf der Friedrich-Wilhelms-Universität zu Berlin im Winter-Semester vom 16. October 1892 bis 15. März 1893 gehalten werden, 22, https://www.digi-hub.de/viewer/toc/DE-11-001799607/1/; W. E. B. Du Bois, "Lecture Notebook," p. 4 [6], Series 10, Du Bois Papers. Du Bois also noted at the time that Wagner was "publishing a new edition of his valuable *Lehrbuch*," of which the *Grundlegung* was the first part: see W. E. B. Du Bois, *The Autobiography of W. E. B. Du Bois: A Soliloquy on Viewing My Life from the Last Decade of Its First Century* (New York: International, 1968), 166.

In a draft essay, probably written in the mid-1890s, Du Bois summarized the position of the lectern socialists, highlighting their "relativity" view:

> The work of the Katheder Socialists, and their followers, has been to formulate an economic theory which would suit the industrial condition of New Germany. That English economic philosophy, which so admirably generalized the economic situation in England during the first half of the century, was, naturally, inadequate, misleading and wrong, when applied without modification to the new and unusual conditions in Germany.

He also explained the position of Schmoller's "younger historical school" with respect to social reform:

> They confine themselves in practice to careful statistical investigation of the history and development of present economic conditions, and social phenomena. From this gradually increasing basis of scientific facts, they attempt to recommend remedies for certain more obvious social ills, but go no further, in such recommendations and generalizations, than a careful interpretation of the facts at hand.

Finally, Du Bois praised this school's support of "socialistic" legislation in Germany—"experiments" which, though not uniformly successful, illustrated the commitment of these economists to "greater social justice."[40] Du Bois, inspired by the Berlin economists, was already connecting legislative experimentation, economic investigation, social reform, and social ethics in the mid-1890s.

But science and social reform would have to wait: although he "begged to be allowed to lecture on sociology," Du Bois ended up teaching only Latin, Greek, English, and German at Wilberforce University in Ohio, where he worked for two years, from 1894 to 1896.[41] Chicago, meanwhile, had become a field site for ethical sociology. Florence Kelley, a resident of Hull House, was leading a team surveying the households of the Nineteenth Ward in May 1893 as part of a Department of Labor project run by Carroll Wright. This survey would eventually result in

---

40. W. E. B. Du Bois, "The Socialism of German Socialists," pp. 21–25, in Series 10, Du Bois Papers; published in W. E. B. Du Bois, "The Socialism of German Socialists," *Central European History* 31 (1998): 194–95. For the dating of this manuscript, see Kenneth D. Barkin, "W. E. B. Du Bois and the Kaiserreich," *Central European History* 31 (1998): 164–65.

41. W. E. B. Du Bois, *Darkwater: Voices from within the Veil* (New York: Harcourt, Brace & Howe, 1920), 18; see also W. E. B. Du Bois, *Dusk of Dawn: An Essay toward an Autobiography of a Race Concept* (New York: Harcourt, 1940), 58.

CHAPTER SIX

the color-coded maps, inspired by those of Charles Booth, that appeared in *Hull-House Maps and Papers*.[42] As Mary Jo Deegan has shown, such research led many observers to view Hull House and related institutions as experimental sociological laboratories. Just a few weeks after Kelley's survey, Robert Archey Woods, who had cofounded a Boston settlement house in 1892, gave a speech in Chicago titled "University Settlements as Laboratories in Social Science." The approach described by Woods was both participatory and experimental:

> A peculiarly important line of social investigation and experiment, which is being undertaken by university settlements, is the discovery of such forms of original organization and co-operation among the people themselves as exist in the neighboring district; and what is equally important, experiment in the way of participating as a local neighbor on the same plane as the rest in such efforts.

Woods specifically cited Hull House as one of the settlements employing this approach, which he thought would "be of the greatest scientific value."[43] At the University of Chicago, the Department of Sociology and Anthropology under Albion Woodbury Small (yet another Berlin-educated American social scientist) had already embraced this laboratory perspective: in February 1893, Charles Worthen Spencer—then a graduate student at the university—gave a presentation titled "The City of Chicago as a Sociological Laboratory" in which he declared that "more social experiments are now in progress in this city than there are sociologists to watch them."[44] Then in May 1894, Daniel Fulcomer—a university extension lecturer at Chicago—invited Julia Lathrop, another of Addams's coresidents, to speak about "Hull House as a Laboratory of Sociological

---

42. Agnes Sinclair Holbrook, "Map Notes and Comments," in *Hull-House Maps and Papers* (New York: Thomas Y. Crowell, 1895), 6–7, 11; Florence Kelley, "I Go to Work," *Survey* 58 (1927): 272; Kathryn Kish Sklar, *Florence Kelley and the Nation's Work: The Rise of Women's Political Culture, 1830–1900* (New Haven, CT: Yale University Press, 1995), 228–29; Kathryn Kish Sklar, "Hull-House Maps and Papers: Social Science as Women's Work in the 1890s," in *The Social Survey in Historical Perspective, 1880–1940*, ed. Martin Bulmer, Kevin Bales, and Kathryn Kish Sklar (Cambridge: Cambridge University Press, 1991), 121–29.

43. Mary Jo Deegan, *Jane Addams and the Men of the Chicago School, 1892–1918* (New Brunswick, NJ: Transaction, 1988), chap. 2; Robert Archey Woods, "University Settlements as Laboratories in Social Science," in *Sociology in Institutions of Learning: Being a Report of the Seventh Section of the International Congress of Charities, Correction and Philanthropy, Chicago, June, 1893*, ed. Amos G. Warner (Baltimore: Johns Hopkins Press, 1894), 36–37.

44. Charles Worthen Spencer, "The City of Chicago as a Sociological Laboratory," *Current Topics* 1 (1893): 273; based on a presentation reported in *Annual Register, July 1, 1892—July 1, 1893, with Announcements for 1893-4* (Chicago: University Press of Chicago, 1893), 230. There is a copy of Spencer's article in Folder 67, Box 2, Charles Worthen Spencer Papers [A1046], Special Collection and University Archives, Colgate University.

Investigation." Although she was not sure about the designation, Lathrop suggested that Kelley's survey maps of the Nineteenth Ward were "a piece of sociological investigation" and that Hull House had "gathered a considerable fund of sociological material." Addams noted in the preface to *Hull-House Maps and Papers* that the residents' energies had "been chiefly directed, not towards sociological investigation, but to constructive work." But she also argued later that same year that their detailed, long-term observations of particular groups of people were contributions to "the study of society which we call sociology." As Matthias Frisch and Joshua Skorburg have recently argued, Hull House, even if not best described as a laboratory, was a site of collaborative experimentation.[45]

Shortly after the publication of *Hull-House Maps and Papers*, Dewey—a frequent visitor to Hull House at the time—began to develop his own field site: a primary school at the University of Chicago that opened early in 1896 under the direction of Clara Isabel Mitchell. In a December 1895 letter to Mitchell, Dewey referred to the school as a "Laboratory of Education."[46] One of the first public descriptions of this laboratory was "A Pedagogical Experiment," an article that Dewey contributed to *Kindergarten Magazine* in June 1896. The goal of the school, said Dewey, was "to keep the theoretical work in touch with the demands of practice"— to be "an experimental station for the testing and development of methods which, when elaborated, may be safely and strongly recommended to other schools." Famously, Dewey insisted "that the child learns most easily . . . when his problems grow out of his practical work, as either involved in it or enrichments of it," and the students thus engaged in cooking and carpentry, among other activities.[47] Mead and Dewey both argued in 1896 that learning by doing was effective because it focused on the children's own interests as natural experimenters: "The simple use of carpenters' tools" and other similar activities, said Mead, naturally

---

45. Julia Lathrop, "Hull House as a Laboratory of Sociological Investigation," in *Proceedings of the National Conference of Charities and Correction, at the First Annual Session Held in Nashville, Tenn., May 23–29, 1894*, ed. Isabel C. Barrows (Boston: Geo. H. Ellis, 1894), 313, 317–18; Jane Addams, "Prefatory Note," in *Hull-House Maps and Papers* (New York: Thomas Y. Crowell, 1895), vii–viii; Jane Addams, "Claim on the College Woman," *Rockford Collegian* 23, no. 6 (1895): 60; Dorothy Ross, "Gendered Social Knowledge: Domestic Dicourse, Jane Addams, and the Possibilities of Social Knowledge," in *Gender and American Social Science: The Formative Years*, ed. Helene Silverberg (Princeton, NJ: Princeton University Press, 1998), 243; Matthias Gross, "Collaborative Experiments: Jane Addams, Hull House and Experimental Social Work," *Social Science Information* 48 (2009); Joshua August Skorburg, "Jane Addams as Experimental Philosopher," *British Journal for the History of Philosophy* 26 (2018). Addams later rejected the laboratory designation in "A Function of the Social Settlement," *Annals of the American Academy of Political and Social Science* 13 (1899): 35.

46. John Dewey to Clara Isabel Mitchell, 24 December 1895, in Dewey, *Correspondence*, no. 00275.

47. John Dewey, "A Pedagogical Experiment," *Kindergarten Magazine* 8 (1896): 739–40.

CHAPTER SIX

stimulates interest because of "their values for life." Dewey told a similar story to his Philosophy of Education class that year:

> A child is an experimenter from the outset because it is only through the movements of his bodily organs, especially of his hands, that he comes to know anything. No child is satisfied with any supposed item of knowledge until he has got the thing to which it refers into his hands and tries to do something with it. Objects to the child are what he can do with them.

At Dewey's laboratory school, experimentation was an activity common to teachers and students, a collaborative approach that mirrored that of Hull House.[48]

Du Bois finally got a chance to do some fieldwork of his own in 1896–97, having been hired by the University of Pennsylvania to study "the social condition of the Colored People of the Seventh Ward of Philadelphia." This research was designed, as the provost of the university wrote in a letter of introduction, "to ascertain every fact which will throw light upon this social problem; and then having this information and these accurate statistics before us, to see to what extent and in what way, proper remedies may be applied."[49] Du Bois's research, like that of Kelley, was connected to a settlement house—in his case, the Philadelphia College Settlement. In an 1897 issue of their paper, the *College Settlement News*, Du Bois published a brief "Program of Social Reform." Echoing his earlier description of Schmoller's approach as well as the language of the provost's letter, Du Bois said that reform aimed to provide scientific "remedies" for social problems, declaring that "ignorance of the cause is the greatest cause" of such problems. As Du Bois noted several years later, social settlements could help provide such remedies.[50] Mary Jo Deegan and others have shown that the social survey approach of Du Bois's Philadelphia research was modeled on that of *Labour and Life of the People* and *Hull-House Maps and Papers*. Isabel Eaton, coauthor of the

---

48. George Herbert Mead, "The Relation of Play to Education," *University Record* (Chicago) 1 (1896): 145; "Philosophy of Education (1896)," in Dewey, *Class Lectures*, 2:76. On collaborative experimentation at Dewey's school, see Charlene Haddock Seigfried, *Pragmatism and Feminism: Reweaving the Social Fabric* (Chicago: University of Chicago Press, 1996), chap. 5; Anne Durst, *Women Educators in the Progressive Era: The Women behind Dewey's Laboratory School* (New York: Palgrave Macmillan, 2010).

49. Charles Custis Harrison to Whom It May Concern, 15 August 1896, Series 1A, Du Bois Papers.

50. W. E. B. Du Bois, "A Program of Social Reform," *College Settlement News* 3, no. 3 (1897); reprinted in W. E. B. Du Bois, "A Program of Social Reform," in *Du Bois on Reform: Periodical-Based Leadership for African Americans*, ed. Brian Johnson (Lanham, MD: AltaMira, 2005), 19; W. E. B. Du Bois, "A Proposed Social Settlement," pp. 3–4, Series 1A, Du Bois Papers.

latter book and a resident of Hull House, even worked in parallel with Du Bois on the 1896–97 project, ultimately contributing a "Special Report on Negro Domestic Service" to Du Bois's *The Philadelphia Negro* that made up about a sixth of the book.[51]

After a brief stint in Farmville, Virginia—preparing a sociological analysis of the black population there for Wright at the Department of Labor—and a few more months in Philadelphia, Du Bois was hired as professor of economics and history at Atlanta University, where he started in 1897–98.[52] As a series of scholars have demonstrated, Du Bois was from 1897 to 1910 the driving force behind the first American school of sociology at Atlanta University, one that preceded but was then overshadowed by the Chicago school.[53] During his first semester at Atlanta, Du Bois addressed the American Academy of Political and Social Science on "The Study of the Negro Problems." He opened his address by arguing that the United States itself was a kind of sociological laboratory:

> The sociologists of few nations have so good an opportunity for observing the growth and evolution of society as those of the United States. The rapid rise of a young country, the vast social changes, the wonderful economic development, the bold political experiments, and the contact of varying moral standards—all these make for American students crucial tests of social action, microcosmic reproductions of long centuries of world history, and rapid—even violent—repetitions of great social problems.

Although targeting a specific social problem, Du Bois kept some strategic distance between social research and social reform, arguing—as Liam

---

51. Mary Jo Deegan, "W. E. B. Du Bois and the Women of Hull-House, 1895–1899," *American Sociologist* 19 (1988): 303–4; Martin Bulmer, "W. E. B. Du Bois as a Social Investigator: *The Philadelphia Negro*, 1899," in *The Social Survey in Historical Perspective, 1880–1940*, ed. Martin Bulmer, Kevin Bales, and Kathryn Kish Sklar (Cambridge: Cambridge University Press, 1991), 175; Samuel McCune Lindsay, introduction to *The Philadelphia Negro: A Social Study*, by W. E. B. Du Bois (Philadelphia: University of Pennsylvania, 1899), x–xi; W. E. B. Du Bois, *The Philadelphia Negro: A Social Study* (Philadelphia: University of Pennsylvania, 1899), 427–509.

52. W. E. B. Du Bois to Carroll Davidson Wright, 5 May 1897, in Du Bois, *Correspondence*, 1:41; W. E. B. Du Bois, "The Negroes of Farmville, Virginia: A Social Study," *Bulletin of the Department of Labor*, no. 14 (1898); *Catalogue of the Officers and Students of Atlanta University, . . . with a Statement of the Courses of Study, Expenses, Etc., 1897–98* (Atlanta: Atlanta University Press, 1898), 4.

53. Shaun L. Gabbidon, "W. E. B. Du Bois and the 'Atlanta School' of Social Scientific Research, 1897–1913," *Journal of Criminal Justice Education* 10 (1999); Earl Wright II, "The Atlanta Sociological Laboratory 1896–1924: A Historical Account of the First American School of Sociology," *Western Journal of Black Studies* 26 (2002); Earl Wright II, "Using the Master's Tools: The Atlanta Sociological Laboratory and American Sociology, 1896–1924," *Sociological Spectrum* 22 (2002); Aldon D. Morris, *The Scholar Denied: W. E. B. Du Bois and the Birth of Modern Sociology* (Berkeley: University of California Press, 2015); Earl Wright II, *The First American School of Sociology: W. E. B. Du Bois and the Atlanta Sociological Laboratory* (New York: Routledge, 2016).

Kofi Bright has discussed—that the latter should only be "the mediate object of a search for truth." He ended his address by declaring that we should not "sneer at the heroism of the laboratory" but instead hold to "the pure ideals of science, and continue to insist that if we would solve a problem we must study it." Thus by 1897, Du Bois had found his "sociological laboratory"—the official title of his sociology seminar room and library at Atlanta University beginning in 1902, but also Atlanta itself, where "the race-hatred of the whites" was "naked and unashamed."[54]

Addams, Dewey, Mead, and Du Bois all developed sites of ethical fieldwork at the end of the nineteenth century: Addams and the other residents experimented at Hull House in Chicago, developing projects in which Dewey and Mead also participated; Dewey and Mead, along with a team of teachers, tested educational and psychological theories in the new laboratory school at the University of Chicago; and Du Bois, partly inspired by Hull House but also drawing on his scientific training in Germany, constructed research programs in Philadelphia and Atlanta that ultimately produced a new school of sociology. All of these field sites connected social science with social reform, and all of the figures involved were aware of one another's work—Du Bois even gave a speech at Hull House in 1907, with Addams returning the favor the next year.[55] These links were not only personal but also theoretical: as I will show in the next section, Du Bois, Dewey, Mead, and Addams all framed their experimental activity in terms of Spencer's organism-environment dichotomy, discussed in chapters 2 and 4.

Organism and Environment in Social Reform

Robert Woods's description of settlements as experimental laboratories for social science was presented in June 1893 at the International Con-

---

54. "November Meeting," *Bulletin of the Academy*, n.s., no. 1 (1897): 1; W. E. B. Du Bois, "The Study of the Negro Problems," *Annals of the American Academy of Political and Social Science* 11 (1898): 1–2, 16, 23; Liam Kofi Bright, "Du Bois' Democratic Defence of the Value Free Ideal," *Synthese* 195 (2018); W. E. B. Du Bois, *Darkwater: Voices from within the Veil* (New York: Harcourt, Brace and Howe, 1920), 21. On Du Bois's "sociological laboratory," see Horace Bumstead to W. E. B. Du Bois, 21 June 1902, Series 1A, Du Bois Papers; W. E. B. Du Bois, ed., *The Negro Artisan: A Social Study* (Atlanta: Atlanta University Press, 1902), 2; *Catalogue of the Officers and Students of Atlanta University, . . . with a Statement of the Courses of Study, Expenses, Etc., 1902–03* (Atlanta: Atlanta University Press, 1903), 15; W. E. B. Du Bois, "The Laboratory in Sociology at Atlanta University," *Annals of the American Academy of Political and Social Science* 21 (1903).

55. Jane Addams to W. E. B. Du Bois, 26 January 1907, Jane Addams Project, Digital Edition, https://digital.janeaddams.ramapo.edu/items/show/1850; W. E. B. Du Bois to Jane Addams, 19 May 1908, Series 1A, Du Bois Papers.

gress of Charities, Correction, and Philanthropy, with Hull House resident Julia Lathrop in attendance. In the oration that closed the opening session of this same congress, held in Chicago in connection with the World's Columbian Exposition, Francis Greenwood Peabody declared that "the new forms of industrial life, the vastly greater social complexity, the increasing wealth, the manifold inventions and the democratic spirit of the last fifty years, have made for us a new social environment, with new problems calling for new rules of conduct." A new environment, said Peabody, demands a new morality. Recalling his earlier essay, quoted in the previous section, he went on to suggest that just as "the scientific mind . . . takes possession of the electric current and harnesses it into the machinery of modern life," it could also intelligently guide the "force of Christian feeling," directing it "along definite economic lines" to address social problems.[56] Peabody's view recalled that of Du Bois's mentor Alexander Crummell, who claimed—as discussed in chapter 4—that the black community was in "need of new ideas and new aims for a new era," with "changed circumstances" demanding "new adjustments in life." Apart from its explicitly Christian perspective, it also echoed Du Bois's teacher Schmoller, who argued—as noted in the previous section—that "we have advanced more quickly in technology than in our ethical views and social institutions."[57] I will demonstrate in this section that Du Bois, Dewey, Mead, and Addams framed such differential progress as a kind of organism-environment mismatch, with a new social environment necessitating new institutions, new legislation, and a new moral framework.

The organism-environment perspective was at the heart of Du Bois's 1897 address "The Study of the Negro Problems." His very definition of a *social problem* was "the failure of an organized social group to realize its group ideals, through the inability to adapt a certain desired line of action to given conditions of life." That is, a social problem is an adaptive mismatch, which could presumably be corrected by altering either the social group or the relevant conditions. Social growth, continued Du Bois, leads inevitably to social problems, which "denote that laborious

---

56. Francis Greenwood Peabody, "The Problem of Charity," in *The Organization of Charities: Being a Report of the Sixth Section of the International Congress of Charities, Corrections, and Philanthropy, Chicago, June, 1893*, ed. Daniel C. Gilman (Baltimore: Johns Hopkins Press, 1894), xxii. For Lathrop's attendance and the timing of Peabody's oration, see *General Exercises of the International Congress of Charities, Correction and Philanthropy, Chicago, June, 1893* (Baltimore: Johns Hopkins Press, 1894), 15, 35.

57. Alexander Crummell, *Africa and America: Addresses and Discourses* (Springfield, MA: Willey, 1891), 20; Gustav Schmoller, *Zur Social- und Gewerbepolitik der Gegenwart* (Leipzig, Ger.: Duncker & Humblot, 1890), 34.

and often baffling adjustment of action and condition which is the essence of progress." According to Du Bois, "the Negro . . . is unusually handicapped" in the United States, despite "the boldness of [this country's] experiments in organized social life," because his problems "are complicated by a peculiar environment"—namely, "the widespread conviction among Americans that no persons of Negro descent should become constituent members of the social body" (less delicately: racist attitudes). Du Bois accused existing studies of the Negro of ignoring his evolution—"his whole reaction against his environment"—and acting as if he "arose from the dead in 1863," resurrected by the Emancipation Proclamation. Finally, in describing his "program of future study," Du Bois divided "the study of the Negro" into "a study of the group" and "a study of the environment," arguing that these categories should be distinguished by the investigator even though they were difficult to disentangle in practice. This division was probably inspired by Richmond Mayo-Smith, who—in a book later assigned by Du Bois at Atlanta University—had divided the material of sociology into (a) demographic, social, and ethnographic classes, (b) the physical environment, (c) the social environment, and (d) forces of social change.[58]

Du Bois realized this program in his 1899 book *The Philadelphia Negro*, based on his 1896–97 research. A complete study of "the Negro problems of Philadelphia," he wrote in its opening pages, "must not confine itself to the group, but must specially notice the environment; the physical environment of city, sections and houses; the far mightier social environment—the surrounding world of custom, wish, whim, and thought which envelops the group and powerfully influences its social development." Echoing the views of Schmoller and other economists, Du Bois insisted that science must precede ethical reform: "A slum is not a simple fact, it is a symptom," and we need to uncover its "removable causes." Later in the book, Du Bois portrayed crime as the result of organism-environment mismatch: "If men are suddenly transported from one environment to another, the result is lack of harmony with the new conditions; lack of harmony with the new physical surroundings leading to disease and death or modification of physique; lack of harmony with social surroundings leading to crime." In other words, crime

---

58. W. E. B. Du Bois, "The Study of the Negro Problems," *Annals of the American Academy of Political and Social Science* 11 (1898): 2, 6, 8, 14, 17–18, 20; Richmond Mayo-Smith, *Statistics and Sociology* (New York: Macmillan, 1896), 6–7; *Catalogue of the Officers and Students of Atlanta University, . . . with a Statement of the Courses of Study, Expenses, Etc., 1898–99* (Atlanta: Atlanta University Press, 1899), 13. Mayo-Smith's book was also cited, along with *Hull-House Maps and Papers*, in W. E. B. Du Bois, *The Philadelphia Negro: A Social Study* (Philadelphia: University of Pennsylvania, 1899), 420.

is a social disease that can be cured by attending to its cause, a lack of harmony between a social group and its environment. "A large number of young Negroes are in such [an] environment," said Du Bois, "that they find it easier to be rogues than honest men." Du Bois also provided a more detailed account of the "peculiar environment," partially created by "color prejudice," to which he had alluded in 1897:

> The influence of homes badly situated and badly managed, with parents untrained for their responsibilities; the influence of social surroundings which by poor laws and inefficient administration leave the bad to be made worse; the influence of economic exclusion which admits Negroes only to those parts of the economic world where it is hardest to retain ambition and self-respect; and finally that indefinable, but real and mighty moral influence that causes men to have a real sense of manhood or leads them to lose aspiration and self-respect.[59]

For Du Bois, the social problems of the Philadelphia Negro resulted from a mismatch between the social environment and the group, and the primary means of correcting this mismatch was to change the environment—although such change could itself be the result of newly evolved group institutions, as I will discuss later.

Dewey, at around the same time, was deploying a modified version of Spencer's organism-environment framework in his sequence of ethics classes at the University of Chicago, even as he rejected the English philosopher's idea of a "completed evolution"—after which, said Dewey, a human being would be beyond all feeling, a machine with "no more adjustments or readjustments to be made."[60] As we saw in chapter 5, Dewey thought that friction and tension were essential to progress and would continue indefinitely. Referring to Baldwin's notion of accommodation as the breaking up of habits (which Peirce had also discussed in his *Monist* series), Dewey glossed habits as "those factors in function which . . . can be depended upon to assist us without further attention . . . in reaching an end." When a problem is encountered—"as function breaks up, or as friction arises in various parts of it"—we start thinking of the environment as "that to which the organism must adapt itself," remembering that this adjustment often also involves, in the human

---

59. Du Bois, *Philadelphia Negro*, 5–6, 235, 254–55, 284–86.
60. "Psychological Ethics (1898)," in Dewey, *Class Lectures*, 1:1216–17. Dewey's ethics sequence in 1897–98 consisted of Philosophy 34–36: Logic of Ethics (Autumn), Psychological Ethics (Winter), and Political Ethics (Spring): see *Annual Register, July, 1896—July, 1897, with Announcements for 1897–8* (Chicago: University of Chicago Press, 1897), 169; *President's Report, July, 1897–July, 1898, with Summaries for 1891–7* (Chicago: University of Chicago Press, 1899), 47.

CHAPTER SIX

case, "the adaptation of the environment to the organism." Dewey described the variation involved in this evolutionary process as "tentative, in so far as it is experimental," but also as "a mediation of function already in existence, . . . a reflection of the activities, of the life habits previously exercised." This variation, he said, should be "interpreted as the accompaniment of tension in the operations of the organism, that tension itself being due to a disturbance of functional unity."[61] To sum up: according to Dewey, progressive change occurs because tension breaks up organism-environment unity, leading to experimental variation and readaptation.

But what does this have to do with ethics and reform? According to Dewey, as for Du Bois, questions of social ethics arise because of a mismatch between institutions and the social environment:

We may say that politics deals with an ethical content in the sense that the values of experience as realized through business life, through family life, through school life, are assumed as there. But now some hitch arises of some sort, some friction, and the question then comes up,—is this value what it purports to be, or is this institution giving us, realizing for us, the values with reference to which it was instituted and with reference to which it ought to function?

Government and legislation, said Dewey, are designed "to locate these points of tension and lay down a general formulation for dealing with them." Just as the German historical school rejected the "absolutism of solutions" in economics, Dewey argued for "the relativity of legislation": government must "adapt its formulation to the needs of the situation, those needs being interpreted along the line of the adjustment of forces which is most problematic at that particular period," and should not act "beyond certain limits, . . . which are determined for the most part experimentally." Thus for Dewey, "the organization of [a] particular type of industrial work and the organization of the structure of a particular type of fishes" must both be placed within the organism-environment framework: in ethics and social reform, just as in biology, what matters—as Dewey put it in his 1898 essay "Evolution and Ethics"—is "the tension between an organ adjusted to a past state and the functioning required by present conditions. And this tension demands reconstruction."[62]

---

61. "Political Ethics (1898)," in Dewey, *Class Lectures*, 1:1653–55, 1658–59. For Baldwin and Peirce on accommodation and habit, see Trevor Pearce, "'Protoplasm Feels': The Role of Physiology in Peirce's Evolutionary Metaphysics," *HOPOS* 8 (2018): 49.

62. "Political Ethics (1898)," in Dewey, *Class Lectures*, 1:1766, 1769, 1799–1800; Dewey, "Evolution and Ethics," *Monist* 8 (1898): 333.

economic tensions—required the "active co-operation" of social classes with diverse habits and points of view:

> What we term social ethics involves so much readjustment of given conditions that the result simply cannot be reached where the stress of that interest falls somewhat exclusively on one side. It is a general reconstruction that we are after; it is a readjustment of social conditions as such, and since that is the case it is perfectly futile for one class to suppose that it can bring about the desired reconstruction.[69]

Addams and Dewey argued that a new social environment—in this case, the modern industrial city with its social, economic, and ethnic diversity—demanded a shift toward a more democratic and social ethics.

Du Bois, Dewey, Mead, and Addams all used the organism-environment dichotomy to frame their analysis of "social questions." They suggested that modern social tensions should be interpreted as a mismatch between present habits, institutions, and codes, on the one hand, and a changed social environment, on the other. This mismatch prompted what they variously termed "readjustment," "readaptation," or "reconstruction"— activities that would in turn, following the organism-environment dialectic of chapter 4, remake the social environment. In the next section, we will see how, according to the second cohort of pragmatists, this reconstructive process—at least when guided by science—could result in social evolutionary progress.

Social Science and Social Evolution

As Barrington Edwards has shown, Du Bois vehemently rejected what was later called *social Darwinism*, and in particular the following view commonly held by whites: "[Negroes] stand on a lower plane of humanity than we, and never have in the past evolved a civilization of their own"; thus they "must in accordance with the universal law of the survival of the fittest yield before the all-conquering Anglo-Saxon."[70] But as

---

69. "Sociology of Ethics (1902–1903)," in Dewey, *Class Lectures*, 1:2378–79, 2439.
70. W. E. B. Du Bois, "The Afro-American [ca. 1894–95]," pp. 10–11, Series 3, Du Bois Papers; published in W. E. B. Du Bois, "The Afro-American," *Journal of Transnational American Studies* 2, no. 1 (2010): 6; see also W. E. B. Du Bois, *Dusk of Dawn: An Essay toward an Autobiography of a Race Concept* (New York: Harcourt, 1940), 98; Barrington S. Edwards, "W. E. B. Du Bois, Empirical Social Research and the Challenge to Race, 1868–1910" (PhD diss., Harvard University, 2001), ProQuest (ID 275863250); Barrington S. Edwards, "W. E. B. Du Bois between Worlds: Berlin, Empirical Social Research, and the Race Question," *Du Bois Review* 3 (2006): 397–98; see also Aldon D. Morris, *The*

CHAPTER SIX

we saw in chapters 3 and 4, he was not opposed to evolutionary ideas more generally and was in fact committed to the organism-environment framework and to evolutionary social progress.

In his sociological research, Du Bois emphasized that the "peculiar environment" of African Americans, and in particular their exclusion from the ordinary institutions of social and political life, had led to the evolution of specialized institutions, most notably "The Negro Church" and "Negro Secret Societies" (black lodges of Odd Fellows, Masons, etc.). As he wrote in the "General Summary" of the Third Conference for the Study of Negro Problems in 1898 at Atlanta University:

> No more interesting example of the growth of organizations within a group could be adduced. Here in a half-century, or at most a century, we have epitomized that intricate specialization of the different human activities, and that adaptation of the thoughts and actions of men to the thoughts and actions about them, which we call advance in civilization. The process here has been hastened, the environment has had unusual features, the action of the group unusual hindrances; and yet we catch here a faint idea of what human progress really means, and how infinitely complicated its methods are.

Although he later criticized the "biological analogy" of "Spencerian Sociologists," here he argued that these new institutions—"slowly evolving organs," as he called them—were readaptations of "African clan life" in a new social environment: "Men seldom invent new ways of social advance, they rather change and adapt old ways to new conditions." Institutions, said Du Bois, acquire new functions in the course of their evolution: hence, after emancipation in 1863,

> the minister added political and economic functions to his religious duties. Next the church itself began to differentiate organizations for different functions; [and] economic and cooperative action became the business of the beneficial society and secret society.... How curious a chapter is this of the adaptation of social methods and ways of thinking to the environment of real life!

Given that "race prejudice" had thrown upon the Negro group "the responsibility of evolving its own methods and organs of civilization," Du Bois argued that efforts toward social reform should be democratic, in the sense endorsed by Dewey and Addams. Benevolent aid must be based on the work of "trained thinkers and observers," but will be most

---

Scholar Denied: W. E. B. Du Bois and the Birth of Modern Sociology (Berkeley: University of California Press, 2015), 129.

effective, said Du Bois, when directed according to "the initiative of the Negroes themselves," with existing social evolutionary trends indicating promising avenues of reform: "If [Negroes] are found striving in new directions, as today toward asylums, homes and hospitals, this is a pretty fair indication of a social want, and judicious aid to such enterprises can be applied usually with gratifying results." In Du Bois's view, evolutionary progress for African Americans had to involve both expert guidance and self-determination, for as he had written a year earlier, "the American Negro . . . simply wishes to make it possible for a man to be both a Negro and an American . . . without losing the opportunity of self-development."[71]

Du Bois returned to the theme of self-development in a 1901 essay, "The Evolution of Negro Leadership," a review of Booker Taliaferro Washington's *Up from Slavery*. When a people face not only the natural environment of "sticks and stones and beasts" but also "an environment of men and ideas," wrote Du Bois, "the attitude of the imprisoned group may take three main forms: a feeling of revolt and revenge; an attempt to adjust all thought and action to the will of the greater group; or, finally, a determined attempt at self-development, self-realization, in spite of environing discouragements and prejudice." Du Bois criticized Washington for taking the second approach—"the old attitude of adjustment to environment"—and instead allied himself with the third group, who sought "that self-development and self-realization in all lines of human endeavor which they believe will eventually place the Negro beside the other races."[72] This embrace of self-realization did not mean that Du Bois thought the social environment irrelevant to race progress; he opposed only the one-sided Spencerian picture (also criticized by James and Dewey) in which a group molds itself to a fixed environment. That is, self-determination was consistent with—to use the example of the previous paragraph—the evolution of new social institutions in response to a peculiar social environment.

As Marilyn Fischer has shown in detail, Addams applied a similar evolutionary framework in "Ethical Survivals in Municipal Corruption" and "The Subtle Problems of Charity," both published in the late 1890s

---

71. W. E. B. Du Bois, "Results of the Investigation," in *Some Efforts of American Negroes for Their Own Social Betterment*, ed. W. E. B. Du Bois (Atlanta: Atlanta University Press, 1898), 42–44; W. E. B. Du Bois, "Strivings of the Negro People," *Atlantic Monthly* 80 (1897): 195. On "the biological analogy," see "Sociology Hesitant" [1904/05], p. 3, Series 3, Du Bois Papers; published in W. E. B. Du Bois, "Sociology Hesitant," *boundary 2* 27 (2000): 39–40; Du Bois, *Dusk of Dawn*, 51.

72. W. E. B. Du Bois, "The Evolution of Negro Leadership," *Dial* 31 (1901): 53–55; also in W. E. B. Du Bois, *The Souls of Black Folk* (Chicago: A. C. McClurg, 1903), 46, 50.

and then included in *Democracy and Social Ethics*.[73] One of the themes of Addams's ethics was the idea that members of the laboring classes were focused on personal character rather than impersonal justice, leading to a conflict between social reformers and the groups they were attempting to help. Citing Wilhelm Wundt's claim that "the idea of morality is at first intimately connected with the person and with personal conduct," she suggested that "the common people" still inhabited this earlier stage of moral evolution:

> Granting . . . that morality develops far earlier in the form of moral fact than in the higher form of moral ideas, it becomes obvious that ideas only operate upon the popular mind through will and character, and that goodness has to be dramatized before it reaches the masses of men. . . . To the common people [ethics as well as political opinions] can only come through example,—through a personality which seizes the popular imagination.

According to Addams, "primitive people, such as the south Italian peasants who live in the Nineteenth Ward, are still in this stage"; thus they tend to elect "a good friend and neighbor" whose public acts create a personal connection with his constituents, even though social reformers would describe the same man as irredeemably corrupt and immoral.[74] Similarly, Addams argued that there was a "striking incongruity" between the ethical standard of the college-educated "charity visitor," which was based on "the industrial virtues," and that of "poor people," which was based on "emotional kindness." "The evolutionists," wrote Addams, "tell us that the instinct to pity, the impulse to aid his fellows, served man at a very early period as a rude rule of right and wrong." This impulse, according to Addams, was still dominant in poor districts: such areas were characterized by "primitive and frontier-like . . . neighborly relations," and shared circumstances made "the ready outflow of sympathy and material assistance the most natural thing in the world." Addams claimed that it was impossible to substitute "a higher ethical standard for the lower one without the intermediate stages of growth,"

---

73. Marilyn Fischer, *Jane Addams's Evolutionary Theorizing: Constructing "Democracy and Social Ethics"* (Chicago: University of Chicago Press, 2019), chap. 4.

74. Jane Addams, "Ethical Survivals in Municipal Corruption," *International Journal of Ethics* 8 (1898): 274, 276, 286; Wilhelm Wundt, *The Facts of the Moral Life*, trans. Julia Gulliver and Edward Bradford Titchener (New York: Macmillan, 1897), 32; also in Jane Addams, *Democracy and Social Ethics* (New York: Macmillan, 1902), 227–29, 254 (Wundt reference removed). Addams's interest in Wundt's book may be partly explained by the fact that Gulliver, its cotranslator, was a professor at Rockford College, Addams's alma mater.

lamenting that "we are singularly slow to apply the evolutionary principle to human affairs in general." Although she seems to insult the laboring classes by placing them at an earlier evolutionary stage, Addams was primarily criticizing those reformers who "ruthlessly force ... conventions and standards" on the poor without attending to the peculiarities of their social environment. Thus, in keeping with her more general commitment to democracy, Addams concluded her essay on charity by suggesting that reform efforts would not be effective without a greater appreciation for the motive power of sympathy, so evident in the daily life of Chicago's Nineteenth Ward. As she put it in *Democracy and Social Ethics*, the charity visitor is ultimately "chagrined to discover that in the actual task of reducing her social scruples to action, her humble beneficiaries are far in advance of her."[75]

In the early 1900s, Dewey—like Addams—invoked the notion of "ethical survivals" in his work on the evolution of morality, also citing Wundt. In a chapter on "Custom and the Moral Life" that Dewey assigned in several of his classes, Wundt had applied the notion of cultural survivals—associated with the anthropologist Edward Burnett Tylor—to the history of morality: "Each phase of our modern life is thus permeated with usages that have survived from long forgotten cults, and which in their power to adapt themselves to the new thoughts that come with changed conditions of life seem to repeat in the mental realm ... the capacity for transformation exhibited in the organic world."[76] This idea of moral practices as adaptations was central to Dewey's programmatic statement of his approach to evolutionary ethics—"The Evolutionary Method as Applied to Morality," published in 1902. The evolutionary method assumes, said Dewey, "that norms and ideals, as well as unreflective customs, arose out of certain situations, in response to the demands of those situations." For example, in an earlier essay, Dewey had claimed that the opposing schools of Hellenistic philosophy were a response to social changes, nicely illustrating the evolutionary approach:

---

75. Jane Addams, "The Subtle Problems of Charity," *Atlantic Monthly* 83 (1899): 164–65, 175, 177–78; also in Addams, *Democracy and Social Ethics*, 17, 19–20, 22, 59, 65–66, 69 (quotation from p. 69).

76. Wilhelm Wundt, *The Facts of the Moral Life*, trans. Julia Gulliver and Edward Bradford Titchener (New York: Macmillan, 1897), 139; see also Edward Burnett Tylor, *Primitive Culture: Researches into the Development of Mythology, Philosophy, Religion, Art, and Custom*, 2 vols. (London: John Murray, 1871), chap. 3. Dewey recommended Wundt's chapter to his students in "Evolution of Morality (1901)" and "Sociology of Ethics (1902–1903)": see Dewey, *Class Lectures*, 1:2062, 2302.

CHAPTER SIX

With the growth of the Macedonian and Roman supremacies, the welfare and customs of the local community came to mean less and less to the individual. He was thrown back upon himself for moral strength and consolation. . . . Both [the Stoic and Epicurean schools] are concerned with the question of how the individual, in an environment which is becoming more and more indifferent to him, can realize satisfaction.[77]

For Dewey, social norms and ethical theories were responses to the social environment.

Dewey's evolutionary analysis was supposed to be of more than merely antiquarian interest: "We are still engaged in forming norms, in setting up ends, in conceiving obligations. If moral science has any constructive value, it must provide standpoints and working instrumentalities for the more adequate performance of these tasks." How does an evolutionary approach to morality accomplish this? According to Dewey, it "reveals to us the conditions under which moral practices and ideas have originated. . . . [And] in seeing where they came from, in what situations they arose, we see their significance." Moral intuitions could thus be subjected to an evolutionary test:

If we can find that the intuition is a legitimate response to enduring and deep-seated conditions, we have some reason to attribute worth to it. If we find that historically the belief has played a part in maintaining the integrity of social life, and in bringing new values to it, our belief in its worth is additionally guaranteed. But if we cannot find such historic origin and functioning, the intuition remains a mere state of consciousness, a hallucination, an illusion, which is not made more worthy by simply multiplying the number of people who have participated in it.

This perspective, which I have elsewhere called *dynamic functionalism*, could also explain moral progress: "It is the lack of adequate functioning in the given adjustments that supplies the conditions which call out a different mode of action; and it is in so far as this is new and different that it gets its standing by transforming or reconstructing the previously existing elements." Moral progress occurs when we demand "that a way of conceiving or interpreting the situation cease to be *mere* idea, and become a practical construction." It is only through "failure from the standpoint of adjustment," and subsequent readjustment, "that history, change

---

77. John Dewey, "The Evolutionary Method as Applied to Morality," *Philosophical Review* 11 (1902): 356; John Dewey, "Moral Philosophy," in *Johnson's Universal Cyclopaedia: A New Edition*, ed. Charles Kendall Adams, vol. 5 (New York: A.J. Johnson, 1894), 881–82.

in quality or values, is made." The winners in this process, according to Dewey, are the values that actually help us resolve our current social problems; the losers are "surds, mere survivals, emotional reactions."[78] For Dewey, ethical practices are adaptations, coevolving with a social environment that is itself constantly in flux.

This evolutionary progress was also experimental: as Dewey told his psychological ethics students at Chicago, "the moral process is an experimental process just like the scientific process"; "the moral life follows the experimental method of science."[79] More or less explicit in the writings of Dewey, Mead, Addams, and Du Bois, this viewpoint stands in direct contrast to Spencer's laissez-faire approach to evolutionary ethics. Dewey presented the following criticism to his students in 1902:

> While Spencer has in one way stimulated more than anyone else the treatment of ethics from the genetic and biological point of view, yet in another sense he has done more than anyone else to keep it back, because his opponents, as well as his followers, have determined to make the case stand or fall with his particular statement of it, instead of going back to criticise the original statement.[80]

Although Beth Eddy has recently suggested that Addams shared Spencer's "unilineal account" of moral evolution, Addams's claim that progress does not require antagonism (discussed in chapter 5) need not imply that she viewed ethics as moving inexorably toward a certain end point. As a social reformer, Addams—like Dewey—emphasized "experiment in the line of industrial amelioration and social advancement," and she indicated in an anecdote the importance of human effort in moral evolution, positively recounting the "bungling" but "suggestive" view of a workingman who analogized workers to fish struggling on the sand to evolve legs, with capitalists sailing comfortably along in the water and ignoring the progressive efforts of those on land.[81]

Dewey moved to Columbia University in 1904, but that did not stop his "Chicago School" from staging a joint attack on Spencer's idea of

---

78. Dewey, "Evolutionary Method as Applied to Morality," 113, 356–57, 367–68, 370. On dynamic functionalism, see Trevor Pearce, "American Pragmatism, Evolution, and Ethics," in *The Cambridge Handbook of Evolutionary Ethics*, ed. Michael Ruse and Robert J. Richards (Cambridge: Cambridge University Press, 2017), 54.
79. "Psychological Ethics (1898)" and "Psychology of Ethics (1901)," in Dewey, *Class Lectures*, 1:1277, 1474.
80. "Sociology of Ethics (1902–1903)," in Dewey, *Class Lectures*, 1:2260.
81. Jane Addams, *Democracy and Social Ethics* (New York: Macmillan, 1902), 164, 173–74; Beth L. Eddy, *Evolutionary Pragmatism and Ethics* (Lanham, MD: Lexington, 2016), 64.

a perfect moral end point in 1908. Mead, in an article in the *International Journal of Ethics*, argued that "moral advance consists not in adapting individual natures to the fixed realities of a moral universe, but in constantly reconstructing and recreating the world as the individuals evolve." The ethicist, said Mead, is like an engineer—"face to face with a real problem." He receives "no plan of procedure . . . as a vision in the mount," but rather "uses working hypotheses" as part of a "process of evolution in which individual and environment mutually determine each other." That same year, Dewey mocked Spencer's view, according to which a "rapid transit system of evolution is carrying us automatically to the goal of a perfect man in a perfect society." Opposing the very existence of a *summum bonum* (highest good), Dewey argued that "the proper business of intelligence is discrimination of multiple and present goods and of the varied immediate means of their realization; not search for the one remote aim." He claimed that "the business of morals . . . is to converge all the instrumentalities of the social arts, of law, education, economics and political science upon the construction of intelligent methods of improving the common lot." That is, moral evolution involves piecemeal progress toward local ends under the guidance of social science. Finally, in a talk at the American Philosophical Association, Addison Webster Moore—who had completed his doctorate under Dewey at Chicago in 1898—presented an elaborate metaphor to make a similar point. According to "the evolutionist" in morals, said Moore, it is not enough to swab the deck; one must also guide the course of the ship:

As a *moral* experience, "this laying the course" means more than running for a harbor already built from all eternity. It means nothing less than that our moral craft carries within her the material and the machinery for the building of *new* shores and ports. And this material is simply the entire world of organized habits and institutions; and the machinery, the method, is thought,—science.

Moore argued that this evolutionary process must occur "whenever the old plans, the old shores and ports become inadequate," and in whatever direction (echoing William James) "promises the largest satisfaction." As for who decides on the direction, and what counts as satisfaction, Moore admitted that conflicting plans were inevitable; yet he maintained that the only available solution was "more planning, more investigation and experiment, [and] more getting together. . . . Are conflicts, as a matter of fact, ever settled in any other way?" Mead, Dewey, and Moore all rejected Spencer's global evolutionary ethics in favor of a local experimental approach—one that would proceed hand in hand with social science and social reform.[82]

Eugenics and Civilization

The pragmatists all rejected the evolutionary racism of Herbert Spencer and others, but their views also illustrate the darker side of social reform efforts in the late nineteenth and early twentieth centuries. Pragmatism's commitment to experimentation in the service of social and moral progress linked it directly to contemporary discussions of eugenics and civilization.

Eugenics, a term that gained popularity at around the same time as *pragmatism*, referred to the improvement of the human population through scientific intervention—"creative evolution become self-conscious," as one practitioner put it.[83] Although primarily associated with the selective modification of what Weismann called the "germ-plasm," the British editors of the *Eugenics Review* emphasized in their inaugural issue that they would "not ignore the importance of Environment." For example, one way of reducing infant mortality in cities might be "to assist 'Nature' by 'Nurture,'" focusing on conditions in hospitals. A few months later in the same journal, Dewey's teacher G. Stanley Hall presented "the basis of the new biological ethics":

Everything is right that makes for the welfare of the yet unborn and all is wrong that injures them, and to do so is the unpardonable sin—the only one nature knows. Just as the soma and all the mortal cells and organs of the body and all their activities throughout our individual lives are only to serve the deathless germ plasm, so every human institution, home, school, state, church and all the rest exist primarily in order to bring children and youth on and up to their highest possible maturity of body and

---

82. George Herbert Mead, "The Philosophical Basis of Ethics," *International Journal of Ethics* 18 (1908): 319–21; John Dewey, *Ethics*, Lectures on Science, Philosophy, and Art, 1907–1908 (New York: Columbia University Press, 1908), 19–21; also in John Dewey, *The Influence of Darwin on Philosophy, and Other Essays in Contemporary Thought* (New York: Henry Holt, 1910), 66–69; Addison Webster Moore, "Absolutism and Teleology," *Philosophical Review* 18 (1909): 315–17; cf. James, "The Moral Philosopher and the Moral Life," *International Journal of Ethics* 1 (1891): 346. For Moore's doctorate, see *University Record* (Chicago, IL) 2 (1898): 410. On the idea of a "Chicago School" of philosophy and psychology at this time, see William James to John Dewey, 11 March 1903, and John Dewey to William James, n.d., in Dewey, *Correspondence*, nos. 00798 and 00797; William James, "The Chicago School," *Psychological Bulletin* 1 (1904).

83. Caleb Williams Saleeby, *The Progress of Eugenics* (New York: Funk & Wagnalls, 1914), [vii].

CHAPTER SIX

soul, and the value not only of all institutions, but of art, science, literature, culture and civilisation itself are ultimately measured and graded by how much they contribute to this supreme end.[84]

For Hall and others, biological and social improvement shared one overarching goal: the welfare of future generations.

Given the experimental-evolutionary approach described in this chapter, it should not surprise us that some of the pragmatists were attracted to eugenics, which was strongly associated with reform movements and "an engineering approach to social problems."[85] As Daylanne English has shown, Du Bois clearly advocated eugenics in the 1920s: in a *Crisis* editorial, for example, he complained that "the Negro has not been breeding for an object." Du Bois had endorsed such views as early as *The Philadelphia Negro* of 1899. Comparing in a table "the size of [Negro] families in the highest and lowest class," Du Bois showed that half of those in "the aristocracy" were having more than three children, in contrast with only a quarter of those in "the 'submerged tenth,'" borrowing the latter phrase from William Booth. He commented, alluding to the popular hypothesis of racial extinction, "this certainly looks like the survival of the fittest, and is hardly an argument for the extinction of the civilized Negro."[86]

This apparently fertile black aristocracy would later feature in Du Bois's 1903 essay "The Talented Tenth":

From the very first it has been the educated and intelligent of the Negro people that have led and elevated the mass, and the sole obstacles that nullified and retarded their efforts were slavery and race prejudice; for what is slavery but the legalized survival of the unfit and the nullification of the work of natural internal leadership? Negro leader-

---

84. "Editorial and Other Notes," *Eugenics Review* 1 (1909): 4; G. Stanley Hall, "Education in Sex-Hygiene," *Eugenics Review* 1 (1910): 242.

85. Nathaniel Comfort, *The Science of Human Perfection: How Genes Became the Heart of American Medicine* (New Haven, CT: Yale University Press, 2012), 46–47; see also Diane B. Paul, "Eugenics and the Left," *Journal of the History of Ideas* 45 (1984); Thomas C. Leonard, *Illiberal Reformers: Race, Eugenics, and American Economics in the Progressive Era* (Princeton, NJ: Princeton University Press, 2016).

86. W. E. B. Du Bois, "Opinion," *Crisis* 24 (1922): 152; quoted in Daylanne K. English, *Unnatural Selections: Eugenics in American Modernism and the Harlem Renaissance* (Chapel Hill: University of North Carolina Press, 2004), 37–38; W. E. B. Du Bois, *The Philadelphia Negro: A Social Study* (Philadelphia: University of Pennsylvania, 1899), 310–11, 319; William Booth, *In Darkest England and the Way Out* (New York: Funk & Wagnalls, 1890), 18–23. On racial extinction, see W. E. B. Du Bois, "Race Traits of the American Negro," *Annals of the American Academy of Political and Social Science* 9 (1897); Patrick Brantlinger, *Dark Vanishings: Discourse on the Extinction of Primitive Races, 1800–1930* (Ithaca, NY: Cornell University Press, 2003).

ship, therefore, sought from the first to rid the race of this awful incubus that it might make way for natural selection and the survival of the fittest.

Such rhetoric also appeared in Du Bois's 1909 address to the National Negro Conference, where he argued for what had recently been called "positive eugenics":

> We cannot ensure the survival of the best blood by the public murder and degradation of unworthy suitors, but we can substitute a civilized human selection of husbands and wives which shall ensure the survival of the fittest. Not the methods of the jungle, not even the careless choices of the drawing room, but the thoughtful selection of the schools and laboratory is the ideal of future marriage.

These views were common among reformers at the time. Although several scholars have claimed Dewey as an opponent of eugenics, the 1932 edition of his textbook *Ethics*—albeit in a section written by James Hayden Tufts—emphasized that "the problem of maintaining the best stocks is one of the most serious that confronts us." Even Alain Locke, a member of the third cohort of pragmatists who usually avoided the biological rhetoric of his predecessors, had no problem in principle with "practical eugenics"—that is, "any really scientific and enlightened policy of population control"—although he argued that it would only be possible after the elimination of racism and bigotry.[87]

The discourse of improvement was also central to anthropological theories of evolutionary progress, with Spencer and others postulating a trajectory from primitive savagery to scientific civilization. Dewey explicitly criticized Spencer for isolating "the psychology of primitive man . . . from the modes of life that are prevalent and [from] environmental conditions," as well as for lumping together diverse groups that "present widely remote cultural resources, varied environments and

---

87. W. E. B. Du Bois, "The Talented Tenth," in *The Negro Problem: A Series of Articles by Representative American Negroes of To-day* (New York: James Pott, 1903), 34–35; see also Joy James, *Transcending the Talented Tenth: Black Leaders and American Intellectuals* (New York: Routledge, 1997), 20; W. E. B. Du Bois, "Evolution of the Race Problem," in *Proceedings of the National Negro Conference, 1909: New York, May 31 and June 1* (n.p., [1909]), 149, 152, https://hdl.handle.net/2027/nyp.33433081797809; also in W. E. B. Du Bois, *John Brown* (Philadelphia: George W. Jacobs, 1909), 384–85; John Dewey and James Hayden Tufts, *Ethics*, rev. ed. (New York: Henry Holt, 1932), 508; Alain Locke, "Darwinian Negro Study," *New York Herald Tribune Books*, July 24, 1938. On "positive eugenics," see Caleb Williams Saleeby, *Parenthood and Race Culture: An Outline of Eugenics* (London: Cassell, 1909), 171–72. On Dewey's opposition to eugenics, see Steven Selden, *Inheriting Shame: The Story of Eugenics and Racism in America* (New York: Teachers College Press, 1999), 113–17; Timothy McCune, "Dewey's Dilemma: Eugenics, Education, and the Art of Living," *Pluralist* 7 (2012).

distinctive institutions." Nevertheless, as Thomas Fallace has shown in detail, Dewey's own endorsement of the primitive-civilized dichotomy was implicitly racist, since the "savage" or "primitive" races were invariably nonwhite.[88] This is certainly not meant as a defense of Dewey, but it is worth pointing out—contrary to Fallace's claims—that even famous antiracists such as Du Bois and Franz Boas were content at the time with the contrast between primitive and civilized cultures. For example, in a speech later noted by Dewey, Boas distinguished "the mind of primitive man" from that of "civilized man," even though he attributed all major differences to the social environment: "The difference in the mode of thought of primitive man and of civilized man seems to consist largely in the difference of character of the new traditional material with which the new perception associates itself. The instruction given to the child of primitive man is not based on centuries of experimentation, but consists of the crude experience of generations."[89] Du Bois, who cited Boas's book, was likewise happy to distinguish "primitive" and "civilized" groups, highlighting the historical contributions of Egypt and other "civilized" nations of Africa and referring to "the reddish dwarfs of the center and the Bushmen of South Africa" as surviving examples of "the primitive Negro."[90]

Addams, as we have seen, blithely spoke of the "primitive people" living around Hull House, placing them at an earlier stage of moral evolution. Newspapers at the time noted this dimension of uplift, depicting reformers like Addams as drivers of social evolution, viewed as a transformation culminating in the civilized American (figure 12). As Denise James notes, "culture (and civilization) were not terms easily divorced from the white supremacist attitudes of the time." Despite her pluralism, Addams still "measured [the cultures of the immigrants] against the

---

88. "Evolution of Morality (1901)," in Dewey, *Class Lectures*, 1:2067; John Dewey, "Interpretation of Savage Mind," *Psychological Review* 9 (1902): 218; cf. Herbert Spencer, *The Principles of Sociology*, vol. 1 (London: Williams & Norgate, 1877), chap. 7; Thomas D. Fallace, *Dewey and the Dilemma of Race: An Intellectual History, 1895–1922* (New York: Teachers College Press, 2011).

89. Franz Boas, "The Mind of Primitive Man," *Science* 13, no. 321 (1901): 286; also in Franz Boas, *The Mind of Primitive Man* (New York: Macmillan, 1911), 203; pace Fallace, *Dewey and the Dilemma of Race*, chap. 5. At the back of his copy of *Mind of Primitive Man*—held in the Special Collections Research Center, Morris Library, Southern Illinois University–Carbondale (call number 010038)—Dewey wrote "202–206," a page range that contains the quoted passage.

90. W. E. B. Du Bois, *The Conservation of Races* (Washington, DC: American Negro Academy, 1897), 9; W. E. B. Du Bois, "The People of Peoples and Their Gifts to Men," *Crisis* 6 (1913): 339; W. E. B. Du Bois, *The Negro* (New York: Henry Holt, 1915), 20–25, 103–16, 138, 245 (quotation from p. 20); see also Adolph L. Reed Jr., *W. E. B. Du Bois and American Political Thought: Fabianism and the Color Line* (Oxford: Oxford University Press, 1997), 122; pace Fallace, *Dewey and the Dilemma of Race*, chap. 5.

Figure 12 "The Evolution of Hull House," a newspaper cartoon depicting an impoverished wretch gradually transforming into a well-dressed gentleman.
From "Noble Work at Hull House," *Chicago Chronicle*, June 9, 1895. This article was cut out and glued into 508* OV Scrapbook v. 3, Series 11, Hull House Collection. Reproduced courtesy of Special Collections and University Archives, University of Illinois at Chicago.

dominant, white US culture of her time and found them important but lacking."[91]

How should historians of philosophy react to these unsettling views? One promising approach might be to look to other aspects of pragmatism for resources. For example, James and Du Bois were both skeptical that science was the key to social progress, arguing that the relevant experimentation could be moral and political rather than scientific—a point I will return to in the conclusion of this volume. Nevertheless, we should not assume that pragmatism as a philosophical approach is easily disassociated from ideas of civilizing progress, just as we should not assume that it is automatically tainted by those same ideas.

The second cohort of pragmatists, despite their differences, shared an experimental approach to moral and social evolution. They constructed new venues for experimental ethical fieldwork, from Hull House and the University Elementary School in Chicago to the Conferences for the Study of the Negro Problems in Atlanta. Despite rejecting Spencer's evolutionary ethics, they all employed a modified version of his influential organism-environment framework—Du Bois in *The Philadelphia Negro*, Dewey in his ethics classes at the University of Chicago, and Mead and Addams in their writings on social reform and social ethics. Finally, they all took an evolutionary approach to social and moral progress: for Du

---

91. Jane Addams, *Democracy and Social Ethics* (New York: Macmillan, 1902), 229; V. Denise James, "Comments on Marilyn Fischer's 'Addams on Cultural Pluralism, European Immigrants, and African Americans,'" *Pluralist* 9 (2014): 68–69; see also Khalil Gibran Muhammad, *The Condemnation of Blackness: Race, Crime, and the Making of Modern Urban America* (Cambridge, MA: Harvard University Press, 2010), 117–27.

Bois, black institutions were adaptations to a peculiar social environment; for Addams, the evolutionary perspective explained persistent conflicts between workers and reformers; and for Dewey and his Chicago school, moral practices were evolved responses to an ever-changing set of social conditions.

Despite the links between pragmatism, eugenics, and civilization, we should not dismiss the pragmatists' biological approach as merely a historical curiosity, inevitably mired in Eurocentrism and racism. As modern proponents have shown, it can still be fruitful. In some ways, the pragmatists anticipated what we now refer to more generally as "evidence-based policy"—which, if we can get beyond the fetishizing of randomized controlled trials, promises to offer effective remedies for moral and social problems through experimental policymaking.[92] Some philosophers have self-consciously attempted to revive pragmatist experimental ethics. Elizabeth Anderson, for example, citing Dewey as inspiration, takes a problem-centered approach to ethics in her book *The Imperative of Integration*, placing social scientific research at the heart of moral inquiry:

> Nonideal theory begins with a diagnosis of the problems and complaints of our society and investigates how to overcome these problems. . . . In nonideal theory, ideals embody imagined solutions to identified problems in a society. They function as *hypotheses*, to be tested in experience. . . . This process is not merely instrumental: it is not a matter of finding better means to a fixed end already fully articulated. Reflection on our experience can give rise to new conceptions of successful conduct.[93]

Taking a similar line, following both Dewey and Du Bois, Paul Taylor argues that racial discourse begins with "problematic situations in need of resolution." He continues:

> Races become visible when we attempt to account for the world in a particular way. They are elements in explanations, interpretations, and readings of the social environment and of our likely experiences in it. Races are, in this sense, social-theoretic posits, deployed in a pragmatic spirit in the attempt to understand and navigate the world more intelligently and productively.[94]

---

92. Nancy Cartwright and Jeremy Hardie, *Evidence-Based Policy: A Practical Guide to Doing It Better* (Oxford: Oxford University Press, 2012).

93. Elizabeth Anderson, *The Imperative of Integration* (Princeton, NJ: Princeton University Press, 2010), 6–7; citing "Logic of Ethics (1900)," in Dewey, *Class Lectures*, 1:1004–14.

94. Paul C. Taylor, "Context and Complaint: On Racial Disorientation," *Graduate Faculty Philosophy Journal* 35 (2014): 346.

For both Anderson and Taylor—although they do not use the biological language of the pragmatists—moral evolution involves working hypotheses as potential adaptations in dynamic social environments. Linking this approach more explicitly to evolution, Philip Kitcher has argued for a neo-Deweyan functionalism in ethics. For all of these philosophers, moral practices have the function of addressing problems posed by the social environment, and moral progress involves "experiments in living."[95]

This chapter has demonstrated that the pragmatists developed an experimental-evolutionary ethics. But pragmatism is best known for its epistemological and metaphysical commitments. In the next chapter, I will suggest that pragmatist views of knowledge and reality should be understood as an experimental-evolutionary approach to logic, paralleling their account of ethics.

---

95. Elizabeth Anderson, "John Stuart Mill and Experiments in Living," *Ethics* 102 (1991); Philip Kitcher, *The Ethical Project* (Cambridge, MA: Harvard University Press, 2011), chap. 3; Taylor, "Context and Complaint," 346; Elizabeth Anderson, *Social Movements, Experiments in Living, and Moral Progress: Case Studies from Britain's Abolition of Slavery*, Lindley Lecture no. 52, February 11, 2014, University of Kansas, KU ScholarWorks, http://hdl.handle.net/1808/14787; see also Ryan Muldoon, "Expanding the Justificatory Framework of Mill's Experiments in Living," *Utilitas* 27 (2015).

SEVEN

Pragmatist Logic: Evolution, Experiment, and Inquiry

Richard Rorty echoed a long line of critics when he accused John Dewey of merely "lifting the vocabulary of the evolutionary biologist out of the laboratory and using it to describe everything that could ever count as 'Knowledge.'" In 1909, William Pepperell Montague—Dewey's colleague at Columbia University and the son of one of the members of the Metaphysical Club—had already criticized what he called "biological pragmatism" for assuming a particular evolutionary story: "When desires, by reason of their complexity, are no longer able to secure immediate and automatic satisfaction, knowledge and thinking are evolved and by natural selection preserved as new and useful instruments of adaptation to environment."[1] Although pragmatism was and still is most famous for its theory of truth, I agree with Isaac Levi that the central insight of the classical pragmatists was their model of inquiry. In this final chapter, I will demonstrate that Rorty and Montague were right to see this model as fundamentally linked to biology: evo-

---

1. Richard Rorty, "Dewey's Metaphysics," in *New Studies in the Philosophy of John Dewey*, ed. Steven M. Cahn (Hanover, NH: University Press of New England, 1977), 66; William Pepperell Montague, "May a Realist Be a Pragmatist?" *Journal of Philosophy, Psychology and Scientific Methods* 6 (1909): 485–86; Cornelis de Waal, "Montague, William Pepperell, Jr. (1873–1953)," in *The Dictionary of Modern American Philosophers*, ed. John R. Shook, 4 vols. (Bristol, UK: Thoemmes Continuum, 2005), 3:1719.

lutionary ideas, and especially Herbert Spencer's organism-environment framework, formed the backdrop to the pragmatists' account of knowledge and scientific inquiry. As I will show, this account was grounded in their psychological and educational research. It also proceeded hand in hand with their analyses of ethics and social reform, as described in chapter 6. In the 1890s and early 1900s, pragmatism (now finally going by that name) presented a model of inquiry—a *logic*, in the terminology of the time—that that was both evolutionary and experimental. As Melvin Rogers has argued, evolutionary ideas inclined the pragmatists toward an interventionist ethos that emphasized both control and contingency.[2]

Those familiar with recent work on pragmatism will immediately think of at least three problems with this thesis: first, James Scott Johnston has argued that "in terms of Dewey's logic, Hegel rather than Darwin emerges as the important precursor"; second, as Vincent Colapietro and Christopher Hookway each have discussed in detail, Charles Sanders Peirce explicitly criticized the "natural history" approach to logic; and third, William James seems to have had no logic to speak of—his enemies were the rationalists who shouted, "Down with psychology, up with logic, in all this question!"[3] In the remainder of this introductory section, I will respond briefly to each of these challenges, and they will be addressed more fully (albeit indirectly) in the body of the chapter.

1. Although I agree with Johnston that Dewey's logic was fundamentally Hegelian, the choice between Hegel and biology is a false one. As shown in chapter 4, Hegel was linked to biological evolution by the very idealists who inspired Dewey: Dewey's logic thus reflected this complex influence. For example, he described Hegel's philosophy using the terms *evolution* and *readjustment*, both of which had biological overtones. The idea that conflict leads to readjustment or adaptation, central to Dewey's logic and ethics, was arguably both biological and Hegelian.[4]

   Johnston's dismissal of the biology connection also depends on his claim that although "Dewey's broad evolutionary accounts are Darwinian in spirit, they are

---

2. Isaac Levi, "Pragmatism and Change of View," in *Pragmatism*, ed. Cheryl Misak (Calgary: University of Calgary Press, 1999), 181; Melvin L. Rogers, *The Undiscovered Dewey: Religion, Morality, and the Ethos of Democracy* (New York: Columbia University Press, 2009), xi.

3. James Scott Johnston, *John Dewey's Earlier Logical Theory* (Albany: State University of New York Press, 2014), 32; Vincent Colapietro, "Experimental Logic: Normative Theory or Natural History?" in *Dewey's Logical Theory: New Studies and Interpretations*, ed. F. Thomas Burke, D. Micah Hester, and Robert B. Talisse (Nashville, TN: Vanderbilt University Press, 2002); Christopher Hookway, "Normative Logic and Psychology: Peirce's Rejection of Psychologism," in *The Pragmatic Maxim: Essays on Peirce and Pragmatism* (Oxford: Oxford University Press, 2012), 102–9; William James, *Pragmatism: A New Name for Some Old Ways of Thinking* (New York: Longmans, Green, 1907), 67.

4. John Dewey, "Movements of Thought in the Nineteenth Century," Lecture 12 (31 November 1891), Edwin Spencer Peck Notebooks (851997 Aa 2), Bentley Historical Library, University of Michigan.

CHAPTER SEVEN

not Darwinian in letter."[5] On this point, I agree: Darwin was a relatively unimportant interlocutor for Dewey in the 1880s and '90s. This fact has been obscured by a unitary focus on Dewey's famous 1909 essay "Darwin's Influence upon Philosophy" and an almost complete neglect of Spencer and the organism-environment dichotomy.[6] Belying the claim that Dewey was "first and foremost a Darwinist," he only rarely discussed Darwin's work prior to the 1909 essay. The one exception was a sustained engagement with *The Expression of the Emotions in Man and Animals*, treated in his 1887 psychology textbook and also in an 1894 article. Thus, although Dewey was at least somewhat familiar with Darwin's *Origin of Species* and *Descent of Man*, including them in several bibliographies, he does not seem to have carefully studied these texts—a stark contrast with members of the first cohort of pragmatists, as described in chapter 1.[7]

Although Johnston is right to downplay Darwin's influence on Dewey, he is wrong to conclude that biological concepts had little role in shaping Dewey's philosophy. As we saw in chapter 3, evolutionary ideas were central to Dewey's education and early career. Spencer was of particular importance: as a college student at Vermont, Dewey checked out Spencer's *Principles of Psychology* from the library more often than any other book (he checked out nothing by Darwin); as a graduate student at Johns Hopkins University, Dewey learned—under the tutelage of George Sylvester Morris—to see Spencer as a key critical target; and as a young professor at Michigan, Dewey taught Philosophy 10: The Philosophy of Herbert Spencer in the late 1880s and also assigned Spencer's *Principles of Sociology* in his early 1890s seminars on the history of political philosophy.[8] Even the famous

---

5. James Scott Johnston, *John Dewey's Earlier Logical Theory* (Albany: State University of New York Press, 2014), 31.

6. John Dewey, "Darwin's Influence upon Philosophy," *Popular Science Monthly* 75 (1909). For this unitary focus, see Raymond D. Boisvert, *Dewey's Metaphysics* (New York: Fordham University Press, 1988), chap. 2; James Campbell, *Understanding John Dewey* (Chicago: Open Court, 1995), chap. 2; Charlene Haddock Seigfried, *Pragmatism and Feminism: Reweaving the Social Fabric* (Chicago: University of Chicago Press, 1996), 177–79; Jerome A. Popp, *Evolution's First Philosopher: John Dewey and the Continuity of Nature* (Albany: State University of New York Press, 2007), chap. 1; Melvin L. Rogers, *The Undiscovered Dewey: Religion, Morality, and the Ethos of Democracy* (New York: Columbia University Press, 2009), chap. 2; Preston Stovall, "Nature, Purpose, and Norm: A Program in American Philosophy," *Journal of the American Philosophical Association* 2 (2016): 621–22.

7. Charles Darwin, *The Expression of the Emotions in Man and Animals* (London: John Murray, 1872); John Dewey, *Psychology* (New York: Harper & Brothers, 1887), 355–56; John Dewey, "The Theory of Emotion. (I.) Emotional Attitudes," *Psychological Review* 1 (1894). For Dewey's bibliographic citations of *Origin* and *Descent*, see Dewey, *Psychology*, 358; John Dewey, *Outlines of a Critical Theory of Ethics* (Ann Arbor, MI: Inland Press, 1891), 78; John Dewey, "Moral Philosophy," in *Johnson's Universal Cyclopaedia: A New Edition*, ed. Charles Kendall Adams, vol. 5 (New York: A. J. Johnson, 1894), 884. For the claim that Dewey was "first and foremost a Darwinist," see Popp, *Evolution's First Philosopher*, 2.

8. Lewis S. Feuer, "John Dewey's Reading at College," *Journal of the History of Ideas* 19 (1958): 419–20; George Sylvester Morris, *British Thought and Thinkers: Introductory Studies, Critical, Biographical and Philosophical* (Chicago: S. C. Griggs, 1880), chap. 12; John Dewey, "Knowledge and the Relativity of Feeling," *Journal of Speculative Philosophy* 17 (1883): 57–58, where Spencer is the im-

thesis of Dewey's 1909 essay on Darwin—namely, that "in treating the forms that had been regarded as types of fixity and perfection as originating and passing away, the 'Origin of Species' introduced a mode of thinking that in the end was bound to transform the logic of knowledge"—was anticipated in his earlier essay on Spencer, written a few months after the philosopher's death: "The transfer from the world of set external facts and of fixed ideal values to the world of free, mobile, self-developing, and self-organizing reality would be unthinkable and impossible were it not for the work of Spencer, which, shot all through as it is with contradictions, thereby all the more effectually served the purpose of a medium of transition from the fixed to the moving." This quotation demonstrates that Dewey viewed Spencer and Darwin as jointly responsible for the transformation of philosophy by evolutionary thought, although Darwin naturally assumed pride of place in a lecture series celebrating his 1909 centennial. It is thus wrong to conclude that biology and evolution were unimportant for Dewey from the fact that he only rarely employed specifically Darwinian terminology.[9]

2. After reading *Studies in Logical Theory*, in which Dewey praised the "natural history" approach of psychology and applied an "evolutionary method" to logic, Peirce complained privately (in 1904) that this method would "substitute for the Normative Science which in my judgment is the greatest need of our age a 'Natural History' of thought or of experience." He then declared publicly in the *Nation* that "calling the new natural history by the name of 'logic'" seemed to be "a way of prejudging the question of whether or not there be a logic which is more than a mere natural history, inasmuch as it would pronounce one proceeding of thought to be sound and valid and another to be otherwise."[10] Colapietro has plausibly suggested that Peirce's criticisms were motivated in part by self-defense: "Peirce was fighting *for* the future of logic and also for his own desperate chance to win a public hearing for technical work," and it was a fight that he lost, since many of his logic manuscripts—including the *Minute Logic*, partially drafted a few years earlier—were not published during his lifetime. But just as Hegel versus biology

---

plicit target; *Calendar of the University of Michigan for 1885–86* (Ann Arbor: University of Michigan, 1886), 55; *Calendar of the University of Michigan for 1887–88* (Ann Arbor: University of Michigan, 1888), 50; *Calendar of the University of Michigan for 1892–93* (Ann Arbor: University of Michigan, 1893), 67; *Calendar of the University of Michigan for 1893–94* (Ann Arbor: University of Michigan, 1894), 68.

9. John Dewey, "Darwin's Influence upon Philosophy," *Popular Science Monthly* 75 (1909): 90; also in John Dewey, *The Influence of Darwin on Philosophy, and Other Essays in Contemporary Thought* (New York: Henry Holt, 1910), 1–2; John Dewey, "The Philosophical Work of Herbert Spencer," *Philosophical Review* 13 (1904): 175. For the centennial lecture series, see *Annual Reports of the President and Treasurer to the Trustees, with Accompanying Documents* (New York: Columbia University, 1909), 154.

10. John Dewey, *Studies in Logical Theory* (Chicago: University of Chicago Press, 1903), 15; Charles Sanders Peirce to John Dewey, 9 June 1904, in Dewey, *Correspondence*, no. 00930; Charles Sanders Peirce, "Logical Lights," *Nation*, September 15, 1904, 220.

CHAPTER SEVEN

is a false choice, so is normative science versus natural history. As Colapietro and Hookway have emphasized, Peirce's logic was consistent with a moderate naturalism and Dewey's logic did not deny the importance of norms.[11] Peirce's approach even contained "natural history" elements, which were especially apparent in his model of scientific inquiry: in the years directly preceding his response to Dewey, for example, Peirce exempted "the fact that the mind struggles to escape from doubt" from his broader critique of psychologism and appealed to biology in his discussions of abduction.[12] Thus, although Peirce and Dewey had their differences, both of their approaches to logic—as I will show in this chapter—are accurately described as evolutionary and experimental.

3. James, like Dewey, made no contributions to exact or formal logic, but his pragmatism—which built on his earlier research in psychology—was relevant to logic in its broader nineteenth-century sense. According to Peirce, logic includes not only speculative grammar (the general theory of signs) and critic (the classification of arguments) but also "methodeutic," which seeks "the general conditions requisite for the attainment of truth" or "the general principles which ought to guide an inquiry." Peirce suggested that scholars look to the philosophy of science for such principles, citing books by Auguste Comte, William Whewell, John Stuart Mill, and William Stanley Jevons, among others. Dewey, likewise, argued that "the distinctions and classifications that have been accumulated in 'formal' logic"—that is, what Peirce called "critic"—demand "interpretation from the standpoint of use." In Dewey's view, logicians had been neglecting "the standpoint of practical deliberation and scientific research"—that is, "all the typical investigatory and verificatory procedures of the various sciences." James's pragmatism, insofar as it aimed to provide a method or model of inquiry, was relevant to logic in this broad sense. Peirce himself declared in 1903, with James in the audience, that pragmatism was "a certain maxim of Logic."[13]

---

11. Vincent Colapietro, "Experimental Logic: Normative Theory or Natural History?" in *Dewey's Logical Theory: New Studies and Interpretations*, ed. F. Thomas Burke, D. Micah Hester, and Robert B. Talisse (Nashville, TN: Vanderbilt University Press, 2002), 59, 65; Christopher Hookway, "Normative Logic and Psychology: Peirce's Rejection of Psychologism," in *The Pragmatic Maxim: Essays on Peirce and Pragmatism* (Oxford: Oxford University Press, 2012), 108n29.

12. James Mark Baldwin, ed., *Dictionary of Philosophy and Psychology*, 2 vols. (New York: Macmillan, 1901–2), 2:22, s.v. "logic"; "Minute Logic, Chapter I. Intended Characters of this Treatise," MS 425 (1902), in Peirce, *Collected Papers*, 2:86. The entry for *logic* in Baldwin's *Dictionary* was authored by Peirce; the initials of Christine Ladd-Franklin signal only her acceptance of its content (Baldwin, *Dictionary*, 1:xii).

13. Baldwin, *Dictionary of Philosophy and Psychology*, 2:21, 75, s.vv. "logic," "method and methodology"; John Dewey, *Studies in Logical Theory* (Chicago: University of Chicago Press, 1903), 6, 8; "The Maxim of Pragmatism [First Harvard Lecture]," MS 301 (26 March 1903), in Peirce, *Essential Peirce*, 2:133; for James's attendance, see Charles Sanders Peirce to William James, 16 March 1903, and William James to Charles Sanders Peirce, 19 March 1903, in James, *Correspondence*, 10:212–13.

This chapter has two parts: in the first, I will describe the pragmatists' "natural history" approach to logic, the roots of which lie in their psychological and educational research; in the second, I will demonstrate that Dewey, Peirce, and James all placed experimentation at the center of their models of inquiry, and that this experimentalism was directly connected with evolutionary ideas. I will conclude with a brief discussion of the relationship between logical and moral inquiry.

The "Natural History" Approach

When William James introduced pragmatism to the world in 1898, he began by describing Charles Sanders Peirce's doubt-belief model of inquiry.[14] Peirce had first presented this model in an 1877–78 series of articles for *Popular Science Monthly*, which was the main American venue for Spencer's work. The series was titled "Illustrations of the Logic of Science," and Peirce positioned the interplay of doubt and belief as the basis of this logic: "The irritation of doubt causes a struggle to attain a state of belief. I shall term this struggle *inquiry*."[15] In the second article of the series (the one quoted by James), Peirce connected his doubt-belief model to what James would later call *pragmatism*:

The essence of belief is the establishment of a habit, and different beliefs are distinguished by the different modes of action to which they give rise. If beliefs do not differ in this respect, if they appease the same doubt by producing the same rule of action, then no mere differences in the manner of consciousness of them can make them different beliefs, any more than playing a tune in different keys is playing different tunes.

That is, if two beliefs are constituted by the same habits and give rise to the same actions, they are not essentially different. Peirce also called belief "the demi-cadence which closes a musical phrase in the symphony of our intellectual life." What he meant by this was that, although belief involves habit, acting according to that habit can lead to renewed doubt, prompting further inquiry (a "demi-cadence," also known as a semicadence or half cadence, is a weak cadence on the dominant—for example, the words *high* and *sky* in "Twinkle Twinkle Little Star"; it is

---

14. William James, "Philosophical Conceptions and Practical Results," *University Chronicle* (Berkeley, CA) 1 (1898): 290.
15. Charles Sanders Peirce, "The Fixation of Belief," *Popular Science Monthly* 12 (1877): 6.

only a temporary stopping point, and the listener expects the song to continue). For Peirce, as James reported in 1898, inquiry is the endless struggle to obtain stable rules for action.[16]

Peirce connected this "struggle to attain belief" with Darwin's "struggle for existence," placing his model of inquiry on an evolutionary basis.[17] This was well before the factors debates discussed in chapter 5, and Peirce referred only to Darwin's preferred factor, natural selection. "Logicality in regard to practical matters," Peirce wrote in the first essay of the series, "is the most useful quality an animal can possess, and might, therefore, result from the action of natural selection."[18] Then in the fifth essay of the series, "The Order of Nature," he broached what was to be a recurring theme in his work—the adapted mind:

> It seems incontestable . . . that the mind of man is strongly adapted to the comprehension of the world. . . . How are we to explain this adaptation? The great utility and indispensableness of the conceptions of time, space, and force, even to the lowest intelligence, are such as to suggest that they are the results of natural selection. Without something like geometrical, kinetical, and mechanical conceptions, no animal could seize his food or do anything which might be necessary for the preservation of the species. . . . As that animal would have an immense advantage in the struggle for life whose mechanical conceptions did not break down in a novel situation . . . , there would be a constant selection in favor of more and more correct ideas of these matters.

That is, "inborn" or "innate" ideas relating to the most basic physical concepts are results of natural selection. Without these ideas, said Peirce, it would be impossible to discover causes: we would "have to hunt among all the events in the world without any scent."[19]

Peirce returned to this theme five years later in "A Theory of Probable Inference"—an essay published in *Studies in Logic*, the 1883 volume collecting the research of his Johns Hopkins students. He began the last

---

16. Charles Sanders Peirce, "How to Make Our Ideas Clear," *Popular Science Monthly* 12 (1878): 291.

17. Darwin popularized the phrase "struggle for existence" to describe the "severe competition" to which all organisms are exposed: see Charles Darwin, *On the Origin of Species by Means of Natural Selection, or the Preservation of Favoured Races in the Struggle for Life* (London: John Murray, 1859), 62. For Peirce's use of the synonymous phrase "struggle for life," see Charles Sanders Peirce, "Fraser's Works of Bishop Berkeley," *North American Review* 113 (1871): 472; Charles Sanders Peirce, "The Order of Nature," *Popular Science Monthly* 13 (1878): 213.

18. Charles Sanders Peirce, "The Fixation of Belief," *Popular Science Monthly* 12 (1877): 3; see also Charles Sanders Peirce, "Fraser's Works of Bishop Berkeley," *North American Review* 113 (1871): 472.

19. Charles Sanders Peirce, "The Order of Nature," *Popular Science Monthly* 13 (1878): 212–14.

section of the essay by asking the reader to imagine a United States Census Report presented to "a being from some remote part of the universe, where the conditions of existence are inconceivably different from ours." Although for us the census is "a mine of valuable inductions," this depends on our ability "to ask intelligent questions not unlikely to furnish the desired key to the problem." The alien being, in contrast, would only be able to check for patterns at random, and even if it found them, would be unable to generate promising hypotheses. Since "nature is a far vaster and less clearly arranged repository of facts than a census report," Peirce continued, our success in science must be partly attributable to "special aptitudes for guessing right," which dramatically narrow the range of plausible explanations. As in the earlier essay, Peirce claimed that these aptitudes were in large part the result of natural selection acting on our evolutionary ancestors:

Not man merely, but all animals derive by inheritance (presumably by natural selection) two classes of ideas which adapt them to their environment. In the first place, they all have from birth some notions, however crude and concrete, of force, matter, space, and time; and in the next place, they have some notion of what sort of objects their fellow-beings are, and of how they will act on given occasions.

These innate ideas, said Peirce, constrain our knowledge by circumscribing the two domains in which we have a tendency to guess right: "Man has thus far not attained to any knowledge that is not in a wide sense either mechanical or anthropological in nature, and it may be reasonably presumed that he never will." Peirce concluded that "all human knowledge, up to the highest flights of science, is but the development of our inborn animal instincts."[20] Thus in 1877–78 and in 1883, Peirce argued that our practical logicality and our scientific knowledge depend on our evolutionary history—on rules of action repeatedly tested in the struggle for existence.

I will argue later in this section that these themes were also present in Peirce's subsequent work, despite his shift to neo-Lamarckism (as discussed in chapter 5) and his criticism of Dewey's "natural history" approach. First, however, we need to take a detour into Dewey's logic and its debt to the psychological research of James and others.

The biological-psychological model of inquiry that Peirce developed in the 1870s was paralleled by James's strategic adoption of what he

---

20. Charles Sanders Peirce, "A Theory of Probable Inference," in *Studies in Logic* (Boston: Little, Brown, 1883), 178–81.

CHAPTER SEVEN

sometimes called the "natural history point of view."[21] As we saw in chapter 2, James's primary critique of Spencer was made from this point of view. He emphasized that the problem was not the organism-environment dichotomy itself but rather Spencer's one-sided treatment of it. "My quarrel with Spencer," James told his publisher in 1878, "is not that he makes much of the environment but that he makes *nothing* of the glaring and patent fact of subjective interests which cooperate with the environment in moulding intelligence." James embraced Spencer's organism-environment framework in *Principles of Psychology* and the work leading up to it, despite his fierce opposition to Spencer, because he found the framework much more fruitful than "old-fashioned 'rational psychology,'" which artificially isolates minds from their environments.[22]

This natural history approach had a direct impact on Dewey, who taught graduate seminars on James's *Principles* at the University of Michigan in 1890–91 and 1891–92, later recalling that he had been "much struck in the 'Psychology' of Professor James with the biological conception."[23] A variety of factors were shaking up Dewey's research program at this time. Upon returning to Michigan as department chair in 1889 following the death of his mentor George Sylvester Morris, Dewey began teaching more classes on ethics and the history of philosophy and was thus exposed to the biological idealist views described in chapter 4. He also started teaching Philosophy 2b: Advanced Psychology, the seminar in which James's book was discussed, and Philosophy 10: Advanced Logic, which covered "the theory of scientific method." Finally, he hired new faculty members—James Hayden Tufts, George Herbert Mead, and Alfred Henry Lloyd—with whom he began to engage in collaborative research.[24]

---

21. William James, "Reflex Action and Theism," *Unitarian Review and Religious Magazine* 16 (1881): 394, 403; William James, "The Moral Philosopher and the Moral Life," *International Journal of Ethics* 1 (1891): 333; also in William James, *The Will to Believe, and Other Essays in Popular Philosophy* (New York: Longmans, Green, 1897), 116, 128, 187.

22. William James to Henry Holt, 22 November 1878, in James, *Correspondence*, 5:24–25; William James, *The Principles of Psychology*, 2 vols. (New York: Henry Holt, 1890), 1:6.

23. *Calendar of the University of Michigan for 1890–91* (Ann Arbor: University of Michigan, 1891), 57; John Dewey to William James, 10 May 1891, in Dewey, *Correspondence*, no. 00459; *Calendar of the University of Michigan for 1891–92* (Ann Arbor: University of Michigan, 1892), 59; John Dewey to Henri Robet, 2 May 1911, in Dewey, *Correspondence*, no. 01991; see also John Dewey, "From Absolutism to Experimentalism," in *Contemporary American Philosophy: Personal Statements*, ed. George Plimpton Adams and William Pepperell Montague, vol. 2 (New York: Russell & Russell, 1930), 23–24.

24. Dewey taught Philosophy 10: Advanced Logic from 1889 to 1892 and Philosophy 2b: Advanced Psychology from 1890 to 1892: see *Calendar of the University of Michigan for 1889–90* (Ann Arbor: University of Michigan, 1890), 55; *Calendar of the University of Michigan for 1890–91* (Ann Arbor: University of Michigan, 1891), 56–57; *Calendar of the University of Michigan for 1891–92* (Ann Arbor:

Tufts, Mead, and Lloyd all pushed Dewey in a more biological and experimental direction—these were the Michigan roots of Dewey's Chicago school. In October 1890, the department secured $110 (equal to about $3,100 in 2019) from the Board of Regents "for the purchase of apparatus to be used in illustrating Physiological Psychology," since Tufts wanted to include a laboratory component in this class. As he recalled in his unpublished autobiography, "no course in physiological psychology had ever been given" at Michigan.[25] The problem was that Tufts had only a reading knowledge of physiology. He got himself up to speed during the summer of 1890 in the home laboratory of Frederick Warren Ellis, a medical doctor who had done graduate work in nervous physiology at Harvard from 1883 to 1886 with Henry Pickering Bowditch but was by then a practicing ophthalmologist in Tufts's hometown of Monson, Massachusetts. In Ellis's lab, "the frog served once more the interests of science," and Tufts studied "the brain, the eye, [and] the distribution of the nerves" with an expert on the subject. By the fall of 1890, Tufts probably had more hands-on experience in physiology than Dewey, who had taken animal physiology with Henry Newell Martin and physiological psychology with Granville Stanley Hall at Johns Hopkins but had skipped the concurrent laboratory work, unlike his fellow students and future colleagues Henry Herbert Donaldson and James McKeen Cattell.[26]

After Tufts left in 1891 to study in Germany, Dewey hired Mead to cover physiological psychology. Mead had just finished two years of study at the University of Berlin, where he worked in Hermann Ebbinghaus's experimental psychology laboratory and attended the physiology

---

University of Michigan, 1892), 59–60. On collaborative research, see John Dewey to James Rowland Angell, 10 May 1893, in Dewey, *Correspondence*, no. 00478.

25. *Proceedings of the Board of Regents, 1886–91* ([Ann Arbor]: [University of Michigan]), 474, https://hdl.handle.net/2027/mdp.35112204232435; James Hayden Tufts, "Some Impressions of the University of Michigan, 1889–91," p. 9, Box 3, Folder 12, James Hayden Tufts Papers (ICU.SPCL.TUFTS), Special Collections Research Center, University of Chicago (thanks to Christopher Green for bringing this document to my attention). Jay Martin misleadingly implies that the Michigan psychology laboratory was founded by Dewey: see Jay Martin, *The Education of John Dewey* (New York: Columbia University Press, 2002), 87. Data and formula for 2019 dollar value taken from Federal Reserve Bank of Minneapolis, "Consumer Price Index (Estimate), 1800–," accessed 24 October 2019, https://www.minneapolisfed.org/community/financial-and-economic-education/cpi-calculator-information/consumer-price-index-1800.

26. Tufts, "Some Impressions of the University of Michigan, 1889–91." For Ellis's graduate work, see *The Harvard University Catalogue, 1883–84* (Cambridge, MA: Harvard University, 1883), 169; Frederick W. Ellis, "Henry Pickering Bowditch and the Development of the Harvard Laboratory of Physiology," *New England Journal of Medicine* 219 (1938): 823–27. For enrollments in classes and laboratories at Johns Hopkins, see "Enumeration of Classes, Second Half-Year, 1882–3," *Johns Hopkins University Circulars* 2 (1883): 90, 93; "Enumeration of Classes, Second Half-Year, 1883–4," *Johns Hopkins University Circulars* 3 (1884): 69.

lectures of Hermann Munk, who had recently published the second edition of his book on the functions of the cerebral cortex.[27] Mead took over Tufts's class in 1891–92 and added another of his own—Advanced Physiological Physiology. As Daniel Huebner has discussed, in April 1892 Dewey told his former student and future Chicago colleague James Rowland Angell that the department was "just getting [its ideas] into shape for some laboratory experimenting." Dewey said that the unifying conception of the "reflex arc" would give more direction to their experiments in physiological psychology and commented that "Lloyd and Mead were both working independently on somewhat the same general conception, Mead on the phys[iological] side & Lloyd on the psy[chological]."[28] As the second semester drew to a close, Mead wrote a letter to his in-laws that described the department's approach. After echoing Dewey's idealist view that "body and soul are but two sides of one thing," he told them that "in Physiological Psychology the especial problem is to recognize that our psychical life can all be read in the functions of our bodies—that it is not the brain that thinks but . . . our organs insofar as they act together in processes of life." He regretted that he could not visit them at their home in Hawaii that summer, as his "new standpoint" in physiological psychology offered "new methods of experiment which must be worked out, and I can't do this if I do not have the summer here for study and the arrangement of the laboratory." In his report at the end of the academic year, President James Burrill Angell (father of James Rowland) highlighted this new direction for the department, stating that the university had "enriched the work in Philosophy by providing courses in Physiological Psychology, a branch in which so important work has of late years been done in Germany and in France."[29]

In 1892–93, his second year as Dewey's colleague at Michigan, Mead taught several more courses from his "new standpoint": Experimental Psychology, which the *Calendar* described as a "statement of psycho-

27. Daniel R. Huebner, *Becoming Mead: The Social Process of Academic Knowledge* (Chicago: University of Chicago Press, 2014), 44–47; Trevor Pearce, "Naturalism and Despair: George Herbert Mead and Evolution in the 1880s," in *The Timeliness of George Herbert Mead*, ed. Hans Joas and Daniel R. Huebner (Chicago: University of Chicago Press, 2016), 130–31; Hermann Munk, *Über die Functionen der Grosshirnrinde: Gesammelte Mittheilungen*, 2nd ed. (Berlin: August Hirschwald, 1890).

28. *Calendar of the University of Michigan for 1891–92* (Ann Arbor: University of Michigan, 1892), 59–60; John Dewey to James Rowland Angell, 25 April 1892, in Dewey, *Correspondence*, no. 00466.

29. George Herbert Mead to Samuel and Mary Castle, 18 June 1892, Box 1, Folder 19, Mead Papers; partially quoted in Daniel R. Huebner, *Becoming Mead: The Social Process of Academic Knowledge* (Chicago: University of Chicago Press, 2014), 49; James Burrill Angell, "The President's Report [for 1891–92]," in *Proceedings of the Board of Regents, 1891–96* ([Ann Arbor]: [University of Michigan]), 80, https://hdl.handle.net/2027/mdp.39015078241778.

logical problems in terms of the organism," including "lectures, demonstrations, and experiments," and a full-year "Seminary" in psychology, the first half involving "investigations into psychical phenomena of living organisms," with "laboratory work and lectures," and the second half concerning "pathological psychology in asylums and hospitals." Around this time, Dewey reportedly told some undergraduate students to take classes with Mead, since "all 'introspective psychology' had come to an end"; at least one of those classes seems to have involved experiments on frogs.[30] Presumably referring to Mead's seminary, the *University Record* stated in February 1893 that "the principal subject of the course in experimental psychology during the next semester will be the nervous system studied from the standpoint of its function in the organism, the results of efforts to localize functions in the brain, and the light which pathological cases throw upon this subject." The *Record* article also noted, in a paragraph likely provided by Mead or Dewey, that research in psychology at Michigan "moves naturally along four lines":

(1) Lecture courses giving the net outcome of the science of physiological psychology, with demonstrations; (2) The study of sense organs in their earliest forms, and especially where they are just being differentiated—the line of biology; (3) The study of the functions of the physical organism, in so far as they correspond to so-called psychical processes, and can be brought within the range of physiological experiments, and therefore united with the investigations of the physiologist, the neurologist, and the pathologist; (4) The study of the intensities of sensations and the effort to bring them to some exact system of measurement—the field of psycho-physics.

At the end of the school year, Dewey reported in another letter to the younger Angell that "the laboratory is beginning to get in shape," with Mead "trying to work out something on sensation on the biological side"—namely, "to see if one could get back of the present qualities and show the sensation as a condensation or precipitation of past organic activities, so that everything which is aesthetic [i.e., sensory] now was once practical or teleological." Mead's treatment of these topics "on the biological side"—including the Spencerian approach taken in his Special Topics in Psychology class the next year, as discussed in chapter 3—probably helped push Dewey's own work in that direction.

---

30. *Calendar of the University of Michigan for 1892–93* (Ann Arbor: University of Michigan, 1893), 67–68; Corliss Lamont, ed., *Dialogue on John Dewey* (New York: Horizon, 1959), 18–20, quoting a letter from Isaac Bernard Lipson, who took a laboratory class with Mead sometime between 1892 and 1894.

CHAPTER SEVEN

Horace Kallen's claim that Dewey abandoned his earlier, more idealistic approach to psychology under Mead's influence thus seems plausible.[31]

Lloyd was hired at the same time as Mead, having just completed two years of graduate work at Harvard and two more in Germany. He taught General Psychology (a lecture course) at Michigan from 1891 to 1894 but turned the subject over to others after Dewey and Mead left for Chicago.[32] Although Lloyd's doctoral thesis at Harvard was in ethics, Dewey told Angell in 1893 that it concerned biology and psychology as well:

> Lloyd made his degree on what seems to me a very suggestive thesis—the title is 'freedom' but might better be called perhaps 'environment.' The idea is to show that from the standpoints both of the best philosophy and of modern physiological psychology the self cannot be conceived as limited by environment, precisely because the self is environment generalized or set free.

Lloyd did not publish anything until the mid-1890s, but in an 1897 article—apparently based on his thesis—he argued that psychology "is more than epistemology; it is biology also." Lloyd's ideas were yet another version of the biological idealism described in chapter 4: "Environment or not-self proves to be the past made present, its qualification always a process of adjustment to the present. In other words, environment as differentially qualified and self as organically free and active develop together."[33] Dewey's letters to Angell in the early 1890s indicate that Lloyd was employing similar ideas at Michigan at that time, contributing to the department's collaborative research program in biology, psychology, and philosophy. In short, Dewey's colleagues at Michigan embraced the "biological conception" that he had found in James's *Principles of Psychology*.

Given all this biological and psychological research by Mead and others at Michigan, along with the organism-environment picture that Dewey found in the work of James and the British idealists, his move to a more biological approach in psychology and philosophy was overdeter-

---

31. "Experimental Psychology," *University Record* (Ann Arbor, MI) 2 (1893): 95; John Dewey to James Rowland Angell, 10 May 1893, in Dewey, *Correspondence*, no. 00478. For Kallen's claim, see Lamont, *Dialogue on John Dewey*, 99. See also Daniel R. Huebner, *Becoming Mead: The Social Process of Academic Knowledge* (Chicago: University of Chicago Press, 2014), 50–51.

32. Arthur Lyon Cross, DeWitt H. Parker, and R. M. Wenley, "Alfred Henry Lloyd, 1864–1927," *Journal of Philosophy* 25 (1928): 126; *Calendar of the University of Michigan for 1891–92* (Ann Arbor: University of Michigan, 1892), 59–60; *Calendar of the University of Michigan for 1892–93* (Ann Arbor: University of Michigan, 1893), 66–67; *Calendar of the University of Michigan for 1893–94* (Ann Arbor: University of Michigan, 1894), 68–69.

33. *The Harvard University Catalogue, 1893–94* (Cambridge, MA: Harvard University, 1893), 508; John Dewey to James Rowland Angell, 10 May 1893, in Dewey, *Correspondence*, no. 00478; Alfred Henry Lloyd, "The Stages of Knowledge," *Psychological Review* 4 (1897): 167, 172.

mined. By the mid-1890s, this approach was playing a prominent role in his psychology articles (as discussed in chapter 5) and in his ethics courses (as discussed in chapter 6). It was also central to his new research program in education at the University of Chicago. The published syllabus for Educational Psychology, which he taught in spring 1896, was filled with references to James and fellow psychologist James Mark Baldwin and opened with the claim that the "primary quality" of the organism is "selection and assimilation of environment."[34] Dewey also applied the "natural history" approach in his fall 1896 class "Philosophy of Education," where he began by defining *knowledge* as

the bridge or connecting link between some difficulty or friction which has arisen in action and a further successful or harmonized activity. Behind it lies the practical difficulty from which the individual is struggling to escape. Ahead of it lies the free or unified activity which he is endeavoring to reach. Knowledge is the path from one form of action to another.

Dewey, perhaps inspired by Mead's evolutionary standpoint, then outlined the biological and psychological foundations of this account of knowledge, which also recalled the connections between friction, readjustment, and evolution in Hegel. First, on the biological side, any "organ of knowledge . . . must have contributed something practical to the furtherance of life in order to gain any leverage for being selected and perpetuated." The brain and nervous system, said Dewey, "add to the control of the organism over the environment and make it possible to gain food or escape an enemy more easily." Second, on the psychological side, "knowledge is a form of attention" or is at least "directly dependent upon attention." This dependence arises because action requires "discovering and selecting a means which will help us reach a certain end"; attention, "directly bound up with interest," is the psychological basis of this selection.[35]

Dewey invoked the organism-environment dichotomy even more directly in his discussion of the child and the curriculum, which he compared to the organism and its environment: "We talk about the adjustment

---

34. "Educational Psychology: Syllabus of a Course of Twelve Lecture-Studies," in John Dewey, *The Early Works, 1882–1898*, ed. Jo Ann Boydston, 5 vols. (Carbondale: Southern Illinois University Press, 1969–72), 5:303–27; see also "The University Extension Division," *Quarterly Calendar* (Chicago) 4 (1896): 143; *Annual Register, July, 1895—July, 1896, with Announcements for 1896-7* (Chicago: University of Chicago Press, 1896), 328.

35. "Philosophy of Education (1896)," in Dewey, *Class Lectures*, 2:75–77; see also *Annual Register, July, 1895—July, 1896*, 54; "Philosophy of Education," *University Record* (Chicago) 1 (1896): 422.

of organism to the environment, about the adaptation to the environment: we talk as Mr. Spencer talks about life being the personal realization of the environment in the individual. What do we mean by all these phrases?" Replying to this question by alluding to his functional psychology, Dewey claimed that "organism and environment are simply the two sides of function. The organism is the method or implement of function. The environment is the supply [of] function." He also nodded to the Hegelian background discussed in chapter 4, claiming that "the actual life process is the only real thing," with *organism* and *environment* merely names for the same life process viewed from two different angles. Applying this general picture to education, Dewey suggested that we should view the teacher not as "the whole environment" but as "the medium through which the environment enters the child." Likewise, said Dewey, what matters is not the curricular framework as such but the "adaptation of it or translation of it into the experience of the child." Dewey's new research in education at Chicago was thus deeply indebted to the biological-psychological approach he developed at Michigan in conversation with James, Mead, and others.[36]

James's own work on education is less well known than Dewey's, but James also explicitly adopted a biological approach. In *Talks to Teachers*, published in 1899, James presented those parts of psychology that might be useful to educators. He began the third lecture by distinguishing two obvious functions of consciousness: "It leads to knowledge, and it leads to action." He then claimed that the "theory of evolution" was responsible for a shift in scientific and philosophical emphasis from the former to the latter:

Man, we now have reason to believe, has been evolved from infra-human ancestors, in whom pure reason hardly existed, if at all, and whose mind, so far as it can have had any function, would appear to have been an organ for adapting their movements to the impressions received from the environment, so as to escape the better from destruction. . . . Deep in our own nature the biological foundations of our consciousness persist, undisguised and undiminished.

James requested that his audience set aside philosophical debates about consciousness and assume the "point of view likely to be of greatest practical use," that of biology: "I shall ask you now . . . to adopt with me, in this course of lectures, the biological conception, . . . and to lay your

---

36. "Philosophy of Education (1896)," in Dewey, *Class Lectures*, 2:94–97.

own emphasis on the fact that man, whatever else he may be, is primarily a practical being, whose mind is given him to aid in adapting him to this world's life." Continuing this theme in the next lecture, James defined education as "the organization of acquired habits of conduct and tendencies to behavior" and argued that since teaching is about inculcating good habits, "the biological conception of the mind"—elsewhere described as "our evolutionary conception of the mind as something instrumental to adaptive behavior"—was the most useful conception for teachers. In the remainder of the series, James introduced his audience to several psychological concepts, including interest and attention. Throughout, he characterized students as "mere walking bundles of habits," with a plasticity that made education possible. James's psychological approach to education, like that of Dewey, was thus constructed on biological foundations.[37]

The methodological approach of *Studies in Logical Theory*, published by Dewey and his Chicago colleagues in 1903, was directly inspired by this psychological and educational research, as well as by Dewey's writings on ethics. In the first chapter of *Studies*, titled "Thought and its Subject-Matter: The General Problem of Logical Theory," Dewey argued that logic had strayed too far from the concrete contexts of "practical deliberation" and "scientific research." He declared that logic should attend to the limits and specificities of these contexts, instead of remaining in the realm of pure abstraction:

Both [deliberation and research] assume that every reflective problem and operation arises with reference to some *specific* situation, and has to subserve a *specific* purpose dependent upon its own occasion. They assume and observe distinct limits—limits from which and to which. There is the limit of origin in the needs of the particular situation which evokes reflection. There is the limit of terminus in successful dealing with the particular problem presented—or in retiring, baffled, to take up some other question.

A logic aligned with the concrete role of reflection "in everyday life and in critical science," said Dewey, would attend to "the natural history of thinking as a life-process." In other words, it would adopt "the evolutionary method," which "in biology and social history" treats organs, cells, and structures as "instrument[s] of adjustment or adaptation to a particular environing situation." Referring to his recently published article on "The Evolutionary Method as Applied to Morality," discussed in chapter 6,

---

37. William James, *Talks to Teachers on Psychology: And to Students on Some of Life's Ideals* (New York: Henry Holt, 1899), 22–25, 29, 31, 36, 65–66, 77.

CHAPTER SEVEN

Dewey claimed that just as we can only see the significance of moral practices and ideas when we attend to the conditions under which they originated, we can only assess the validity of a particular verification procedure or mode of reasoning "by reference to its efficiency in meeting its problems," both historical and contemporary. For Dewey and his Chicago colleagues, logic—like morality—was fundamentally "a mode of adaptation."[38]

James and Peirce were both excited by Dewey's manifesto, but in very different ways: James was enthusiastic and positive, whereas Peirce was skeptical and negative. In March 1903, after reading "with almost absurd pleasure" an essay by Dewey's colleague and former student Addison Webster Moore, James told Dewey that he saw "an entirely new 'school of thought' forming" at Chicago—like *Studies*, Moore's essay had appeared in the university's *Decennial Publications* series. Dewey replied that their approach was not new but went back to his years with Lloyd and Mead at Michigan, with Mead still working "mainly in biological terms" at Chicago. Dewey said that *Principles of Psychology* had been "the spiritual progenitor of the whole industry" and asked James whether he was willing to be the dedicatee of *Studies*, then in proofs. A week later, Dewey told him that the Chicago research "all go[es] back to certain ideas of life activity, of growth, and of adjustment."[39] As Moore wrote in the essay that prompted James's initial letter, psychologists—influenced by "the conceptions of biological evolution"—understood habit as "a co-ordination of activities" that is constantly being interrupted, leading to reorganization or reconstruction. Moore called this "the evolutionary character of experience." The logical implication of this view was that ideas only appear in response to such interruptions, since they do the work of reorganization. Its metaphysical implication was that reality is dynamic: "it is a reality of activity, of development, whose own very ongoing is ever creating a demand for new purposings, new thought, new effort."[40]

---

38. John Dewey, *Studies in Logical Theory* (Chicago: University of Chicago Press, 1903), 4–5, 13, 15–16; citing, on p. 15, John Dewey, "The Evolutionary Method as Applied to Morality," *Philosophical Review* 11 (1902). See also James Rowland Angell, "The Relations of Structural and Functional Psychology to Philosophy," in *The Decennial Publications*, 1st ser., vol. 3, pt. 2 (Chicago: University of Chicago Press, 1903), 62–66.

39. William James to John Dewey, 11 March 1903, John Dewey to William James, 20 March 1903, and John Dewey to William James, 27 March 1903, in James, *Correspondence*, 10:210, 214–15, 219; see also John Dewey, "Notes upon Logical Topics," *Journal of Philosophy, Psychology and Scientific Methods* 1 (1904): 60–61.

40. Addison Webster Moore, "Existence, Meaning, and Reality in Locke's Essay and in Present Epistemology," in *The Decennial Publications*, 1st ser., vol. 3, pt. 2 (Chicago: University of Chicago Press, 1903), 41–43, 50–51.

James became even more enthusiastic after reading *Studies in Logical Theory* later that year. He proclaimed to artist Sarah Wyman Whitman that "Chicago University has during the past 6 months given birth to the fruit of its 10 years gestation under John Dewey. The result is wonderful—a *real School*, and *real Thought*. Important thought, too! Did you ever hear of such a City or such a University? Here we have thought, but no school. At Yale a school but no thought. Chicago has both." A few days later, on November 2, James gave an informal presentation to Royce and other Harvard colleagues on "the Chicago school of Thought," which was soon expanded and published as "The Chicago School," a review essay that opened the first issue of Baldwin's new journal *Psychological Bulletin*.[41] Although James, along with Moore and Dewey, had previously noted the Hegelian flavor of the Chicago approach, he now highlighted its biological aspects:

Like Spencer's philosophy, Dewey's is an evolutionism. . . . Like Spencer, again, Dewey makes biology and psychology continuous. "Life," or "experience," is the fundamental conception; and whether you take it physically or mentally, it involves an adjustment between terms. Dewey's favorite word is "situation." A situation implies at least two factors, each of which is both an independent variable and a function of the other variable. Call them E (environment) and O (organism) for simplicity's sake. They interact and develop each other without end; for each action of E upon O changes O, whose reaction in turn upon E changes E, so that E's new action upon O gets different, eliciting a new reaction, and so on indefinitely. The situation gets perpetually "reconstructed," to use another of Professor Dewey's favorite words, and this reconstruction is the process of which all reality consists.[42]

For James, this dialectical model of organism-environment interaction (discussed in chapter 4) was the key to the Chicago perspective: not only do organisms change their environments and vice versa, but it is only the reconstructive process itself that is real. According to James, then, one of the strengths of Dewey's logic was its natural history approach.

---

41. William James to Sarah Wyman Whitman, 29 October 1903, in James, *Correspondence*, 10:324; William James, "Appendix IV: Notes for a Report on the Chicago School," in *Essays in Philosophy*, ed. Frederick H. Burkhardt, Fredson Bowers, and Ignas K. Skrupskelis (Cambridge, MA: Harvard University Press, 1978); William James, "The Chicago School," *Psychological Bulletin* 1 (1904).

42. James, "Chicago School," 2. For the Hegel connection, see Addison Webster Moore, "Existence, Meaning, and Reality in Locke's Essay and in Present Epistemology," in *The Decennial Publications*, 1st ser., vol. 3, pt. 2 (Chicago: University of Chicago Press, 1903), 16; William James to John Dewey, 23 March 1903, and John Dewey to William James, 27 March 1903, in James, *Correspondence*, 10:217, 220.

For Peirce, however, this approach was its main weakness. Dewey sent him a copy of *Studies in Logical Theory* in January 1904, and in a reply written in June (but perhaps never sent), Peirce sketched an argument against logic as natural history:

1. The two branches of natural history are physiology and anatomy.
2. The Chicago School must be engaged in "the Anatomy of Thought," since physiology has not been "revolutionized by conceptions of evolution" (paraphrasing Dewey's claim about natural history).
3. "Experiential diversity & absence of most possible forms . . . renders the kind of study called anatomy possible."
4. Logic is a normative science.
5. Normative science is concerned with "pure possibilities," not with "particular and variable facts," and "pure possibilities vary and diverge from one another . . . in every possible way."
6. Therefore, anatomy cannot be a normative science.
7. Therefore, there can be no "Anatomy of Thought" and logic cannot be natural history.

The fourth and fifth premises (among others) could of course be challenged, although Peirce insisted that if there were "a 'Natural History' . . . of thought," it would not be "the merely *possible* thought that Normative Science studies, but thought as it presents itself in an *apparently* inexplicable & irrational experience." At the end of his *Nation* review of *Studies*, published in September 1904, Peirce implied that such a natural history was not even deserving of the name "logic," as it could not "pronounce one proceeding of thought to be sound and valid and another to be otherwise."[43]

What explains this negative response? I agree with Larry Hickman's suggestion that Peirce—always eager to take offense—was personally hurt by *Studies*. Although Dewey had signaled his "indebtedness" to Peirce in the 1904 letter accompanying the book, Peirce's name did not appear in its pages, and this despite Dewey's contact with Peirce in graduate school and his reading of the older philosopher's *Monist* series. Notwithstanding the structural similarity between Dewey's biological idealist model of inquiry and Peirce's earlier doubt-belief model, there

---

43. John Dewey to Charles Sanders Peirce, 11 January 1904, and Charles Sanders Peirce to John Dewey, 9 June 1904, in Dewey, *Correspondence*, nos. 00929–30; paraphrasing John Dewey, *Studies in Logical Theory* (Chicago: University of Chicago Press, 1903), 15; Charles Sanders Peirce, "Logical Lights," *Nation*, September 15, 1904, 220.

appears to have been a real lack of engagement on Dewey's part at the time: in March 1903, by which time his logic was mostly developed, Dewey told James that he was just beginning to get more out of Peirce, who a few years earlier had been "mostly a sealed book." Dewey seems not to have read anything from Peirce's 1877–78 series, which appeared when he was still in college, until after James cited "How to Make Our Ideas Clear" in 1898. Even Dewey's 1900 essay "Some Stages of Logical Thought," with its claim that "thought is to be interpreted as a doubt-inquiry function, conducted for the purpose of arriving at that mental equilibrium known as assurance or knowledge," failed to cite Peirce. Nevertheless, Dewey did see Peirce's work as consonant with his own: around the time of the January 1904 letter, he suggested that Peirce's mathematical logic could potentially "transcend . . . the limitations of mere formalism and become a potent instrumentality in developing a system which has inherent reference to the pursuit of truth and the validation of belief."[44]

Beyond this biographical point, and as Jean-Marie Chevalier has argued in detail, Peirce's views of naturalism, normativity, logic, and psychology were much more complicated than his June 1904 letter indicated.[45] In this letter and in the *Nation* review, Peirce emphasized the part of logic he called "critic." As defined in Peirce's entry on "Logic" for Baldwin's *Dictionary of Philosophy and Psychology*, which also featured several entries by Dewey, *critic* is that branch of logic which produces a "classification of arguments, so that all those that are bad are thrown into one division, and those which are good into another." But much of Peirce's work on the logic of science straddled the line between this branch and another, "methodeutic," which "teaches the general

---

44. Larry Hickman, "Why Peirce Didn't Like Dewey's Logic," *Southwest Philosophy Review* 3 (1986): 179; Dewey to Peirce, 11 January 1904, and John Dewey to William James, 27 March 1903, in Dewey, *Correspondence*, nos. 00929 and 00800; John Dewey, "Some Stages of Logical Thought," *Philosophical Review* 9 (1900): 486; Dewey, "Notes upon Logical Topics," *Journal of Philosophy, Psychology and Scientific Methods* 1 (1904): 60. Although Dewey borrowed many journals from his college library, *Popular Science Monthly* was not among them: see Lewis S. Feuer, "John Dewey's Reading at College," *Journal of the History of Ideas* 19 (1958). On Dewey's early contact with Peirce, see John Dewey to H. A. P. Torrey, 5 October 1882, and John Dewey to William Torrey Harris, 17 January 1884, in Dewey, *Correspondence*, nos. 00415 and 00429. On Dewey's reading of Peirce's *Monist* series, see John Dewey, "The Superstition of Necessity," *Monist* 3 (1893): 362n; James Mark Baldwin, ed. *Dictionary of Philosophy and Psychology*, 2 vols. (New York: Macmillan, 1901–2), 2:721, s.v. "tychism"; John Dewey, "Logical Conditions of a Scientific Treatment of Morality," in *The Decennial Publications*, 1st ser., vol. 3, pt. 2 (Chicago: University of Chicago Press, 1903), 126n.

45. Jean-Marie Chevalier, "Why Ought We to Be Logical? Peirce's Naturalism on Norms and Rational Requirements," in *Liber Amicorum Pascal Engel*, ed. Julien Dutant, Davide Fassio, and Anne Meylan (Geneva: University of Geneva, 2014); see also Claudine Tiercelin, "Was Peirce a Genuine Anti-Psychologist in Logic?" *European Journal of Pragmatism and American Philosophy* 9, no. 1 (2017).

principles which ought to guide an inquiry."[46] As Hookway has discussed, in the early 1900s Peirce frequently criticized appeals to psychology by logicians such as Christoph Sigwart. But where did that leave the doubt-belief model, which, as we have seen, was explicitly psychological and biological? In Baldwin's *Dictionary*, Peirce treated this model as prior to and unaffected by psychological research:

> Under an appeal to psychology is not meant every appeal to any fact relating to the mind. For it is, for logical purposes, important to discriminate between facts of that description which are supposed to be ascertained by the systematic study of the mind, and facts the knowledge of which altogether antecedes such study, and is not in the least affected by it; such as the fact that there is such a state of mind as doubt, and the fact that the mind struggles to escape from doubt.

Although Peirce granted that such facts must be "carefully examined by the logician before he uses them as the basis of his doctrine," he believed that his doubt-belief model had survived such examination; thus, he could still treat it, along with its evolutionary underpinnings, as the basis of the logic of science.[47]

Just as it retained the doubt-belief model, Peirce's later logic also embraced the other "natural history" elements of his earlier work. In 1878 and 1883, as detailed earlier in this section, Peirce argued that natural selection had adapted the human mind to the natural and social world, resulting in our instinctive "special aptitudes for guessing right." In 1898, he made the same point in his Cambridge Conferences lectures:

> The instincts connected with the need of nutrition have furnished all animals with some virtual knowledge of space and of force, and made them applied physicists. The instincts connected with sexual reproduction have furnished all animals at all like ourselves with some virtual comprehension of the minds of other animals of their kind, so that they are applied psychists.[48]

---

46. James Mark Baldwin, ed., *Dictionary of Philosophy and Psychology*, 2 vols. (New York: Macmillan, 1901–2), 2:21, 75, s.vv. "logic," "method and methodology."

47. Christopher Hookway, "Normative Logic and Psychology: Peirce's Rejection of Psychologism," in *The Pragmatic Maxim: Essays on Peirce and Pragmatism* (Oxford: Oxford University Press, 2012); Baldwin, *Dictionary of Philosophy and Psychology*, 2:22, s.v. "logic."

48. "The First Rule of Logic [Fourth Cambridge Conferences Lecture]," MS 442 (21 February 1898), in Peirce, *Essential Peirce*, 2:51; see also "How to Theorize [Eighth Lowell Lecture]," MS 475 (17 December 1903), pp. 30–40, in Peirce, *Collected Papers*, 5:590–92; "Guessing," MS 687 (1907), in *Hound & Horn* 2 (1929): 268–69 and in *Collected Papers*, 7:38–40.

According to Peirce, evolution had given us an innate knowledge of nature, including human nature.

A few years later, in Baldwin's *Dictionary*, Peirce suggested that the adapted mind is what makes abduction possible—abduction being that form of probable reasoning in which "the reasoner . . . notices some remarkable character or relation among [the features of a phenomenon], which he at once recognizes as being characteristic of some conception with which his mind is already stored, so that a theory is suggested which would *explain* (that is, render necessary) that which is surprising." Peirce claimed, again recalling his earlier writings, that humans' success at abduction—a vital part of the logic of science—is due to "an affinity between [the reasoner's] ideas and nature's ways."[49] He put it as follows in an unpublished 1901 manuscript:

> It is a primary hypothesis underlying all abduction that the human mind is akin to the truth in the sense that in a finite number of guesses it will light upon the correct hypothesis. . . . Science will cease to progress if ever we reach the point where there is no longer an infinite saving of expense in experimentation to be effected by care that our hypotheses are such as naturally recommend themselves to the mind. . . . For the existence of a natural instinct for truth is, after all, the sheet anchor of science.

Whence this instinct? Why are we so good at abduction? In one of his 1903 lectures at the Lowell Institute in Boston, Peirce called it "a natural adaptation," and in one of his lectures that same year at Harvard, he hinted at a similar answer: "You may say that evolution accounts for the thing. I don't doubt it is evolution. But as for explaining evolution by chance, there has not been time enough." Setting aside the oblique criticism of "evolution by chance," which I will discuss in the next section, Peirce's implication was that we are all abduction engines, built by evolution. He even argued that our perceptual judgments—the rough accuracy of which is also presumably a result of evolution—should be "regarded as an extreme case of abductive inference." Thus, early and late, Peirce claimed that our evolutionary history underpins at least one key component of the logic of science.[50]

---

49. James Mark Baldwin, ed., *Dictionary of Philosophy and Psychology*, 2 vols. (New York: Macmillan, 1901–2), 2:427, s.v. "reasoning."

50. "On the Logic of Drawing History from Ancient Documents," MS 690 (1901); "The Nature of Meaning [Sixth Harvard Lecture]," MSS 314 and 316 (7 May 1903); and "Pragmatism as the Logic of Abduction [Seventh Harvard Lecture]," MS 315 (15 May 1903), in Peirce, *Essential Peirce*, 2:108, 217, 227; "How to Theorize [Eighth Lowell Lecture]," MS 475 (17 December 1903), p. 38, in Peirce,

Whether or not he received Peirce's 1904 letter, Dewey had already addressed similar criticisms. In the "Philosophy" entry for Baldwin's *Dictionary*, he described how in the nineteenth century, "historical method has had so profound an influence upon philosophic thought, that it is not yet possible to comprehend it, or to state its limits." Mentioning both Hegel and Spencer, Dewey suggested that "the radical distinction between questions of genesis, dealing with *how* things came to be, and questions of analysis, dealing with *what* they are," might be merely "a survival of an age which had not the historical point of view."[51] In his pamphlet *The Child and the Curriculum*, published in 1902, Dewey drew a related distinction between "the logical and psychological aspects of experience," comparing them to "the notes which an explorer makes in a new country, blazing a trail and finding his way along as best he may, and the finished map that is constructed after the country has been thoroughly explored." What is the point of the map? What is the use of logical theory? According to Dewey:

> The map, a summary, an arranged and orderly view of previous experiences, serves as a guide to future experience; it gives direction; it facilitates control; it economizes effort, preventing useless wandering, and pointing out the paths which lead most quickly and most certainly to a desired result. . . . That which we call a science or study puts the net product of past experience in the form which makes it most available for the future.

In short, logical theory offers control, guiding future experience by summarizing past experience.[52]

Although Dewey never discussed Peirce's distinction between *logica utens*, the "more or less conscious . . . classification of arguments, antecedent to any systematic study of [logic]," and *logica docens*, the improved classification that results from this study, his own account of science—and of logical theory more specifically—relied on a similar distinction. He put it this way in his presidential address before the American Philosophical Association in December 1905:

> The whole procedure of thinking as developed in those extensive and intensive inquiries which constitute the sciences, is but rendering into a systematic technique, into

---

Collected Papers, 5:591. For more on the instinctive aspects of Peircean abduction, see Sami Paavola, "Peircean Abduction: Instinct or Inference?" *Semiotica* 153 (2005).

51. James Mark Baldwin, ed., *Dictionary of Philosophy and Psychology*, 2 vols. (New York: Macmillan, 1901–2), 2:295–96, s.v. "philosophy."

52. John Dewey, *The Child and the Curriculum* (Chicago: University of Chicago Press, 1902), 25–27; see also John Dewey, *How We Think* (Boston: D. C. Heath, 1910), chap. 5.

an art deliberately and delightfully pursued, the rougher and cruder means by which practical human beings have in all ages worked out the implications of their beliefs, tested them and endeavored in the interests of economy, efficiency, and freedom, to render them coherent with one another.

Peirce was angry at Dewey for ignoring critic, but Dewey could have replied that he was actually engaged in methodeutic—a search for the general principles that should guide inquiry. For Dewey (as he wrote in *Studies*), the difference between "the methods of science and those of the plain man" could be summed up in one word: *control*. Just as science has given us greater control over our natural and social environments, logic promises to give us greater control over the process of inquiry itself, helping us identify the right problems and choose the right concepts. Dewey and Peirce disagreed about the importance of critic and the role of truth in inquiry, but they were both committed to guidance and control. It was this shared commitment that prompted each of them to seek the biological roots of inquiry. As Dewey put it in his 1905 address, "The testimony of biology is unambiguous to the effect that the organic instruments of the whole intellectual life, the sense-organs and brain and their connections, have been developed on a definitely practical basis and for practical aims, for the purpose of such control over conditions as will sustain and vary the meanings of life." For Dewey, and arguably even for Peirce, the instruments of logic are grounded in these practical aims of biology.[53]

Evolutionary Experimentalism

The pragmatists' focus on control was accompanied by a championing of experiment. In this final section, I will show that Charles Sanders Peirce, William James, and John Dewey all viewed pragmatism—and logic more generally—as closely connected with experiment. Each of them also linked repeated experimentation to evolutionary progress. But because of their divergent responses to the factors of evolution debates, described in chapter 5, they disagreed about the nature of this progress: for Dewey and James, it was open-ended and under our direct control; for Peirce, it was part of a cosmic tendency toward reasonableness.

---

53. James Mark Baldwin, ed., *Dictionary of Philosophy and Psychology*, 2 vols. (New York: Macmillan, 1901–2), 2:21, s.v. "logic"; John Dewey, "Beliefs and Realities," *Philosophical Review* 15 (1906): 123–25; John Dewey, *Studies in Logical Theory* (Chicago: University of Chicago Press, 1903), 9–10.

The link between pragmatism and the experimental attitude is well known. Peirce lamented in Baldwin's *Dictionary* that "for the last three centuries thought has been conducted in laboratories, in the field, or otherwise in the face of the facts, while chairs of logic have been filled by men who breathe the atmosphere of the seminary." He continued this line in the opening passage of his 1905 essay "What Pragmatism Is," published in the *Monist*:

> Every physicist, and every chemist, and, in short, every master in any department of experimental science, has had his mind moulded by his life in the laboratory to a degree that is little suspected. . . . Excepting perhaps upon topics where his mind is trammelled by personal feeling or by his bringing up, his disposition is to think of everything just as everything is thought of in the laboratory, that is, as a question of experimentation.

Peirce, who had engaged in astronomical, geodetic, and psychological research, saw himself as exemplifying this "experimentalist type" and pointed to the laboratory attitude as the key to pragmatism: "Whatever assertion you may make to [the typical experimentalist], he will either understand as meaning that if a given prescription for an experiment ever can be and ever is carried out in act, an experience of a given description will result, or else he will see no sense at all in what you say." For Peirce, pragmatism was fundamentally experimental.[54]

Dewey was delighted with "What Pragmatism Is": he wrote to Peirce in April 1905, telling him that the essay would "go far in clearing up and away a lot of current misconceptions." In an ungrateful reply, which he may never have sent, Peirce continued his attack on Dewey's genetic or evolutionary approach—*genetic* meaning relating to origination or development. Accusing Dewey of assuming "that any non-genetic logic will reach no conclusions that have any meaning in their real applications," Peirce reasserted his scientific authority:

> All my studies are conducted in full view of actual scientific memoirs and other records of scientific inquiry, in which they lead to denials of conclusions to which bad logic has led their authors; and some of my non-genetical studies have led directly to discoveries in mathematics and others to instituting experimental researches about the reality, if not the solidity, of which there can be no question.[55]

---

54. Charles Sanders Peirce, "What Pragmatism Is," *Monist* 15 (1905): 161–62.
55. John Dewey to Charles Sanders Peirce, 11 April 1905, and Charles Sanders Peirce to John Dewey, n.d., in Dewey, *Correspondence*, nos. 01007 and 00806; see also John Dewey, "What Does Pragmatism Mean by Practical?" *Journal of Philosophy, Psychology and Scientific Methods* 5 (1908): 86.

But this frustration at Dewey's marginalization of symbolic logic concealed areas of broad agreement. In particular, both Dewey and Peirce pointed to experimentation and control as important aspects of logic.

In "Issues of Pragmaticism," also published in the *Monist* in 1905, Peirce claimed that intellectual "self-control" is the goal of logic and that one of the "essential ingredients" of "the machinery of logical self-control" is "the formation of habits under imaginary action." A pragmatist approach to logic, said Peirce, which judges the content of a belief "by the conduct that it determines," can give us conscious access to that which normally "hides in the depths of our nature." Such access matters because "to say that an operation of the mind is controlled is to say that it is, in a special sense, a conscious operation; and this no doubt is the consciousness of reasoning." But what counts as reasoning? "In reasoning," said Peirce, "we should be conscious, not only of the conclusion, and of our deliberate approval of it, but also of its being the result of the premiss [*sic*] from which it does result, and furthermore that the inference is one of a possible class of inferences which conform to one guiding principle." If the reasoner is conscious of this guiding principle, the inference rises to the level of "logical argumentation." The true aim of logic, in Peirce's view, is conscious control over the habits and rules of one's conduct. Thus, in "What Pragmatism Is," he had described the pragmatist outlook as experimental and forward-looking. The meaning of a proposition is "the general description of all the experimental phenomena which the assertion of the proposition virtually predicts." Why? Because "future conduct is the only conduct that is subject to self-control."[56]

According to Peirce, all three of the modes of reasoning that make up the logic of science—deduction, induction, and abduction—are in some sense experimental. Take deduction, the least plausible case. In his Cambridge Conferences lectures of 1898, Peirce gave the audience a brief introduction to his "Existential Graphs," a multitier diagrammatic system for propositional, predicate, and even modal logic. These graphs demonstrated, he argued, that deductive reasoning involves experimentation: once we have laid out the premises of an argument in a single diagrammatic proposition,

> we proceed attentively to observe the graph. It is just as much an operation of *observation* as is the observation of bees. This observation leads us to make an *experiment*

---

56. Charles Sanders Peirce, "Issues of Pragmaticism," *Monist* 15 (1905): 482–83; Charles Sanders Peirce, "What Pragmatism Is," *Monist* 15 (1905): 174.

upon the graph. Namely, we first, duplicate portions of it; and then we erase portions of it, that is, we put out of sight part of the assertion in order to see what the rest of it is. We observe the result of this experiment, and that is our deductive conclusion.

Although Peirce admitted that this process involved only one experiment, he claimed that this was also the case in those physical sciences that study relatively uniform phenomena: an expert chemist, said Peirce, "contents himself with a single experiment to establish any qualitative fact." When he published another version of his system of existential graphs in the October 1906 issue of the *Monist*, Peirce again highlighted the experimental aspect of diagrammatic deduction:

> One can make exact experiments upon uniform diagrams; and when one does so, one must keep a bright lookout for unintended and unexpected changes thereby brought about in the relations of different significant parts of the diagram to one another. Such operations upon diagrams, whether external or imaginary, take the place of the experiments upon real things that one performs in chemical and physical research. Chemists have ere now, I need not say, described experimentation as the putting of questions to Nature. Just so, experiments upon diagrams are questions put to the Nature of the relations concerned.

According to Peirce, the logic of science—including even its deductive phase—is experimental.[57]

Dewey agreed: in a series of papers published between 1906 and 1908, he connected logic to biology, pragmatism, control, and experimentation. This focus on experiment distinguished his naturalism from that of his Columbia colleagues: as Wendell Bush reportedly said, "[Frederick] Woodbridge, he's looking backwards; Woodbridge should have been a bishop. Dewey—Dewey, he lives in a laboratory."[58] The laboratory attitude was prominent in Dewey's work. In a 1906 essay in *Mind*, for example, he boasted that "the experimental or pragmatic theory of knowledge explains the dominating importance of science" by interpreting meanings as "the instruments upon which fulfilment depends *so far as that is controlled* or other than accidental." For Dewey, as for Peirce, things become interesting when habit or expectation is frustrated: for instance,

---

57. "The First Rule of Logic [Fourth Cambridge Conferences Lecture]," MS 442 (1898), in Peirce, *Essential Peirce*, 2:45; Charles Sanders Peirce, "Prolegomena to an Apology for Pragmaticism," *Monist* 16 (1906): 493. Peirce also published part of his system of existential graphs in James Mark Baldwin, ed., *Dictionary of Philosophy and Psychology*, 2 vols. (New York: Macmillan, 1901–2), 2:645–650, s.v. "symbolic logic."

58. Corliss Lamont, ed., *Dialogue on John Dewey* (New York: Horizon, 1959), 96.

if we smell a rose but are unable to locate the source of the smell. At this point, according to Dewey, there are two options:

> By reason of disappointment, the person may turn epistemologist. He may then take the discrepancy, the failure of the smell to execute its own intended meaning, . . . as evidence of a contrast in general between things meaning and things meant. . . . One may then say: Woe is me; smells are only *my* smells, subjective states existing in an order of being made out of consciousness, while roses exist in another order made out of a radically different sort of stuff.

Or instead, "observing the futility of such a method, one may turn scientist, and then epistemologist only as logician, only, that is, as reflecting upon the nature and implications of the scientific process." For the scientist, the discrepancy is "evidence of the need of a more cautious and thorough inspection of odours and execution of operations indicated by them":

> One might, that is, observe the cases in which odours mean other things than just roses, might voluntarily produce new cases for the sake of further inspection, and thus come to account for the cases where meanings had been falsified in the issue; to discriminate more carefully the peculiarities of those meanings which the event verified, and thus to safeguard and bulwark to some extent the employ of similar meanings in the future.

Dewey argued that this logic of experiment should replace traditional epistemology, thus avoiding the radical gap between knowledge and reality that had so exercised philosophers.[59]

Dewey also linked experiment, environment, and guidance in his three-part 1907 essay "The Control of Ideas by Facts," a response to critics who claimed that pragmatism denied the external constraints of reality. John Edward Russell, one of these critics, introduced the case of a "lost sojourner in the Adirondacks" trying to find his way home. Russell claimed that the pragmatist ignored the "objective conditions" of the traveler's situation—that is, "the environment to which his action must be adjusted if it [is] to have a successful result."[60] Dewey replied

---

59. John Dewey, "The Experimental Theory of Knowledge," *Mind* 15 (1906): 303–4, 306. This essay was substantially revised in 1910, but the new version made the same basic points: see John Dewey, *The Influence of Darwin on Philosophy, and Other Essays in Contemporary Thought* (New York: Henry Holt, 1910), 99, 102, 109.

60. John E. Russell, "The Pragmatist's Meaning of Truth," *Journal of Philosophy, Psychology and Scientific Methods* 3 (1906): 600.

that according to his "experimental theory of logic," an idea was "not some little psychical entity or piece of consciousness-stuff" matching "the actually visible environment" but rather an "interpretation of the locally present environment in reference to its absent portions." The relevant agreement was not between our thoughts and reality but "between purpose, plan, and its own execution, fulfillment; between a map of a course constructed for the sake of guiding behavior and the result attained in acting upon the indications of the map." For Dewey, "the function of observation is to define the facts that describe the problem of a situation," and this always involves a specific purpose that "prescribes the selective determination of a constitution of the 'given' facts. The environment varies, in intellectual definition, as the organism, character or agent varies." Ongoing experimental verification was at the heart of Dewey's account:

> If by acting in accordance with the experimental definition of facts, *viz.*, as obstacles and conditions, and the experimental definition of the end or intent, *viz.*, as plan and method of action, a harmonized situation effectually presents itself, we have the adequate and the only conceivable verification of the intellectual factors. If the action indicated be carried out and the disordered or disturbed situation persists, then we have not merely confuted the tentative positions of intelligence, but we have in the very process of acting introduced new data and eliminated some of the old ones, and thus afforded a fresh opportunity for the resurvey of the facts and the revision of the plan of action.

For Dewey, what mattered was repeated experimentation guiding future action. As he stated at the beginning of the essay, this logic of verification had a biological basis—namely, "the interests of intelligence with all that intelligence imports in the exercise of the life functions."[61]

Expanding on the meaning of *life functions*, Dewey's 1908 essay "Does Reality Possess Practical Character?" summarized the biological outlook of pragmatism:

> The organism has its appropriate functions. To maintain, to expand adequate functioning is its business. This functioning does not occur *in vacuo* [in a vacuum]. It involves co-operative and readjusted changes in the cosmic medium. Hence the appropriate

---

61. John Dewey, "The Control of Ideas by Facts," *Journal of Philosophy, Psychology and Scientific Methods* 4 (1907): 201–2, 311, 313–14; also in John Dewey, *Essays in Experimental Logic* (Chicago: Chicago University Press, 1916), 237–41. The passage discussing "the function of observation" and "the 'given' facts" was removed from the 1916 version.

subject-matter of awareness is not reality at large, a metaphysical heaven to be mimeographed at many removes upon a badly constructed mental carbon paper which yields at best only fragmentary, blurred, and erroneous copies. Its proper and legitimate object is that relationship of organism and environment in which functioning is most amply and effectively attained; or by which, in case of obstruction and consequent needed experimentation, its later eventual free course is most facilitated.

He made a similar point later that year in the *Progressive Journal of Education*:

Pragmatism holds that all the higher achievements of individual organic life result from the stress and strain of the problem of maintaining the functions of life. For life can be kept going only as the organism "*makes* its living," by proper manipulation of the environment and adjustment of the latter to its own vital ends. Reduced to [its] simplest terms, the biological problem of the individual . . . is to subordinate the materials and forces of the natural environment so that they shall be rendered tributary to life-functions.

For Dewey, experimentation on, manipulation of, and adjustment to the environment were the biological functions that grounded pragmatist inquiry.[62]

These two 1908 essays were written in the wake of James's book *Pragmatism: A New Name for Some Old Ways of Thinking*, originally presented as lectures to Dewey and hundreds of others at Columbia University in early 1907. After deciding to give the Columbia lectures, which he had already presented at the Lowell Institute in Boston, James jokingly assumed the natural history point of view in a letter to Dewey's colleague James McKeen Cattell: "My organism can stand *that* extra strain, surely, before taking its eternal repose from the lecturing function."[63] As we have seen, James had adopted this point of view more seriously in his writings on psychology and education. It was also apparent in his 1904–5 essays on radical empiricism, which featured the notion of "pure experience"—that is, "the original flux of life before reflexion has categorized it." James argued that human experience was never pure but always translated into "a more intellectualized form, filling it with ever

---

62. John Dewey, "Does Reality Possess Practical Character?" in *Essays Philosophical and Psychological, in Honor of William James* (New York: Longmans, Green, 1908), 70–71; John Dewey, "The Bearings of Pragmatism upon Education: First Paper," *Progressive Journal of Education* 1, no. 2 (1908): 1.

63. William James, *Pragmatism: A New Name for Some Old Ways of Thinking* (New York: Longmans, Green, 1907), vii; William James to James McKeen Cattell, 1 January 1907, and William James to Alice Howe Gibbens James, 1 February 1907, in James, *Correspondence*, 11:297, 310–11.

more abounding verbalized distinctions." Why was this translation necessary? According to James, "the pragmatic answer" to this question "is that the environment kills as well as sustains us, and that the tendency of raw experience to extinguish the experient himself is lessened just in the degree in which the elements in it that have a practical bearing upon life are analyzed out of the continuum and verbally fixed and coupled together, so that we may know what is in the wind for us and get ready to react in time."[64] Continuing this line of thought in *Pragmatism*, James argued that our commonsense categories "are discoveries of exceedingly remote ancestors, which have been able to preserve themselves throughout the experience of all subsequent time." For instance, the idea of *kind* is merely a "colossally useful *denkmittel* [instrument of thought]," which helps straighten "the tangle of our experience's immediate flux." James's pragmatism thus echoed Dewey's logic: the theories of both science and common sense, said James, "are mental modes of *adaptation* to reality."[65]

This story about our knowledge of the world grew from James's earlier focus, in his critique of Spencer's psychology as discussed in chapter 2, on the importance of interest and attention. Our sensations, he wrote in *Pragmatism*, are "undoubtedly beyond our control; but *which* we attend to, note, and make emphatic in our conclusions depends on our own interests."[66] As in his experimental approach to ethics, described briefly in chapter 6, James saw reality as essentially open. We could, James suggested, imagine an account of reality "which it proves impossible to better or alter" and view the permanence of this impossibility as constituting the truth of that account. But in the end, what is primary is our own active role in shaping experience and reality: "We plunge forward into the field of fresh experience with the beliefs our ancestors and we have made already; these determine what we notice; what we notice determines what we do; what we do again determines what we experience." According to James, this open-endedness was what distinguished pragmatism from its competitors: "for rationalism reality is ready-made and complete from all eternity, while for pragmatism it is still in the

---

64. William James, "The Thing and Its Relations," *Journal of Philosophy, Psychology and Scientific Methods* 2 (1905): 30–31; also in William James, *A Pluralistic Universe: Hibbert Lectures at Manchester College on the Present Situation in Philosophy* (New York: Longmans, Green, 1909), 350. In the later version, James changed "the pragmatic answer" to "the naturalist answer."

65. William James, *Pragmatism: A New Name for Some Old Ways of Thinking* (New York: Longmans, Green, 1907), 170, 178–79, 194.

66. James, *Pragmatism*, 245; echoing William James, *The Principles of Psychology*, 2 vols. (New York: Henry Holt, 1890), 1:402.

making." James's pragmatism was directed toward an open-ended future and embraced a kind of evolutionary metaphysics.[67]

This evolutionary-experimental outlook was also emphasized by Dewey and others in their own overviews of pragmatism. For example, in the syllabus for "The Pragmatic Movement of Contemporary Thought," a class he taught at Columbia in the summer of 1909, Dewey provided the following as "historical background": "On the negative side, the pragmatic movement is developed by various deadlocks into which modern thought has run, thereby necessitating a reconsideration of fundamental premises. On its positive side, it grows out of the development of experimental methods and of genetic and evolutionary conceptions in science."[68] In another overview at John Hopkins early in 1910, which identified the same negative and positive "motives for philosophic revision," Dewey devoted an entire lecture to "The Biological Foundations," summarized in the syllabus as follows: "The problem of control of the environment. The function of sense-organs; of the central organs. Adjustment (habit) and adjusting (attention); reflection as readjusting. The novel, prospective, and precarious factor. Needs, experiments and success (satisfaction)." His next lecture reported that these biological foundations—which involved control of the environment, adjustment, and experimentation—had "equivalents in logical theory."[69] Dewey thus framed pragmatism as experimental and biological not only in his more technical articles but also in his overviews of the movement.

That same year, Addison Webster Moore—whose 1903 essay had so impressed James and whom Mead later called "after Mr. Dewey the most important and most authoritative member of the so-called Chicago school"—published *Pragmatism and Its Critics*, dedicated to Dewey. The first five chapters, which were based on a series of public lectures given at the University of Chicago in 1908, provided yet another summary of the pragmatic movement. They covered "some phases of the movement" that Moore deemed neglected in the general discussion of pragmatism, including its "historical background" and "the central rôle of the conception of evolution."[70] He suggested that the pragmatists'

---

67. James, *Pragmatism*, 250, 255, 257 (italics removed).
68. "Syllabus: The Pragmatic Movement of Contemporary Thought," in John Dewey, *The Middle Works, 1899–1924*, ed. Jo Ann Boydston, 15 vols. (Carbondale: Southern Illinois University Press, 1976–83), 4:253; "Department of Philosophy," *Columbia University Quarterly* 11 (1909): 385–86.
69. "Syllabus of Six Lectures on 'Aspects of the Pragmatic Movement of Modern Philosophy,'" in Dewey, *Middle Works*, 6:175–76; "Notes and News," *Journal of Philosophy, Psychology and Scientific Methods* 7 (1910): 83.
70. George Herbert Mead, "Doctor Moore's Philosophy," *University Record* (Chicago), n.s., 17 (1931): 48; Moore, *Pragmatism and Its Critics* (Chicago: University of Chicago Press, 1910), [v]–vii;

contribution to the factors of evolution debates had been to highlight "purposive, ideational control" of biological variation, alluding to Peirce's and Dewey's sympathy for directed variation in evolution, as discussed in chapter 5:

> Does not the simple recognition of the variation of types open the way for any type of variation that may be efficient, and therefore possibly of the ideational, purposive type of variation? That is, must not variation in species admit variation in species of variation as well? The admission of variation of types, with the limitation of the type of variation to merely "natural," that is, non-purposive selection, seems dogmatic, to say nothing of the facts. Variation in species implies at least the possibility of a purposive species of variation.[71]

Moore may have been recalling Dewey's ethics sequence at the University of Chicago, which he probably attended as a graduate student in in the 1890s. Lecture notes from the 1898 version of Dewey's Political Ethics class include the following passage:

> There is much discussion among speculative biologists as to whether the variations are determinate or indeterminate. If we realize that the variation is a variation, it is indeterminate, that is, in so far as the variation is tentative, in so far as it is experimental. If we look at it the moment it occurs with relation to its future development it would be indeterminate, but if we remember that every variation must be a mediation of function already in existence, that that variation cannot break in arbitrarily from the outside, nor break loose arbitrarily from the inside, but that it represents simply a reflection of the activities, of the life habits previously exercised, we would think of the variation as determinate. It is mere mythology to say that there is nothing at all which controls it. Of course it is controlled all the time by the function of which it is after all simply a modification, simply a mediation.[72]

That is, according to Dewey and Moore, variation is tentative and experimental but is also purposive or determinate insofar as it mediates existing habits. This purposive character is even more obvious in logical theory: for Moore, the idea of the "working hypothesis," which "marks

---

"The Summer Quarter," *Chicago Alumni Magazine* 2 (1908): 75. Evolution and biology were also central to H. Heath Bawden, *The Principles of Pragmatism: A Philosophical Interpretation of Experience* (Boston: Houghton Mifflin, 1910).

71. Addison Webster Moore, *Pragmatism and Its Critics* (Chicago: University of Chicago Press, 1910), 77; see also Horace M. Kallen, "John Dewey and the Spirit of Pragmatism," in *John Dewey: Philosopher of Science and Freedom*, ed. Sidney Hook (New York: Dial, 1950), 13.

72. "Political Ethics (1898)," in Dewey, *Class Lectures*, 1:1658–59.

the appearance of the conception of evolution in logic," implies "that thought is an actual manipulation of our 'spontaneous' experiences," leading "to the control of these spontaneous variations, and to the introduction of new variations." Directed variation had its analogue in logical control.[73]

Some readers may be wondering at this point about the status of Rorty's criticism, which opened this chapter. Is "biological pragmatism" a fundamentally flawed approach to logic? I will return to the question in the conclusion of this book, but it is worth stating here that some of Dewey's and James's earliest critics drew on evolutionary ideas to criticize their dismissal of formal logic. Grace and Theodore de Laguna, writing in 1910, admitted that "pragmatism [was] the first whole-hearted attempt at an appreciation of the significance of Darwinism for logical theory." But the Lagunas, who both taught at Bryn Mawr College, argued that the pragmatists had "not carried their evolutionism far enough," criticizing them for failing to distinguish between survival value and emotional satisfaction and suggesting that the rapidity of social evolution made it "increasingly independent" of the control of natural selection. They also claimed that Dewey's attack on logical abstraction was undermined by ordinary cases of hypothesis-testing, in which a failed test almost never calls into question the logical validity of the inference procedure. The Lagunas proposed that the pragmatists (i.e., Dewey and James) give up their official opposition to formal logic and acknowledge that there is "a specific interest attaching to the logical situation as such," independent of our more concrete practical interests. All of these criticisms were made from an evolutionary point of view, since this was the one aspect of pragmatism that the Lagunas enthusiastically embraced. They even presented an evolutionary interpretation of the a priori, suggesting that the pragmatists were too quick to dismiss "fundamental categories of thought":

When a succession of concepts appears, each of which has arisen as a modification of the preceding complex, a certain relative stability belongs to the earlier members. Not as if temporal priority gave a logical priority in the ordinary sense of the term; for the later does not come as a mere accretion to the earlier, but as a modification of it which goes to the formation of a more complex unity. But the earlier has nevertheless this preference: that, as the further revision of the complex becomes necessary, this takes place, as far as possible, in the later elements; and only such portion of the correction

---

73. Addison Webster Moore, *Pragmatism and Its Critics* (Chicago: University of Chicago Press, 1910), 78–79.

as cannot be made here is passed back farther and farther, until the disturbing conditions are satisfied. This, indeed, appears to be a general characteristic of all evolution, and forms a part, at least, of what is commonly alluded to as the "continuity" of the process. It may, therefore, naturally be expected, that among our concepts there are certain ones which are not observably affected in the course of ordinary experience, and thus stand to the whole of our thought as nearly as possible in the relation of an *a priori* ground.[74]

Although many of pragmatism's opponents were opposed to the use of biological ideas in philosophy, the example of the Lagunas—like that of Peirce—shows that even those in favor of a broader evolutionary naturalism could challenge the pragmatists' approach to logic.

James, Dewey, and Peirce all linked pragmatism to both experimentation and evolution. According to James and Dewey, we have control over the direction of epistemic evolution: just as there is no *summum bonum* (highest good) in ethics, there is no *summa veritas* (highest truth) in logic; experimentalism rules out any guiding teleology. Moreover, although Dewey was attracted to the idea of directed variation in evolution, he still—like James—emphasized individual and social purposes. According to Peirce, on the other hand, the overall direction of the evolution of thought transcends human interests and cannot be fully explained by the operation of chance or the force of logic. His neo-Lamarckism, as we saw in chapter 5, implied that we are inevitably swayed "by an immediate attraction for the idea itself, whose nature is divined before the mind possesses it, by the power of sympathy." This mental evolution also corresponded to a broader cosmic evolution. Peirce argued in the first essay of his early 1890s *Monist* series that the broader process, guided by a "generalizing tendency," had an ultimate goal—namely, "an absolutely perfect, rational, and symmetrical system, in which mind is at last crystallized in the infinitely distant future."[75]

In the early 1900s, Peirce frequently invoked this evolutionary goal in his discussions of logic and pragmatism. His entry on *pragmatism* in Baldwin's *Dictionary of Philosophy and Psychology* claimed that "the only ultimate good which the practical facts to which [the pragmatic maxim]

---

74. Grace de Laguna and Theodore de Laguna, *Dogmatism and Evolution: Studies in Modern Philosophy* (New York: Macmillan, 1910), 123, 137–38, 148, 208–10, 214–15. On the conservation of earlier members of a complex as "a general characteristic of all evolution," see Jeffrey C. Schank and William C. Wimsatt, "Generative Entrenchment and Evolution," in *PSA: Proceedings of the Biennial Meeting of the Philosophy of Science Association* 2 (1986).

75. Charles Sanders Peirce, "Evolutionary Love," *Monist* 3 (1893): 191; Charles Sanders Peirce, "The Architecture of Theories," *Monist* 1 (1891): 176.

directs attention can subserve is to further the development of concrete reasonableness," and he stated that "almost everybody will now agree that the ultimate good lies in the evolutionary process in some way." This "evolutionary process," which he glossed as "the growth of reasonableness," involved "the coalescence, the becoming continuous, the becoming governed by laws, the becoming instinct with general ideas." The phrase "becoming instinct with" is obscure to us, but Peirce's own *Century Dictionary* definition of *instinct* makes his usage clear: "urged or animated from within; moved inwardly; infused or filled with some active principle." For Peirce, the evolutionary growth of reasonableness thus amounts to the universe becoming more and more infused with and animated by general ideas and laws. Discussing the same topic in his Baldwin's *Dictionary* entry on *uniformity*, Peirce argued (as in his *Monist* series) that microcosm mirrored macrocosm: "All laws are results of evolution. . . . Underlying all other laws is the only tendency which can grow by its own virtue, the tendency of all things to take habits. Now since this same tendency is the one sole fundamental law of mind, it follows that the physical evolution works towards ends in the same way that mental action works toward ends." For Peirce, chance and natural selection were not sufficient to explain this evolution "from difformity to uniformity," and he concluded optimistically that "all this, according to the writer, constitutes a hypothesis capable of being tested by experiment."[76]

Although Peirce was critical of Spencer in the *Monist* series and in Baldwin's *Dictionary*, he shared the English philosopher's view that natural selection of fortuitous variations was not the primary factor in evolution. This neo-Lamarckian commitment explains Peirce's comment at Harvard in 1903 (quoted earlier), since he associated chance variation with natural selection and neo-Darwinism: "As for explaining evolution by chance, there has not been time enough." As he had noted the year before, "The neo-Darwinians seem to wish to make reproduction and variation as mechanical as they can. This is a praiseworthy effort, because it must inevitably eventuate in making the truth more plain that they are not mechanical."[77] In his *Minute Logic* manuscripts,

---

76. James Mark Baldwin, ed., *Dictionary of Philosophy and Psychology*, 2 vols. (New York: Macmillan, 1901–2), 2:322, 2:731, s.vv. "pragmatism," "uniformity"; William Dwight Whitney, ed., *The Century Dictionary: An Encyclopedic Lexicon of the English Language*, 6 vols. (New York: Century, 1889–91), 3:3123, s.v. "instinct." Thanks to Wayne Myrvold for alerting me to this earlier usage of *instinct*.

77. "The Nature of Meaning [Sixth Harvard Lecture]," MSS 314 and 316 (7 May 1903), in Peirce, *Essential Peirce*, 2:217; "Carnegie Institution Correspondence," L75 (1902), in Peirce, *New Elements*, 4:66. For Peirce's criticisms of Spencer's definition of evolution, see Charles Sanders Peirce, "The

also written in 1902, Peirce connected his neo-Lamarckism with both experimentation and abduction. Scientists and engineers, said Peirce, "proceed by experimentation" and make steady improvements to human knowledge and inventions. He argued that evolutionary progress was a parallel case, as "the theory of natural selection is that nature proceeds by similar experimentation to adapt a stock of animals or plants precisely to its environment." But chance variation and natural selection are not sufficient to explain any of these progressive changes, according to Peirce. Both logical abduction and biological variation involve a kind of adaptive tendency: "It is no light question how it is that a stock in some degree out of adjustment with its environment immediately begins to sport, and that not wildly but in ways having some sort of relation to the change needed. Still more remarkable is the fact that a man before whom a scientific problem is placed immediately begins to make guesses, not wildly remote from the true guess." In Peirce's view, "this marked, though excessively imperfect, divinatory power of guessing right on the part of the man and on the part of the organic stock" could be explained in one of two ways: either reason knows "how Reason will act" and "Nature is ruled by a Reasonable Power," or "the tendency to guess nearly right is itself the result of a similar experimental procedure." Peirce indicated that each of these explanations of the adapted mind had its attractions, although it is unclear which he supported. This much is clear, however: Peirce believed that the logic of science was both evolutionary and experimental, mirroring nature's own bias toward reasonableness.[78]

For all of the "classical" pragmatists, so named because they were the most prominent American defenders of pragmatism in the years immediately following its introduction in 1898, logic was tied to evolution and organism-environment interaction. In the 1870s and '80s, well before pragmatism got its name, Peirce grounded his logic in the struggle to escape doubt and attain belief, tracing this struggle to its evolutionary source. At around the same time, James adopted the natural history point of view—based on Spencer's organism-environment dichotomy—in much of his psychological work. This work, along with the

---

Architecture of Theories," *Monist* 1 (1891): 165; Baldwin, *Dictionary of Philosophy and Psychology*, 2:731, s.v. "uniformity." For Spencer's criticism of natural selection, see Herbert Spencer, "The Inadequacy of 'Natural Selection,'" *Contemporary Review* 63 (1893).

78. "Minute Logic, Chapter I. Intended Characters of this Treatise," MS 425 (1902), in Peirce, *Collected Papers*, 2:86; see also "Guessing," MS 687 (1907), *Hound and Horn* 2 (1929): 268–69, also in Peirce, *Collected Papers*, 7:38–40.

experimental research of Tufts and Mead and the biological idealism of the Oxford Hegelians, pushed Dewey in the direction of biology in the early 1890s. The late 1890s educational writings of both Dewey and James employed the organism-environment framework, which was also central to Dewey's approach to ethics at the time, as discussed in chapter 6. In *Studies in Logical Theory*, Dewey and his Chicago colleagues endorsed an "evolutionary method" in logic, which impressed James and horrified Peirce. But although Peirce attacked the Chicagoans' "natural history" approach, he still linked certain aspects of his own logic to evolution—namely, the doubt-belief model and abduction. Moreover, Peirce, Dewey, and James all connected logic—and pragmatism more generally—to experimentation, which was also tied to evolution. For Peirce, the logic of science and the pragmatic maxim were essentially experimental, and adaptive variation was the norm in both scientific and evolutionary progress. Dewey traced his "experimental logic" to the "life-functions" that allow an organism to adapt to and manipulate its environment, and James treated our basic concepts and categories as products of evolution. Although Dewey and Peirce were both sympathetic to directed variation in evolution, Dewey and his students linked it to human purposes, whereas Peirce viewed it as part of a cosmic "generalizing tendency."

As the parallels between this chapter and chapter 6 suggest, the pragmatists saw both ethics and logic, along with moral and epistemic progress, as evolutionary and experimental. James described the framework of the Chicago school as one in which organism and environment "interact and develop each other without end," and as we have seen over the course of chapters 5 and 6, the Chicagoans applied this framework in both their ethical and logical theories. When Dewey endorsed an "evolutionary method" in logic, he cited "The Evolutionary Method as Applied to Morality," and he may also have been thinking of Addams's evolutionary approach to social ethics. For Dewey and his colleagues, inquiry as organism-environment interaction was addressed to both moral and scientific problems.[79] Peirce likewise emphasized parallels between logic and ethics throughout his career. In his 1877–78 series, he argued that the logician and probabilistic reasoner is forced to adopt the perspective of a community extending "to all races of beings with whom we can come into immediate or mediate intellectual relation," reaching

---

79. William James, "The Chicago School," *Psychological Bulletin* 1 (1904): 2; John Dewey, *Studies in Logical Theory* (Chicago: University of Chicago Press, 1903), 15; John Dewey, "The Evolutionary Method as Applied to Morality," *Philosophical Review* 11 (1902).

"beyond this geological epoch, beyond all bounds." He declared that "he who would not sacrifice his own soul to save the whole world, is, as it seems to me, illogical in all his inferences, collectively. Logic is rooted in the social principle." In later work, Peirce continued to highlight the parallels between logic and morals: both were normative sciences based on self-control.[80]

As Paul Forster has shown, Peirce viewed the pursuit of truth as itself a moral ideal.[81] Like Dewey, he saw both scientific and moral inquiry as in some sense evolutionary. But instead of organism-environment interaction, Peirce highlighted his aforementioned cosmic telos—reasonableness. In 1901, Peirce contrasted his own evolutionism with that of the English statistician and biologist Karl Pearson, whose book *The Grammar of Science* had claimed on its very first page that "a stable and efficient society" is the result of "the inertness, nay, rather active hostility, with which human societies receive all new ideas." This hostility, said Pearson, is the social analogue of natural selection: "It is the crucible in which the dross is separated from the genuine metal, and which saves the body-social from a succession of unprofitable and possibly injurious experimental variations."[82] Criticizing the first chapter of *Grammar of Science*, Peirce implicitly opposed his neo-Lamarckism to Pearson's neo-Darwinism. According to Peirce, "the man of science" has not been primarily motivated by interest in individual happiness or social stability, but rather by

> a deep impression of the majesty of truth, as that to which sooner or later, every knee must bow. He has further found that his own mind is sufficiently akin to that truth, to enable him, on condition of submissive observation, to interpret it in some measure. . . . The very being of law, general truth, reason—call it what you will—consists in its expressing itself in a cosmos and in intellects which reflect it, and in doing this progressively; and that which makes progressive creation worth doing—so the researcher comes to feel—is precisely the reason, the law, the general truth for the sake of which it takes place.

The foregoing analyses of Peirce's writings on abduction and cosmic evolution point to a simple interpretation of this passage: the minds of scientific researchers are evolutionarily prepared to tap into the progres-

---

80. Charles Sanders Peirce, "The Doctrine of Chances," *Popular Science Monthly* 12 (1878): 610–11; "The Three Normative Sciences [Fifth Harvard Lecture]," MS 312 (30 April 1903), in Peirce, *Essential Peirce*, 2:196–207; Charles Sanders Peirce, "Issues of Pragmaticism," *Monist* 15 (1905): 482.

81. Paul Forster, *Peirce and the Threat of Nominalism* (Cambridge, MA: Cambridge University Press, 2011), chap. 11.

82. Karl Pearson, *The Grammar of Science*, 2nd ed. (London: Adam & Charles Black, 1900), 1. Although Peirce was reviewing the second edition, this passage also opened the first edition of 1892.

sive growth of cosmic reasonableness. He went on to propose that something like the motive of the man of science is also the *summum bonum* in ethics: "The only desirable object which is quite satisfactory in itself without any ulterior reason for desiring it, is the reasonable itself." Peirce concluded his criticism by suggesting that the evolutionary ethics of Leslie Stephen (editor, critic, mountaineer, freethinker) had revealed the inadequacy of any utilitarianism as crude as Pearson's.[83]

Stephen's own vision of moral progress helps underline the differences between Peirce and Dewey, who also cited the English critic's influence:

> Moral progress involves a constant laying down of new problems. Old evils are avoided, old hostilities reconciled, the whole life is fuller and more vigorous; but the process implies at the same time that the new capacities and sensibilities developed constantly bring with them new evils or difficulties which again require to be reconsidered. . . . To improve, whether for the race or the individual, whether in knowledge or sympathy, is to be put in a position where a new set of experiments has to be tried, and experience to be bought at the price of pain. And as this seems to be esssentially [sic] implied in all progress that we can imagine, I see no reason to suppose that pain will be eliminated. . . . From the scientific point of view we may hold that evolution implies progress—progress at any rate to a point beyond our present achievements; and, further, progress implies a solution of many discords, and an extirpation of many evils; but I can at least see no reason for supposing that it implies the extirpation of evil in general or the definitive substitution of harmony for discord.

For Stephen, as for Dewey, progress will forever bring new difficulties, and continued experimentation will always be necessary—this is why both of them were opposed to what James referred to as Spencer's "milk-and-water paradise." For Peirce, on the other hand, progress has "the reasonable itself" as its ultimate goal. His view of scientific and moral evolution was thus distinct from that of Dewey and James, who were opposed to any global teleology.[84]

A key experimentalist concept that linked pragmatist logic and pragmatist ethics was the *working hypothesis*, an idea that was used in the last third of the nineteenth century to refer to a position strategically adopted for its experimental tractability and its potential to be fruitful

---

83. Charles Sanders Peirce, "Pearson's Grammar of Science: Annotations on the First Three Chapters," *Popular Science Monthly* 58 (1901): 296–300; citing, on p. 300, Leslie Stephen, *The Science of Ethics* (London: Smith, Elder, 1882).

84. Stephen, *Science of Ethics*, 445–46; William James, "Herbert Spencer's Data of Ethics," *Nation*, September 11, 1879, 179. Dewey acknowledged his deep obligation to Stephen's book in John Dewey, *Outlines of a Critical Theory of Ethics* (Ann Arbor, MI: Inland Press, 1891), vii.

even if ultimately proven false.[85] The concept was often employed by those associated with the Metaphysical Club of the early 1870s. Chauncey Wright said of natural selection, "[It is in] its value and use as a working hypothesis, that its principal claim to respect consists." James, in an early version of his will-to-believe doctrine, claimed that *"faith* and *working hypothesis"* differ only in "the time required for verification": "A certain hypothesis, in physics, will be verified after half an hour. A hypothesis like that of transformism [i.e., evolution] will require more than a generation to solidly establish itself, and hypotheses of a universal order, such as those we are talking about [i.e., the problems of philosophy], may remain subject to doubt for many more centuries." Later, in the abridged version of his *Psychology*—alluding to the views of Thomas Henry Huxley—James assumed "the uniform correlation of mind-states with brain-states," calling it "the 'working hypothesis' which underlies all the 'physiological psychology' of recent years." In a mid-1890s advertisement for his never-published work *The Principles of Philosophy*, Peirce admitted that his "theory of universal evolution, which supposes matter and its laws to be the result of evolution, . . . is to be regarded for the present as no more than a working hypothesis." In 1898, he suggested that reality itself is only an abduction, "a working hypothesis which we try, our one desperate forlorn hope of knowing anything."[86]

James and Peirce thus designated many of their philosophical views as working hypotheses, consistent with their more general experimental attitude. The Chicago pragmatists also gave the notion a prominent place in their work. In his 1894 book *The Study of Ethics*, Dewey argued that ideals in ethics are akin to the hypotheses that guide scientific research: "True ideals are the *working hypotheses* of action; they are the

---

85. Benjamin E. Smith, ed., *The Century Dictionary Supplement*, 2 vols. (New York: Century, 1909), 1:616, s.v. "working hypothesis." Since Peirce wrote the entry for *hypothesis* in the original dictionary, this later entry may also have been written by Peirce. It is relatively similar to his earlier definition of the phrase in Charles Sanders Peirce, "Ritchie's Darwin and Hegel," *Nation*, November 23, 1893, 394.

86. Chauncey Wright, "The Uses and Origin of the Arrangement of Leaves in Plants," *Memoirs of the American Academy of Arts and Sciences* 9 (1873): 379; William James, "Quelques considérations sur la méthode subjective," *Critique Philosophique* 6 (1878): 411–12 ("working hypothesis" was left in English); see also William James, "Rationality, Activity and Faith," *Princeton Review* 2 (1882): 73–74; William James, *Psychology: Briefer Course* (New York: Henry Holt, 1892), 6; Thomas Henry Huxley, "On Sensation and the Unity of Structure of Sensiferous Organs," *Nineteenth Century* 5 (1879): 606; [Advertisement for] *The Principles of Philosophy*, reprinted in Charles Sanders Peirce, *Reasoning and the Logic of Things: The Cambridge Conferences Lectures of 1898*, ed. Kenneth Laine Ketner (Cambridge, MA: Harvard University Press, 1992), 14; Charles Sanders Peirce, "The Logic of Relatives [Third Cambridge Conferences Lecture]," MS 439 (1898), in Peirce, *Reasoning and the Logic of Things*, 161. Huxley seems to have been one of the first to use the phrase "working hypothesis": see Thomas Henry Huxley, "Science," *Westminster Review* 63 (1855): 251.

best comprehension we can get of the value of our acts; their use is that they mark our consciousness of what we are doing, not that they set up remote goals. Ideals are like the stars; we steer by them, not towards them."[87] As we saw in chapter 6, Mead argued for the importance of working hypotheses in social reform, claiming that just as the natural sciences assume "that the world is as a whole governed by laws that involve the interaction of all its forces," social reform assumes "that human society is governed by laws that involve its solidarity." The Chicago pragmatists also invoked the concept in their discussions of logic. Recall that according to Moore, both ideas in logic and variations in biology are examples of working hypotheses under some form of purposive control: "The working hypothesis, as employed in modern science, marks the appearance of the conception of evolution in logic. As it is the logical expression of mutation of species, so, on the other hand, it marks the appearance of ideas in the process of evolution. As it is an evolutionizing of logic, of science, so it is a logicizing of evolution."[88] For the pragmatists, ideals in ethical inquiry and rules in logical inquiry are working hypotheses, adopted provisionally and subject to experimental revision. Evolution requires variation, and moral and scientific progress require experimentation.

---

87. John Dewey, *The Study of Ethics: A Syllabus* (Ann Arbor, MI: Inland, 1894), 41; see also John Dewey, "Self-Realization as the Moral Ideal," *Philosophical Review* 2 (1893): 664.
88. Addison Webster Moore, *Pragmatism and Its Critics* (Chicago: University of Chicago Press, 1910), 78.

Conclusion

I have shown in this book that the early American pragmatists were enthusiastic participants in conversations about biology and evolution in the late nineteenth century, an interest that culminated in the development of an experimental-evolutionary approach to moral and scientific inquiry in the late 1890s and early 1900s.

Returning to the questions posed in the introduction, we are now in a position to provide some answers. As shown in chapter 2, Herbert Spencer was at least as important as Charles Darwin for the first-cohort pragmatists. Chauncey Wright, Charles Sanders Peirce, and William James defined their own philosophical projects in opposition to those of Spencer, who—as James emphasized—popularized the notion of life or experience as the interaction of an organism and its environment.[1] The English philosopher was also arguably the most important scientific interlocutor for the older members of the second cohort of pragmatists, most of whom read Spencerian periodicals such as *Popular Science Monthly* in college and eventually taught courses on Spencer's philosophy, passing his ideas on to their younger students. When we repeatedly encounter the term *environment* in the early sociological writings of W. E. B. Du Bois, or in anything written by George Herbert Mead or John Dewey, this is evidence that they were part of a broadly Spencerian

---

1. William James, "Herbert Spencer Dead," *Evening Post* (New York), December 8, 1903; William James, "The Chicago School," *Psychological Bulletin* 1 (1904): 2; see also Trevor Pearce, "From 'Circumstances' to 'Environment': Herbert Spencer and the Origins of the Idea of Organism-Environment Interaction," *Studies in History and Philosophy of Biological and Biomedical Sciences* 41 (2010).

tradition in philosophy and the social sciences, even as they opposed many of the views of Spencer himself.[2] The celebrated title essay of Dewey's *The Influence of Darwin on Philosophy*, written for Darwin's centenary, has obscured the fact that Dewey did not study Darwin's work in any detail. Even when noted, this fact has misled some scholars into thinking that Dewey failed to understand evolution or that biological ideas played a superficial role in his philosophy.[3] But as this book has shown, evolutionary ideas in the late nineteenth century were not synonymous with those of Darwin.

At least in earlier scholarship, there has likewise been a tendency to assume that evolution is monolithic: commentators link passages discussing evolution from texts written decades apart, as if the scientific context remained the same.[4] As the cohort approach helps demonstrate, however, the pragmatists engaged biological ideas in a wide variety of contexts. In the first two chapters, I described how the first cohort of pragmatists joined the debates over evolution right after finishing college, in the immediate wake of Darwin's *Origin of Species* and Spencer's *First Principles*. Discussions of evolution at this time, the 1860s and '70s, usually focused on whether evolution was the correct account of the history of life or on whether it undercut Christian theology. When the second cohort of pragmatists started college, however, the first of these questions was settled. As discussed in chapter 3, their natural history and religion teachers assumed the fact of evolution, giving a different tenor to the second question: it is one thing to, in James's words, "contemplate the possibility of our ape descent now and then"; it is another to assume that descent and grapple with the theological and ethical consequences.[5] By the 1890s, as shown in chapter 5, scientific debates about evolution were primarily over the relative importance of its causes. These "factors" debates were the immediate context for Dewey's allusions to Weismannism in the 1890s as well as for Peirce's "Evolutionary Love," which was part of the neo-Lamarckian reaction to August Weismann's work.

---

2. Mark Francis, "The Reforming Spencerians: William James, Josiah Royce, and John Dewey," in *Global Spencerism: The Communication and Appropriation of a British Evolutionist*, ed. Bernard Lightman (Leiden, Neth.: Brill, 2016); see also Ferhat Taylan, *Mésopolitique: Connaître, théoriser et gouverner les milieux de vie (1750–1900)* (Paris: Éditions de la Sorbonne, 2018), 238–45.
3. Jennifer Welchman, *Dewey's Ethical Thought* (Ithaca, NY: Cornell University Press, 1995), 121; James Scott Johnston, *John Dewey's Earlier Logical Theory* (Albany: State University of New York Press, 2014), 31.
4. For example, Philip P. Wiener, *Evolution and the Founders of Pragmatism* (Cambridge, MA: Harvard University Press, 1949), chap. 4.
5. William James, "Huxley's Comparative Anatomy," *North American Review* 100 (1865): 291.

CONCLUSION

Inspired by Cornel West and others, I have looked at a somewhat broader cast of characters than the so-called classical pragmatists.[6] The conceptual parallels between the social scientific work of Jane Addams, Mead, and Du Bois and the more explicitly philosophical teachings of the other second-cohort pragmatists illustrate how our modern disciplinary boundaries can mislead us: sociology and ethics were closely linked in the late nineteenth century, as chapter 6 demonstrated. These parallels reflect another benefit of the cohort approach: it groups thinkers based on when they experienced some broader social event rather than on their occupation. For instance, positivists such as Wright, John Fiske, and Francis Ellingwood Abbot—as discussed in chapters 1 and 2—were key conversation partners for the first-cohort pragmatists, even though they were not professional philosophers. It is my hope that the cohort tables presented in the introduction will spur interest in a greater diversity of philosophical voices, highlighting the fact that neglected and canonical voices were usually part of the same conversation.

The implications of my historical story vary, depending on the reader. The main benefit for historians of pragmatism and those already working in the pragmatist tradition is arguably the contextual details themselves. For example, if you were studying Dewey's famous essay "The Reflex Arc Concept in Psychology" and came across the footnote referring to "the whole controversy in biology regarding the source of variation, represented by Weismann and Spencer respectively," it would be difficult to understand what Dewey was getting at without the background presented in chapter 5.[7] The same is true for Peirce's "Evolutionary Love," James's *Pragmatism*, and the early works of Addams and Du Bois. Although my story officially ends in 1910, it can also illuminate later pragmatist works. As books such as *Experience and Nature*, *Art as Experience*, and *Logic: The Theory of Inquiry* make clear, Dewey never abandoned the organism-environment framework. Therefore, to understand these books, we need to understand where the language of organism and environment came from, how Dewey's deployment of these and other concepts differed from that of Spencer, and so on.[8] Du Bois, in contrast,

---

6. Cornel West, *The American Evasion of Philosophy: A Genealogy of Pragmatism* (Madison: University of Wisconsin Press, 1989); Charlene Haddock Seigfried, *Pragmatism and Feminism: Reweaving the Social Fabric* (Chicago: University of Chicago Press, 1996); Paul C. Taylor, "What's the Use of Calling Du Bois a Pragmatist?" *Metaphilosophy* 35 (2004).

7. John Dewey, "The Reflex Arc Concept in Psychology," *Psychological Review* 3 (1896): 360n2.

8. John Dewey, *Experience and Nature* (Chicago: Open Court, 1925), chaps. 6–8; John Dewey, *Art as Experience* (New York: Milton, Balch, 1934), 13–19; John Dewey, *Logic: The Theory of Inquiry* (New

used biological language only sparingly in his later writings. Chapter 6 thus presents us with new research questions: Why did Du Bois move away from the biological framework in texts such as *Darkwater*? Did this shift represent new worries about his previous approach, or was it simply a change of focus away from urban sociology and toward global politics and colonialism?[9] With this book in hand, historians of pragmatism can thus ask new questions of long-studied texts.

For historians of biology and the social sciences, another set of implications is salient. First, I have confirmed and extended my earlier claim that Spencer's organism-environment perspective played a central role in late nineteenth-century scientific and philosophical discussions.[10] As shown in chapter 5, the 1890s debates about the nature and origin of evolutionary variation were focused on Weismann's work, but Spencer's contributions were just as important, and he should not be dismissed simply because his neo-Lamarckian views were ultimately rejected. Second, as chapter 3 demonstrated, biological ideas were of particular interest to philosophically minded students in the 1870s and '80s, who were bombarded by these ideas from all sides—in the classroom, in the books and journals they read, and in their college newspapers. None of the early second-cohort pragmatists pursued a career in biology, but they all transformed and deployed ideas from the life sciences in their research and teaching. Perhaps most surprisingly from our modern point of view, Spencer's books were required reading in philosophy departments across the United States, with whole courses devoted to his work around the turn of the twentieth century at Harvard, Michigan, Yale, Western Reserve, Minnesota, and the University of the South.[11] His books were also paired with those of canonical philosophers: Royce's Philosophy of

---

York: Henry Holt, 1938), chap. 2. For pioneering work on the Dewey-Spencer contrast, see Peter Godfrey-Smith, *Complexity and the Function of Mind in Nature* (Cambridge: Cambridge University Press, 1996), chaps. 2–4.

9. W. E. B. Du Bois, *Darkwater: Voices from within the Veil* (New York: Harcourt, Brace & Howe, 1920). For some early skepticism about the biological approach, see "Sociology Hesitant" [1904/05], p. 3, Series 3, Du Bois Papers; published in W. E. B. Du Bois, "Sociology Hesitant," *boundary 2* 27 (2000): 39–40.

10. See Trevor Pearce, "The Origins and Development of the Idea of Organism-Environment Interaction," in *Entangled Life: Organism and Environment in the Biological and Social Sciences*, ed. Gillian Barker, Eric Desjardins, and Trevor Pearce (Dordrecht, Neth.: Springer, 2014).

11. *The Harvard University Catalogue, 1879–80* (Cambridge, MA: Harvard University, 1879), 84; *Calendar of the University of Michigan for 1885–86* (Ann Arbor: University of Michigan 1886), 55; *Catalogue of Yale University, 1893–94* (New Haven, CT: Tuttle, Morehouse & Taylor, 1893), 46; *The Western Reserve University Catalogue, 1894–95* (Cleveland: Winn & Judson, 1895), 69; *Catalogue for the Year 1902–1903* (Minneapolis: University of Minnesota, 1903), 100; *Catalogue and Announcement, 1903–1904* (Sewanee, TN: University of the South, 1904), 72.

Nature class assigned Baruch Spinoza's *Ethics* and Spencer's *First Principles*; Dewey's History of Political Philosophy class assigned Plato's *Republic*, the first part of Immanuel Kant's *Metaphysics of Morals*, and Spencer's *Principles of Sociology*.[12]

For philosophers of biology, as well as for ethicists and epistemologists, the most interesting implications relate to the last two chapters of this book, which described the pragmatists' own experimental-evolutionary program. As mentioned at the end of chapter 6, there are at least some ethicists today who have self-consciously adopted a pragmatist approach to moral and social problems. But there are also striking differences between proponents: for example, Philip Kitcher provides an explicitly evolutionary ethics whereas Elizabeth Anderson says almost nothing about biology.[13] It is thus an open question whether a specifically biological pragmatism is still relevant in ethics. Another potentially fruitful area of research might be to compare the evolutionary functionalism of the pragmatists to more recent versions of functionalism, which is an uncommon position in metaethics despite some prominent defenders.[14]

The pragmatist account of logic presented in chapter 7 promises to be even more controversial. As William Pepperell Montague and other early realist critics of pragmatism pointed out, if scientific inquiry is mere adaptation or adjustment, this seems to undermine the possibility of objective truth. Perhaps later pragmatists including C. I. Lewis and W. V. O. Quine left the organism-environment framework behind in response to this internal problem—and not because of external factors, as I hypothesized in the introduction.[15] It should be noted that some later pragmatists retained some sympathy for the biological approach, even as they moved away from its vocabulary: Quine himself followed Peirce in providing an evolutionary explanation for abduction.[16] But even if the third- and fourth-cohort pragmatists did move away from biological pragmatism because they were worried about its implications for objectivity and truth, I think these worries are to a certain extent overblown. It is difficult to articulate precisely how the naturalistic approach of the early

---

12. *The Harvard University Catalogue, 1886–87* (Cambridge, MA: Harvard University, 1886), 106; *Calendar of the University of Michigan for 1892–93* (Ann Arbor: University of Michigan, 1893), 67.

13. Philip Kitcher, *The Ethical Project* (Harvard: Harvard University Press, 2011); Elizabeth Anderson, *The Imperative of Integration* (Princeton, NJ: Princeton University Press, 2010).

14. David Copp, *Morality, Normativity, and Society* (Oxford: Oxford University Press, 1995); Frank Jackson, *From Metaphysics to Ethics: A Defence of Conceptual Analysis* (Oxford: Clarendon Press, 1998); David Wong, *Natural Moralities: A Defense of Pluralistic Relativism* (Oxford: Oxford University Press, 2006).

15. I am grateful to an anonymous reviewer for making this suggestion.

16. Willard Van Orman Quine, *Ontological Relativity and Other Essays* (New York: Columbia University Press, 1969), 126–28. Thanks to Jay Odenbaugh for reminding me of this passage.

pragmatists relates to normative force in ethics and epistemology, but this is a research question rather than a reason to dismiss pragmatism.[17]

Setting historical questions to one side, does the pragmatist account of scientific inquiry really undermine objectivity? I am somewhat skeptical, if only because philosophers of science have made a series of arguments in the last few decades that have brought us closer to the pragmatists. Much of the confusion over pragmatism and objectivity is due to Richard Rorty himself, who characterized pragmatism in his 1979 presidential address to the American Philosophical Association as "the doctrine that there are no constraints on inquiry save conversational ones—no wholesale constraints derived from the nature of the objects, or of the mind, or of language, but only those retail constraints provided by the remarks of our fellow-inquirers."[18] As is well known, however, this doctrine does not accurately describe the position of the first- and second-cohort pragmatists, who consistently cited all sorts of nondiscursive constraints. For example, as discussed in chapter 7, Dewey denied that his logic ignored "objective conditions": experimental verification essentially involves the possibility that one's hypothesis be "confuted" by "new data," a fact also noted by James.[19] The pragmatists did reject the notion that we should view scientific theories as attempts to reflect the structure of some ultimate reality, as do many philosophers of science today. But these philosophers of science, like the pragmatists, also maintain that there are objective constraints on our theories. Pragmatism does not entail the rejection of objectivity.[20]

Philosophy of science as a whole has recently taken on a more pragmatist flavor, even if pragmatism is rarely named specifically. Presidential

---

17. Peter Godfrey-Smith, "Dewey, Continuity, and McDowell," in *Naturalism and Normativity*, ed. Mario de Caro and David MacArthur (New York: Columbia University Press, 2010); Philip Kitcher, "Afterthoughts: Reply to Comments," *Analyse & Kritik* 34 (2012): 185–88; Jean-Marie Chevalier, "Why Ought We to Be Logical? Peirce's Naturalism on Norms and Rational Requirements," in *Liber Amicorum Pascal Engel*, ed. Julien Dutant, Davide Fassio, and Anne Meylan (Geneva: University of Geneva, 2014).

18. Richard Rorty, "Pragmatism, Relativism, and Irrationalism," *Proceedings and Addresses of the American Philosophical Association* 53 (1980): 726; see also Steven Levine, *Pragmatism, Objectivity, and Experience* (Cambridge: Cambridge University Press, 2019).

19. John Dewey, "The Control of Ideas by Facts," *Journal of Philosophy, Psychology and Scientific Methods* 4 (1907): 314; William James, *Pragmatism: A New Name for Some Old Ways of Thinking* (New York: Longmans, Green, 1907), 201–2.

20. Helen Longino, *Science as Social Knowledge: Values and Objectivity in Scientific Inquiry* (Princeton, NJ: Princeton University Press, 1990), chap. 4; Heather Douglas, *Science, Policy, and the Value-Free Ideal* (Pittsburgh: University of Pittsburgh Press, 2009), chap. 6; Matthew Slater, "Natural Kindness," *British Journal for the Philosophy of Science* 66 (2015); Philip Kitcher, "Pragmatism and Progress," *Transactions of the Charles S. Peirce Society* 51 (2015).

addresses to the Philosophy of Science Association during the 2010s exemplify this turn to practice, with Nancy Cartwright discussing evidence-based policy, James Woodward analyzing causal thinking in terms of its purposes, Helen Longino and C. Kenneth Waters each explicitly defending "practice centrism" in the philosophy of science, and Sandra Mitchell arguing that "diversity and pluralism are required for making science an effective epistemic enterprise."[21] Some of these philosophers have made the connection explicit: Waters recently urged that we adopt "a form of pragmatism" in our analyses of scientific practice. Others are only a few genealogical steps away from the characters in this book: Cartwright recently noted the influence of her former Stanford colleague Patrick Suppes, who studied Dewey's logic with Ernest Nagel at Columbia in 1947.[22]

Finally, the work of pragmatists like Dewey and Addams often comes across as naively optimistic—"Progress for whom?" as Beth Eddy asks.[23] Leonard Harris argued in 2002 that the "social engineering" approach of the pragmatists contained a fundamental flaw: the method of "evaluating processes, means, [and] ends" does not lead to progress if the social engineer is racist or elitist. This possible flaw becomes even more worrisome when we recall that the evolutionary-experimental ethics of the pragmatists did not prevent them from endorsing pernicious cultural hierarchies, as discussed in chapter 6. According to Harris, "even if one is committed to an evolutionary view of change, there is no history of evolution without the history of insurrections, revolts, and revolutions." The latter, he claimed, require full commitment even when the chance of success is vanishingly small; the pragmatist approach—based on ex-

21. Nancy Cartwright, "Will This Policy Work for You? Predicting Effectiveness Better: How Philosophy Helps," *Philosophy of Science* 79 (2012); James Woodward, "A Functional Account of Causation; or, A Defense of the Legitimacy of Causal Thinking by Reference to the Only Standard That Matters—Usefulness (as Opposed to Metaphysics or Agreement with Intuitive Judgment)," *Philosophy of Science* 81 (2014); Helen Longino, "Foregrounding the Background," *Philosophy of Science* 83 (2016); C. Kenneth Waters, "An Epistemology of Scientific Practice," *Philosophy of Science* 86 (2019); Sandra Mitchell, "Through the Fractured Looking Glass," Presidential Address, Twenty-Sixth Biennial Meeting of the Philosophy of Science Association, Seattle, WA, November 3, 2018, video, 42:02, quotation at 2:18–24, https://spark.adobe.com/page/AiXfAUmLTaEbB/.

22. C. Kenneth Waters, "Ask Not 'What *Is* an Individual?'" in *Individuation, Process, and Scientific Practices*, ed. Otávio Bueno, Ruey-Lin Chen, and Melinda B. Fagan (New York: Oxford University Press, 2018), 99; Nancy Cartwright, "The Philosophy of Social Technology: Get On Board," *Proceedings and Addresses of the American Philosophical Association* 89 (2015), 106–8; Patrick Suppes, "Nagel's Lectures on Dewey's Logic," in *Philosophy, Science, and Method: Essays in Honor of Ernest Nagel*, ed. Sidney Morgenbesser, Patrick Suppes, and Morton White (New York: St. Martin's, 1969), 2.

23. Beth L. Eddy, *Evolutionary Pragmatism and Ethics* (Lanham, MD: Lexington, 2016), 54.

perimental evidence and rational expectation—would thus fail to support them.²⁴

As I mentioned in passing in chapter 6, however, at least some of the pragmatists suggested that revolutionary action, even if not adaptive or rational in the short term, could contribute to social and moral progress. James, in his lectures on "The Value of Saintliness," focused on the progressive impulse given to social evolution by saintly behavior. Citing Spencer's *Data of Ethics*, James pointed out that "saintly conduct would be the most perfect conduct conceivable in an environment where all were saints already," but that in the present environment, "where few are saints, and many the exact reverse of saints, [such conduct] must be ill adapted." Nevertheless, James continued, saints have a "vital and essential" function in "social evolution": despite "their impracticability and non-adaptation to present environmental conditions, . . . they help to break the edge of the general reign of hardness, and are slow leavens of a better order." Saints, like the "socialists and anarchists" who promote "Utopian dreams of social justice," are

> the tip of the wedge, the clearers of the darkness. Like the single drops which sparkle in the sun as they are flung far ahead of the advancing edge of a wave-crest or of a flood, they show the way and are forerunners. The world is not yet with them, . . . yet they are impregnators of the world, vivifiers and animaters of potentialities of goodness which but for them would lie forever dormant.²⁵

According to James, both saints and socialists were pushing evolution forward, in their dreams and their actions alike.

Du Bois, although he was trained as an economist and founded a school of sociology, also eventually decided that science in the service of reform was not enough for social evolutionary progress. As he later recalled, referring to his editorship (starting in 1910) of the *Crisis*, the monthly magazine of the National Association for the Advancement of Colored People (NAACP), "my career as a scientist was to be swallowed up in my role as master of propaganda." Just as Addams had determined "to test the value of human knowledge by action," Du Bois now asked, "What with all my dreaming, studying, and teaching was I going to *do*

---

24. Leonard Harris, "Insurrectionist Ethics: Advocacy, Moral Psychology, and Pragmatism," in *Ethical Issues for a New Millennium*, ed. John Howie (Carbondale: Southern Illinois University Press, 2002), 202–3, 206. See also Lee A. McBride III, ed., "Symposium on Insurrectionist Ethics," *Transactions of the Charles S. Peirce Society* 49 (2013): 27–111.

25. William James, *The Varieties of Religious Experience: A Study in Human Nature* (New York: Longmans, Green, 1902), 355–60.

CONCLUSION

in this fierce fight?"[26] His view of the importance of activism for progress was on display in his response, in a 1914 issue of the *Crisis*, to a letter from the Unitarian minister Charles Fletcher Dole. In his letter, Dole had complained about the antagonistic tone of the magazine: "Please do the least possible to arouse resentment of bitterness, which is sure to react upon those who stir it. Please do more of what you are doing every month, to show the growth of a kindly good will among all kinds of people. For good will is the only irresistible power in the universe." In his scathing reply—perhaps alluding to Dole's book *The Ethics of Progress*, which had described its author's philosophy as one "of evolution or growth" and confidently asserted "that the world is growing better and not worse"—Du Bois declared that social evolutionary progress requires active struggle. One cannot not just sit on the sidelines, as some scholars have accused Dewey of doing when it came to questions of race:

Humanity is progressing toward an ideal; but not, please God, solely by help of men who sit in cloistered ease, hesitate from action and seek sweetness and light; rather we progress today, as in the past, by the soul-torn strength of those who can never sit still and silent while the disinherited and the damned clog our gutters and gasp their lives out on our front porches. These are the men who . . . make this world so damned uncomfortable with its nasty burden of evil that it tries to get good and does get better. Evolution is evolving the millennium, but one of the unescapable factors in evolution are the men who hate wickedness and oppression with perfect hatred, who will not equivocate, will not excuse, and will be heard.[27]

We can make evolutionary progress, but not unless we stop making excuses and start making the people in power uncomfortable.

It is impossible to fully understand the works of the American pragmatists without attending to developments in the life sciences in the late nineteenth century, in particular the ideas of evolution, adaptation, and environment. *Evolution* evokes an image of slow and steady biological

---

26. W. E. B. Du Bois, *Dusk of Dawn: An Essay toward an Autobiography of a Race Concept* (New York: Harcourt, 1940), 94; Jane Addams, "A Function of the Social Settlement," *Annals of the American Academy of Political and Social Science* 13 (1899): 36; W. E. B. Du Bois, *Darkwater: Voices from within the Veil* (New York: Harcourt, Brace and Howe, 1920), 21.

27. Charles Fletcher Dole, "A Question of Policy," *Crisis* 8 (1914): 24; Charles Fletcher Dole, *The Ethics of Progress, or, The Theory and the Practice by which Civilization Proceeds* (New York: Thomas Y. Crowell, 1909), 382; W. E. B. Du Bois, "The Philosophy of Mr. Dole," *Crisis* 8 (1914): 26. For Dewey's silence on race, see Shannon Sullivan, "(Re)construction Zone: Beware of Falling Statues," in *In Dewey's Wake: Unfinished Work of Pragmatic Reconstruction*, ed. William J. Gavin (Albany: State University of New York Press, 2003); Paul C. Taylor, "Silence and Sympathy: Dewey's Whiteness," in *What White Looks Like*, ed. George Yancy (New York: Routledge, 2004).

change. As I have argued, however, in the years around 1900 it encompassed much more, including the active modification of social norms and institutions. For the pragmatists, moral and scientific progress was synonymous with experimentally guided evolution. But since political action is—in Du Bois's words—"one of the unescapable factors in evolution," experimenters must ask not only "What would happen?" but also "What should happen?" and "Can we imagine?" As Du Bois argued in *Dusk of Dawn*, progressive social change requires more than just empirical evidence and rational persuasion: "It needs carefully planned scientific propaganda; the vision of a world of intelligent men [and women] with sufficient income to live decently and with the will to build a beautiful world."[28]

---

28. W. E. B. Du Bois, *Dusk of Dawn: An Essay toward an Autobiography of a Race Concept* (New York: Harcourt, 1940), 172.

Acknowledgments

This book has been an object lesson in gradualist evolution. Initial spontaneous variation took place during my doctoral work in the Committee on Conceptual and Historical Studies of Science at the University of Chicago, where a small group of graduate students—including Chris DiTeresi, Bill Sterner, Katie Tabb, and Cecelia Watson—began meeting in 2006 to discuss James's *Pragmatism*. This "Pragmatism Reading Group" was reconvened the next year by Beckett Sterner, with variations up to 2014 featuring *The Essential Peirce*, Dewey's *Logic*, and Rorty's *Philosophy and the Mirror of Nature*. I am deeply grateful to everyone who participated in this group over the years, including (in later versions) Michael Pettit, Joyce Havstad, and others. With pragmatism mostly absent from syllabi at Chicago—an irony given its history—I owe almost all my initial knowledge of the pragmatists to careful group reading and conversations with fellow graduate students.

Although pragmatism didn't make it into my writing at Chicago, this book nonetheless bears the imprint of my co-supervisors there: Robert Richards, whose first book covered many of the themes taken up in these pages, and William Wimsatt, who has defended a pragmatist or "engineering" approach to the philosophy of biology.[1] My historical methodology also owes a great deal to Alan Richardson, one of

---

1. Robert J. Richards, *Darwin and the Emergence of Evolutionary Theories of Mind and Behavior* (Chicago: University of Chicago Press, 1987); William C. Wimsatt, *Re-Engineering Philosophy for Limited Beings: Piecewise Approximations to Reality* (Cambridge, MA: Harvard University Press, 2007).

my undergraduate mentors at the University of British Columbia. Most of my knowledge of evolutionary biology stems from classes and conversations at Chicago with Leigh Van Valen, Michael LaBarbera, and David Jablonski, all of whom supported my work in the history and philosophy of biology.

I started writing this book during two postdoctoral fellowships, the first at the Rotman Institute of Philosophy at Western University and the second at the Center for Humanities at the University of Wisconsin–Madison. Without the generous travel funding provided during these years by the Rotman Institute, the Social Sciences and Humanities Council of Canada, and the Andrew W. Mellon Foundation, I could never have undertaken the extensive archival research that grounds the book. Essential conversations and criticisms were provided by my successive supervisors, Gillian Barker and Elliott Sober, along with the wonderful faculty and students in philosophy at Western and Wisconsin. I also benefited from conversations at Wisconsin with Lynn Nyhart, Jennifer Ratner-Rosenhagen, Gregg Mitman, and members of the "Intellectual History Group," as well as from a workshop with the visiting Peter Godfrey-Smith and my fellow Center for Humanities postdocs.

I completed the book while on faculty in the Department of Philosophy at the University of North Carolina at Charlotte. UNC Charlotte provided me with two Faculty Research Grants that supported archival research at the Huntington Library (which also provided funding in the form of a short-term fellowship) and elsewhere. Even more important, the university has a very generous parental leave policy, which allowed me to avoid writing during the dreaded "fourth trimester" of both my children. My colleagues here, including many who study pragmatism, have been incredibly supportive of the project. Finally, I'm not sure I would ever have finished the book without a Scholars Award from the National Science Foundation, which provided a full year of research leave.

Many people have read and commented on earlier drafts of this material: although I'm certainly forgetting many people, I'm grateful to Peter Godfrey-Smith, Elliott Sober, Quayshawn Spencer, Greg Radick, Gregg Mitman, David Depew, Marilyn Fischer, Jay Odenbaugh, Daniel Liu, Alexander Klein, Jeremy Dunham, Sarah Woolwine, several anonymous reviewers, my fellow postdocs at Wisconsin, and my faculty colleagues at UNC Charlotte. Peter Kupfer and Ryan Dahn helped with some of the trickier German translations. I'm also thankful for comments from audiences at the Max Planck Institute for the History of Science, York University, the University of Chicago, the University of San Francisco, Washington State University, the University of Sheffield, the University of Toronto, the University of Oregon, the New School for Social Research, the University of Milan, and Boston University, as well as at meetings of the International Society for the History, Philosophy, and Social Studies of Biology, the International Society for History of Philosophy of Science, the History of Science Society, the Society for the Advancement of American Philosophy, and the Josiah Royce Society. Helpful feedback on the overall project was provided by the audience of a "book-in-progress" workshop at the Summer Institute in American Philosophy at the University of Dayton in 2018.

The book depended on access to archival material, which was generously provided by the American Museum of Natural History, the American Philosophical Society, Harvard University, the Huntington Library, Indiana University–Purdue University Indianapolis, the New-York Historical Society, Oberlin College, the Oxford University Museum of Natural History, the Pike County Historical Society, Southern Illinois University Carbondale, the University of Chicago, the University of Illinois at Chicago, the University of Manchester, and the University of Michigan. I am also grateful to several archivists and librarians for scanning and sending materials, including Joanna Bares (Rockford University), Robin Carlaw (Harvard University), Alhaji Conteh (Howard University), Nicholas Guardiano (SIU Carbondale), Erin Patterson (Colgate University), and James Stimpert (Johns Hopkins University). Thanks are also due to the David Graham Du Bois Trust for permission to quote from W. E. B. Du Bois's unpublished papers, held at the University of Massachusetts Amherst and freely available online.

Some of the material in the book first appeared in other venues. Chapters 1, 2, and 7 draw from "James and Evolution," in *The Oxford Handbook of William James*, ed. Alexander Klein (Oxford: Oxford University Press, forthcoming). Chapter 3 draws from "Naturalism and Despair: George Herbert Mead and Evolution in the 1880s," in *The Timeliness of George Herbert Mead*, ed. Hans Joas and Daniel R. Huebner (Chicago: University of Chicago Press, 2016), and from "The Origins and Development of the Idea of Organism-Environment Interaction," in *Entangled Life: Organism and Environment in the Biological and Social Sciences*, ed. Gillian Barker, Eric Desjardins, and Trevor Pearce (Dordrecht: Springer, 2014). Chapter 4 draws from "The Dialectical Biologist, circa 1890: John Dewey and the Oxford Hegelians," *Journal of the History of Philosophy* 52 (2014). Chapter 5 draws from "'Protoplasm Feels': The Role of Physiology in Peirce's Evolutionary Metaphysics," *HOPOS* 8 (2018); from "American Pragmatism, Evolution, and Ethics," in *The Cambridge Handbook of Evolutionary Ethics*, ed. Michael Ruse and Robert J. Richards (Cambridge: Cambridge University

Press, 2017); and from "Dewey, le darwinisme et la variation dirigée," in *Penser l'évolution: Nietzsche, Bergson, Dewey*, ed. Antoine Daratos and Paul Walter (Paris: Vrin, 2019).

The editing and publishing process was uniformly smooth thanks to the efficient work of Karen Darling, Tristan Bates, and Tamara Ghattas at the University of Chicago Press, as well as Lori Meek Schuldt, who copyedited the manuscript on behalf of the press.

Last but not least, I'd like to thank my family: my parents and sister, who fostered my love of reading and reflection, and more recently my wife, Elise, who has been the best partner and mother I can imagine. I should also thank my children, Asa and Tovin, so they can see their names in print. Although they didn't exactly speed up the book's completion, they did at least give me a personal perspective on Dewey and Mead's image of children as enthusiastic experimenters.

Index

Page numbers in italics refer to figures.

AAAS (American Association for the Advancement of Science), 27, 30

Abbot, Francis Ellingwood: overview of role in Darwin debates, 25–26, 55–56, 100; criticism of Spencer's theories, 59, 70–71; education, 29, 34–35; endorsement of *Origin of Species*, 30–31, 36; Metaphysical Club colleagues, 24; religious views about evolution, 24–25, 36

abduction mode, Peirce's, 311, 315, 326

Adams, Elizabeth, 11, 12

Adams, Henry Carter, 258, 259

Adams, John, 250–51

adaptation processes. *See* organism-environment *entries*

adapted mind, Peirce's theorizing, 296, 311

adaptive characters, in natural selection arguments, 52

Addams, Jane: Applied Ethics lectures, 258; Baltimore year, 135–37; in Dewey's teaching, 274–75; graduate-level studies, 129–30; Hull House programs, 149–50, *151*; on moral progress, 277–79; Pullman labor strike, 239–40, 241–42; social reform significance, 3; social reform writings, 273, 274; Spencer's importance, 59; undergraduate education, 104–5, 107, 120–21; white culture context, 286–87

agapastic evolution hypothesis, Peirce's, 221

Agassiz, Elizabeth Cabot, 29, 43

Agassiz, Louis: at AAAS meeting, 30; Brace's argument against, 45–46; Brazil expedition, 42–45; Coast Survey expedition, 36–38, 39; creation arguments, 26, 27, 29, 43–44; Fiske's criticism of, 47–48; influence of, 26–27, 116; James's criticisms of, 42–43, 45, 47; LeConte's defense of, 116; Methods of Study lectures, 40–41; in B. Peirce's criticism of *Origin of Species*, 38; in C. Peirce's remarks about evolution debate, 56; in *Popular Science Monthly*, 116–17; publications of, 26, 27–29, 103; in Saturday Club, 26–27; Spencer's criticisms of, 46, 49; student connections, 25, 26, 29, 39, 101; Wright's criticism of, 29, 47

Agassiz Natural History Society, 37–38

agnosticism, 23, 60, 122–23, 124, 132. *See also* theological perspectives *entries*

agricultural biology, University of Vermont, 106–7

INDEX

Alexander, Samuel, 161, 164–66, 177–78, 192
Allen, Grant, 58, 87–88, 89–90, 91, 95, 200
"All-Sufficiency of Natural Selection, The" (Weismann), 226–27
Amazon area expedition, 42–45
American Academy of Political and Social Science, 267–68
American Association for the Advancement of Science (AAAS), 27, 30
American Philosophical Association, 282, 312–13, 337
Ames, Edward, 11, 12
anarchism and socialism, implications of natural selection arguments, 231–35
ancestral choice idea, James's, 84–85, 97–98
Anderson, Elizabeth, 288, 336
Andover Review, 256
anesthetic invention example, in Peirce's arguments, 223
Angell, James Burrill, 300
Angell, James Rowland, 10, 11, 12, 156, 300, 302
Appiah, Kwame Anthony, 186–87
archaesthetism hypothesis, Cope's, 214–16
"Architecture of Theories, The" (Peirce), 212–14, 215–16, 218–19, 224, 324
"Are Acquired Characteristics Hereditary?" (Weismann), 202
"Are We Automata?" (James), 84
Argyll, Duke of (Campbell, George), 197, 202, 204
Arnold, Matthew, 121
astronomy, Wright's discussions, 65
astronomy class, Addams's, 121
Atlanta University, Du Bois's teaching, 153–54, 187, 267
Atlantic Monthly, 46, 116
attention, function of, 74–77, 84, 152–53, 320, 321
Auburn Theological Seminary, 110
Auxier, Randall, 16
Axtelle, George, 15
Ayres, Clarence Edwin, 13, 16

BAAS (British Association for the Advancement of Science), 202, 203
Bache, Alexander Dallas, 27
Backe, Andrew, 179
Bagehot, Walter, 88–89

Bain, Alexander, 72, 82, 119
Bair, Barbara, 120
Baldwin, James Mark, 238, 239, 244, 245, 270, 303, 307
Balliol College, 162
Barnett, Henrietta, 255
Barnett, Samuel, 255
Barrows, Samuel, 207–8
Bascom, John, 125
Bateson, William, 136
Bawden, Henry Heath, 11
bees, cell-making instincts, 30–34
Beiser, Frederick, 162, 194–95
belief-doubt model, Peirce's, 295–96, 310, 326
Bellamy, Edward, 232
Bentley, Arthur, 11
Berlin. *See* Germany
Bernard, Claude, 68
Bernstein, Richard Jacob, 4, 16, 159
biological evolution. *See* evolution theory entries
biological idealism tradition. *See* organism-environment relationship, dialectical approach
biological-psychological model. *See* logic model of inquiry, pragmatism's; pragmatism-biology relationship, overview
biological variation. *See* variation phenomenon, causes debate
Biology and Philosophy journal, 6
biology courses, 135–36, 149–50, *151*, 155
Blumer, Herbert, 16
Bluntschli, Johann Caspar, 252
Boas, Franz, 286
Bode, Boyd Henry, 11, 12
Bois-Reymond, Emil du, 45
Boodin, John Elof, 10
Booth, Charles, 261, 264
Booth, William, 284
botany courses, 103
Bowditch, Henry Pickering, 299
Bowen, Francis, 30–32, 34, 184
Bowler, Peter, 196
Boydston, Jo Ann, 16
Boyle, Robert, 248
Brace, Charles, 45–46
Bradley, Francis Herbert, 178
Brazil expedition, Agassiz's, 42–45
Brent, Joseph, 39

348

Bright, Liam Kofi, 267–68
British Association for the Advancement of Science (BAAS), 202, 203
"British Logicians" (Peirce), 40
British Quarterly Review, 115
British Thought and Thinkers (Morris), 132
Brooks, Amelia Katherine Schultz, 135
Brooks, John Graham, 260
Brooks, William Keith, 135–36, 155
"Brute and Human Intellect" (James), 80–81, 86, 120
Buckle, Henry Thomas, 63–64
Bunting, Percy William, 225
Bush, Wendell, 316
Butler, Samuel, 204

Caird, Edward, 161.192, 162, 163–64, 173, 179
Calendar of the University of Michigan, 300–301
California, University of, 102–5, 106, 114–17, 143
Calkins, Mary Whiton, 10, 11–12
Campbell, Donald Thomas, 6
Campbell, George (Duke of Argyll), 197, 202, 204
capitalism, 218, 239–41, 256–57
Carpenter, William, 75
Cartwright, Nancy, 338
Carus, Paul, 211–12
Cassirer, Ernst, 14
Castle, Henry Northrup, 109–10, *111*, 114, 122–23, 140
"Catagenesis" (Cope), 214
catagenesis hypothesis, Cope's, 215–16
Cattell, James McKeen, 134, 299, 319
causes vs. classification, dispute tradition, 40. *See also* organism-environment *entries*
cell-making instincts, natural selection arguments, 30–34
Century Dictionary (Peirce), 211, 215, 250–51, 325
chance factor, 207–8, 212, 217, 218–19, 311, 325–26
charity arguments, Addams's, 278–79
Chase, Frederick Augustus, 110
chemistry analogy, B. Peirce's, 38–39
chemistry courses, 39, 40, 41
Chevalier, Jean-Marie, 309
Chicago, Nineteenth Ward survey, 263–65

Chicago, University of. *See* University of Chicago
"Chicago School, The" (James), 307
Chicago Times, 239, 240
Child, Charles Manning, 156
Child and the Curriculum, The (Dewey), 312
Childs, John, 13
chordate relationships, Bateson's research, 136
Christianity: in Peabody's argument, 18–19; in Peirce's arguments, 218, 221, 222, 223; in social reform experimentation, 256, 269. *See also* theological perspectives *entries*
Christian Register symposium, 207–8, 214
churches, Du Bois's argument, 276
Churchman, Charles West, 16
City College of New York, 15
Civil War hypothesis, 9
Clapp, Elsie Ripley, 13
Clark University, 155
Class Book of Botany (Wood), 102
classification vs. causes, dispute tradition, 40. *See also* organism-environment *entries*
Clifford, William Kingdon, 152
climate factor, 88, 221
coadaptation argument, Spencer's, 206, 226–28
Coast Survey expedition, 27, 36–38, 39
Coffin, Alfred Oscar, 125–26
Coffin, Samuel Allen, 126–27
Cohen, Morris Raphael, 13, 15
cohort groupings, overview: comparisons of evolution debate experience, 157–58, 333; comparisons of organism-environment perspectives, 16–17; members in, 10–16; methodology for, 8–10; research advantages, 333–34
Colapietro, Vincent, 291, 293–94
College Settlement News, 266
"College Woman and the Family Claim, The" (Addams), 273
color variation, Wallace's theory, 89, 135, 136
color vision theory, Ladd-Franklin's, 18
Columbia University, 2, 11, 13, 15, 18, 319, 321
Common Law, The (Holmes), 18
Comte, Auguste, 81, 124, 141, 175–76, 185
consciousness: Clifford's arguments, 152; Cope's theorizing, 214–17; Dewey's

349

consciousness (*cont.*)
    arguments, 147–49, 235; Hegel's argument, 186; Helmholtz's argument, 76; Hodgson's theorizing, 74–75; Huxley's arguments, 119; James's arguments, 75, 83–85, 99, 304; Murray's theorizing, 146; Peirce's arguments, 215–17, 295, 317; Spencer's statement, 83; White's argument, 129; Wundt's arguments, 76–77
"Consciousness in Evolution" (Cope), 214
"Conservation of Races, The" (Du Bois), 189, 191
Contemporary Review, 225–26
Contemporary Socialism (Rae), 261
Contributions to the Natural History of the United States (Agassiz), 26, 27
control: Dewey's arguments, 312, 313, 317–18, 321, 322, 324; Moore's argument, 322–23; in Peirce's logic theory, 315
"Control of Ideas by Facts, The" (Dewey), 317–18
conversation model, Hutton's, 7
Cook, Joseph, 134
Cooley, Charles, 10, 11
Coolidge, Mary, 13
Cooper, Anna Julia, 189–90
Cope, Edward Drinker: Dewey's criticism, 244; evolution theory of, 214–17; immortality symposium argument, 207, 208; influence of, 213; on natural selection hypothesis, 197; on origin of the fittest, 85, 95; in Osborn-Poulton friendship, 203; Peirce's comments about, 211, 213; on plasticity, 213; in variation causes debate, 204; and White, 129
corals, Holmes's thesis, 108, *109*
Cornell University, 11, 259
cosmological evolution, 121, 216
Cravath, Erastus Milo, 125
creation arguments. *See* theological perspectives *entries*
Crisis magazine, 156, 339–40
critic, in Peirce's logic, 294, 309–10, 313
Critical Philosophy of Immanuel Kant, The (Caird), 176
Critique of Pure Reason (Kant), 72 183
Critique Philosophique journal, 72
Croce, Paul, 47
Crummell, Alexander, 189–91, 269
Cummings, Edward, 260–61

Cunningham, Holly Estil, 13
Cunningham, Suzanne, 17
Curry, Tommy, 9

Dana, James Dwight, 101, 102, 104–5, 110, 113–14, 198
Darwin, Charles: overview of defenders, 55–56; on Agassiz's influence, 27; and Dewey's thinking, 291–92, 293, 333; on his philosophy perspective, 25–26; on instincts, 205; in James's arguments about organism changes, 68–69; meeting with Fiske, 49; meeting with Wright, 56; Peirce's praise of, 40; Spencer reputation compared, 67; on stability of species, 52; on variation question, 197, 206; on Weismann's theory, 199; and Wright's mathematics, 34; and Wright's review of Mivart's *Genesis of Species*, 55. *See also* evolution theory *entries*
Darwin, Erasmus, 51
"Darwin and Hegel" (Ritchie), 166
Darwinism (Wallace), 137, 203–4
Darwinism (Wright), 55
"Darwin's Successor at Home" (Kidd), 200
Data of Ethics, The (Spencer), 143, 233, 234
Davidson, Donald, 16
Davis, Bradley Moore, 149–50, *151*
Dawkins, Richard, 6
deduction mode, Peirce's argument, 315–16
Deegan, Mary Jo, 264, 266
deer antlers example, in Weismann's argument, 228
Delboeuf, Joseph, 135
Deledalle, Gérard, 4
Democracy and Social Ethics (Addams), 252, 273, 274, 278
Descartes, René, 119
"Design and Chance" (Peirce), 207
development theory. *See* evolution theory *entries*
Dewey, Alice, 239, 240
Dewey, John: and cohort groupings, 10, 11–12, 13, 15, 18; on conflict's role, 240–42, 271, 291; Darwin's influence, 291–92, 293, 333; in eugenics arguments, 285–86; on experimental approach, 249, 251, 253–54, 271–72, 316–19, 321, 322–23, 324, 337; functional psychology model, 179–80, 194, 242–43, 271–72, 280–81, 304; graduate-level studies,

130, 132–35, 252, 292; on habit, 237–38, 241–44, 271–72, 316–17, 321; and Hegel's publications, 172, 173; hiring of faculty members, 151, 298, 299; on his philosophical influences, 159; on intelligence, 148–49, 235, 237, 247, 318; James's excitement about, 306, 307; and James's thinking, 253–54, 298; Johnston's criticisms, 291–93; Lagunas's criticisms, 323; in logic model arguments, 291–94, 305–6; morality theorizing, 178–79, 233, 240–42, 252, 279–81, 282, 330–31; in organism-environment dialectic, 169–70, 178–79, 192, 194, 334; Peirce's criticisms, 293–94, 308–13, 314–15; on Pullman labor strike, 239–41; and research colleagues, 299–302; reviews of social evolution books, 230, 231, 236, 240; Rorty's criticisms, 290; scholarship contributions summarized, 3; scholarship perspectives on, 160; school laboratory, 265–66; on social evolution, 1–2; on social readjustment, 274–75; social reform arguments, 271–72, 274–75; on source of philosophers' problems, 17; on Spencer's significance, 59, 62; on Spencer's theories, 60, 62, 235, 282, 285–86, 293, 2811; teaching positions and topics, 146–49, 166–68, 179–80, 239, 249, 253, 274, 281, 292; undergraduate education, 103–5, 106–7, 117–20, 292; in variation causes debate, 236–37, 244, 322, 324; on working hypothesis, 330–31

Dialectical Biologist, The (Levins and Lewontin), 169, 192–93

Dictionary of Philosophy and Psychology (Baldwin), 309, 310, 311, 312, 314, 324–25

Dilthey, Wilhelm, 139, 140, 184–85

"Does Reality Possess Practical Character?" (Dewey), 318–19

dog example, in James's arguments, 81, 85

Dole, Charles Fletcher, 340

Donaldson, Herbert, 299

double-consciousness concept, Du Bois's, 186

doubt-belief model, Peirce's, 295–97, 310, 326

Draper, John William, 67

dual aspects arguments, in organism-environment dialectic, 70–71, 169–70, 172–74, 176–77, 178–80, 193–95

Du Bois, W. E. B.: and cohort groupings, 10, 11, 18; and colleague Turner, 155; editing of NAACP's magazine, 156; on experimental approach, 249–50; graduate-level studies, 141–42, 183–85, 186–87, 249–50, 251, 256–57, 258–61; in organism-environment dialectic, 161, 192, 334–35; Philadelphia survey project, 266–67; scholarship contributions summarized, 3; scholarship perspectives on, 160, 183, 185, 186–88; Spencer's importance to, 59; teaching positions and topics, 153–54, 187, 263, 267; undergraduate education, 103–6, 110–13, 114, 124–26, 143, 144, 198–99, 205; on Weismann's influence, 245–46

Du Bois, W. E. B. (writing/speech topics): eugenics, 284–85; lectern socialists, 263; primitive man, 286; race conservation, 189; social evolution, 1, 185–86, 191–92, 275–77, 339–40; social problem causes, 269–71; social reform goals, 265, 276–77, 341; social research opportunities, 267–68; working environments, 257

Dusk of Dawn (Du Bois), 341

dynamic functionalism perspective, 280–81

Eames, Elizabeth Ramsden, 16

Eames, Samuel Morris, 15, 16

Eaton, Isabel, 266–67

Ebbinghaus, Hermann, 139, 299

Eclipse of Darwinism, The (Bowler), 196

Economics (Hadley), 153

economics courses, 141–42, 153, 256–57, 258–62

economics-ethics connection, Peabody's argument, 256, 258. *See also* social reform, experimental approach

Eddy, Beth, 241, 281, 338

Edel, Abraham, 15

educational psychology, 303–5, 312. *See also* Dewey, John

Edwards, Barrington, 275

Eisele, Carolyn, 15

Elements of Comparative Anatomy (Huxley), 41

Elements of Geology (LeConte), 105

Elements of Zoology (Tenney), 103–4

Eliot, Charles William, 78

INDEX

Elkus, Savilla, 11
Ellis, Frederick Warren, 299
Ellis, John Millot, 122
Elmendorf, John Jay, 124
Ely, Richard Theodore, 252–53, 261
Emerson, Ralph Waldo, 26
Encyclopedia Britannica, 211
Encyclopedia Logic (Hegel), 172
energy, in Cope's catagenesis hypothesis, 215–16
Engels, Friedrich, 193
English, Daylanne, 284
Entwicklung concept, Hegel's, 162, 163–64
environmental influences. *See* organism-environment *entries*
Essays (Spencer), 219
Essays in Philosophical Criticism (Seth and R. Haldane, eds.), 173–75
Essays on Heredity (Weismann), 203, 205, 209
ethical fieldwork sites. *See* social reform, experimental approach
"Ethical Survivals in Municipal Corruption" (Addams), 277–78
ethics: scholarship opportunities, 335–36; Spencer's system, 232–34, 235, 253, 281, 282–83. *See also* moral progress; social reform *entries*
Ethics (Dewey), 285
Ethics (Spinoza), 144
ethics courses: at Harvard College, 112–13, 143, 256; at Hull House, 258; at University of Chicago, 242–43, 271, 274, 322; at University of Michigan, 298
Ethics of Progress, The (Dole), 340
eugenics, 283–87
Evening Post, 65
Evidence as to Man's Place in Nature (Huxley), 41–42
evidence-based policy, modern, 288–89. *See also* experimental *entries*
"Evolution; Pro and Con" (Coffin), 126–27
"Evolution and Ethics" (Dewey), 272
Evolution and the Founders of Pragmatism (Wiener), 4
"Evolutionary Love" (Peirce), 211, 218–25
"Evolutionary Method as Applied to Morality, The" (Dewey), 279–80, 305–6
evolutionary philosophy, Spencer's: Abbot's criticisms, 59, 69–70, 81; in Addams's college era essays, 121; Alexander's arguments, 165, 177; basic tenets of, 60, 62, 63, 170–71, 201, 232–33; Caird's arguments, 163–64; consciousness's function, 83; Crummell's argument, 190–91; and Dewey's graduate studies, 132–33; Dewey's remarks on, 60, 62, 285–86; Dewey's teaching of, 147–48, 253, 282, 292–93; Dilthey's remarks on, 185, 282; and Du Bois's graduate studies, 141–42; in Fiske's review of Buckle's *History*, 63–64; Green's criticisms, 162–63, 171–73; Holmes's agreement with, 69; James's remarks on, 60, 62; Mead's teaching of, 151–52, 167–68; Morris's criticism, 132–33; Osborn's arguments, 209–10; Peirce's arguments, 62, 207–8, 208–9, 210–13, 219, 220–21; in *Popular Science Monthly*, 115; Renouvier's criticism, 80, 81; Royce's arguments, 115–16, 143–44, 183, 253; in variation causes debate, 206–7, 226–27; Ward's disagreement with, 69–70; Wright's arguments, 64, 66. *See also* James, William (and Spencer's theories); organism-environment *entries*; Spencer, Herbert
"Evolution of Negro Leadership, The" (Du Bois), 277
evolution theory, Darwin's: AAAS discussions about, 30; Agassiz's arguments against, 26, 40; Alexander's argument, 164–65; Bowen's criticisms, 30–32; Brace's defense of, 45–46; Caird's argument, 163–64; cohort experiences compared, 157–58, 333; Dewey's arguments, 147, 250, 291–92; as experimental process, 250; Green's acceptance of, 162–63; Hegel linkage, 162–69, 180–85, 192–93, 291–93; Hegel's opposition to, 161; Hull House lectures about, 149–50; in James's discussions, 41–42, 46–47, 69, 102; in James's psychology course, 78; in Johnston's criticism of Dewey, 291–92; Mivart's arguments, 50–54; B. Peirce's criticisms, 38–39; C. Peirce's arguments, 207, 212, 218–20, 222; scientific acceptance, 196–98; Spencer's defense of, 45–46; Stirling's argument, 165–66; striving/struggle idea, 188; Wright's arguments for, 29–30, 32–34, 47. *See also* natural selection hypothesis
evolution theory, in graduate-level studies: at German universities, 127–29, 137–42;

at Johns Hopkins University, 130–31, 136; at Peabody Institute, 136–37; at Women's Medical College, 129–30
evolution theory, in undergraduate education: astronomy class, 121; ethics course, 112–13; mind science class, 125; natural history courses, 102–7, 109–11, 112, 113–14, 120–21; periodical articles, 115–20, 121, 123–24, 126–27; philosophy and theology courses, 112–13, 122, 124; physiology courses, 105–6, 118–19. *See also* theological perspectives, in 19th century education
"Existential Graphs" (Peirce), 315–16
experience, James's arguments, 77, 97, 319–21, 324. *See also* experimental *entries*
experimental approach: overviews, 248–55, 287–89, 313; Dewey's arguments, 316–19, 321, 322–23, 324, 337; James's theorizing, 319–21; Moore's discussions, 321–23; Peirce's arguments, 314–16, 326; as pragmatism focus, 313–14, 327, 329–31; scholarship opportunities, 337–39. *See also* social reform, experimental approach
experimental philosophy era, 248–49
experimental psychology: Dewey's school laboratory, 265–66; at Johns Hopkins, 133–34, 135; rise of, 248–49; at University of Chicago, 265–66, 307; at University of Michigan, 299–302, 306

factors of evolution. *See* variation phenomenon, causes debate
Factors of Organic Evolution (Spencer), 201, 202, 206, 219
Fairchild, James Harris, 123
Fallace, Thomas, 286
family ethics, Addams's arguments, 273
Farmville, Virginia, 267
Fay, Harriet Melusina (later Peirce), 29
feeling hypothesis, Peirce's, 216
feelings of pleasure, Spencer's hypothesis, 75
Ferrin, Allan Conant, 124
Ferris, William Henry, 10, 19
fifth cohort members, overview, 10–11, 16
first cohort members, overview, 10–11
First Principles (Spencer), 62, 63, 64, 66–67, 69, 70–71, 125, 144, 147
Fisch, Max, 15
Fischer, Marilyn, 277

Fisk College, Du Bois's education, 103–6, 110, 124–26
Fiske, Abby, 48–49
Fiske, John: overview of role in Darwin debates, 25–26, 55–56, 100; on Agassiz's *Contributions*, 29; criticism of James, 90–91; criticism of Paley, 113; criticisms of Agassiz, 47–49; endorsement of *Origin of Species*, 30, 34; on Gray-Agassiz debates, 30; meeting with Darwin, 49; Metaphysical Club colleagues, 24; on mind-environment relationship, 81; review of Buckle's *History*, 63–64; review of Mivart's *Genesis of Species*, 50, 53, 54, 55; Royce's interest in, 116; social change arguments, 90–91, 92–93, 95; on sociology's concerns, 95; Spencer defender role, 57, 59, 62–64, 73–74, 90–91, 100; on spontaneous generation, 116; theological perspectives, 24–25, 35–36, 74; on Wright's influence, 50
Fisk Herald, 126
Flint, Austin, 119
Flower, Elizabeth, 15, 16
Follett, Mary Parker, 11
Fontaine, William, 16
Forster, Paul, 328
fossil record, *28*, *51*, *54*, 204–5
Foster, Michael, 129
Foundations of Political Economy (Wagner), 141–42
Fourier, Charles, 234
fourth cohort members, overview, 14–16
Frank, Jerome, 13
"Freedman's Bureau, The" (Du Bois), 155
freedom arguments, 1, 72, 73, 234–35, 246, 313
Frege, Gottlob, 17
Frisch, Matthias, 265
Fulcomer, Daniel, 264
functional psychology, Dewey's, 179–80, 194, 242–43, 271–72, 280–81, 304

Galton, Francis, 51, 53
Gammon Theological Seminary, 156
Garrison, Jim, 242
Geiger, Josef Roy, 13
General History of Civilization in Europe (Guizot), 189
genetic characters, in natural selection arguments, 52

geology courses: Fisk College, 198; Harvard College, 111–12, 114, 198–99; Rockford College, 104–5, 120–21, 198; University of California, 106, 114; University of Vermont, 106–7, 198; Vassar College, 101

George, Henry, 232, 233–34

German philosophers, in Royce's Harvard courses, 181–82, 184

Germany: economic debates, 252–53, 259–60; for graduate studies, 127–29, 137–42, 184–85, 258–62, 299–300; James's studies in, 45–46

germinal selection theory, Weismann's, 229

germ-plasm: in Hall's biological ethics argument, 283–84; in Weismann's heredity theory, 199, 201–2, 220, 227, 228, 238

Germ-Plasm, The (Weismann), 238

germ theory example, in Peirce's theorizing, 224

Gilman, Charlotte Perkins, 10, 19

Girel, Mathias, 80

Godfrey-Smith, Peter, 73, 160, 169, 170, 193

Gooding-Williams, Robert, 160, 183

Goodman, Nelson, 14–15, 16

Goodsell, Willystine, 13

Gordon, Kate, 13, 17

Gothic architecture example, in Peirce's arguments, 223

Göttingen. *See* Germany

Goudge, Thomas, 6

Gould, Augustus Addison, 27–29

government's role, 259–60, 272. *See also* social reform, experimental approach

Grammar of Science, The (Pearson), 328

Gray, Asa, 30, 34, 38, 42, 102, 196

Gray, John Henry, 259–60

"great man" theories, 87–89, 90, 95. *See also* social change debate

Green, Nicholas St. John, 24, 27

Green, Thomas Hill, 162–63, 164, 171–73, 192

Grothe, Hermann, 249–50

Grundzüge der physiologischen Psychologie (Wundt), 76–77, 133

Guizot, François, 189

Gunton, George, 257

Guyot, Arnold, 35

habit: Cope's argument, 214; Dewey's arguments, 237–38, 241–44, 271–72, 316–17, 321; James's arguments, 206, 238, 305; Moore's argument, 306; Peirce's arguments, 215–16, 217, 219, 221, 224, 295–96, 315

Hadley, Arthur Twining, 153, 187, 259

Haeckel, Ernst, 109–10, 123, 127

Haldane, John Scott, 165, 173–74

Haldeman, George Bowman, 135–36, 155

Hall, Alexander Wilford, 123

Hall, Edwin, 110

Hall, G. Stanley, 132, 133–34, 283–84, 299

Hamlin, Fletcher, 123–24

Handbook of Psychology (Murray), 146

Hardie, Thomas Melville, 149

Harris, Leonard, 338

Harvard College: cohort groupings, 9, 10–11, 13, 16; ethics courses, 112–13, 143, 256; geology courses, 111–12, 114, 198–99; logic classes, 249; natural history classes, 27–29; philosophy courses, 73–74, 78–79, 86–87, 143–45, 180–84, 199, 205, 255–56; psychology classes, 78, 86–87, 249; writing classes, 143, 249

Harvard Philosophical Club, 144

"Has America a Race Problem?" (Cooper), 189

Hegel (Caird), 173

Hegel, G. W. F.: on agency of organism, 172; in Dewey's teaching, 166–67; and Dewey's thinking, 159, 291–92; on dual aspects challenge, 194–95; in Du Bois's race worldview, 185–86; on evolution of state, 187; natural sciences investigation, 162; on natural vs. spiritual development, 186; opposition to evolution theory, 161; in Royce's Harvard course, 181–82; Santayana's opposition to, 184

Hegel-Darwin linkage, 162–69, 180–85, 192–93, 291–93

"Hegel's Conception of Nature" (Alexander), 164–65

Helmholtz, Hermann von, 75–76, 79

Henke, Frederick, 11, 12

Herder, Johann Gottfried, 167, 188, 189

heredity theory, Weismann's: and Baldwin's theorizing, 238, 239; basic principles, 201–2; introduction of, 197–98, 199–200; Morgan's remarks on, 238; Osborn's remarks on, 209. *See also* Weismann, August

Herrick, Clarence Luther, 155

Hickman, Larry, 308
"History and Problems in Philosophy" (Royce), 183
history approach, in Peirce's variable causes arguments, 221–25
History of Civilization (Buckle), 63–64
History of Philosophy (Ueberweg), 131
Hocking, William Ernest, 13
Hodgson, Shadworth, 74–75
Holmes, Mary Emilie, 107–8, *109*, 120
Holmes, Oliver Wendell, Jr., 18, 24, 59, 69
Holmes, Oliver Wendell, Sr., 26–27
Holt, Henry, 98–99
honeycomb structure, natural selection arguments, 30–34
Hook, Sidney, 15
Hooker, Joseph, 27
Hookway, Christopher, 6, 291, 310
horn corals, Holmes's thesis, 108, *109*
Huebner, Daniel, 139, 300
Hull House, 149–50, *151*, 240, 255, 258, 263–65. *See also* Addams, Jane
Hull-House Maps and Papers, 252, 264, 265
Human Physiology (Draper), 67
Husserl, Edmund, 17
Hutchinson, Joseph Chrisman, 105
Hutton, Sarah, 7
Huxley, Thomas Henry, 41–42, 48–49, 105, 118–19, 121, 211, 231, 243–44

idealism: Dewey's studies and teaching, 132–33, 134, 146–49; in Du Bois's mental science class, 125; influences on Dewey's shift from, 159–60; Murray's argument, 146; Peirce's, 216; Royce's argument, 145–46; in second cohort, 157. *See also* Hegel *entries*; organism-environment relationship, dialectical approach
"Illustrations of the Logic of Science" (Peirce), 295–96
immortality symposium, *Christian Register*'s, 207–8
Imperative of Integration, The (Anderson), 288
"Inadequacy of Natural Selection, The" (Spencer), 226
Index magazine, 36, 56
induction mode, Peirce's argument, 311, 315
inductive ethics, Peabody's, 256

Influence of Darwin on Philosophy, The (Dewey), 2, 333
insect examples, 156, 227, 255
instincts, 30–34, 98, 205–6, 297, 310–11, 325
intelligence: Dewey's arguments, 148–49, 235, 237, 247, 318; Fiske's arguments, 74, 81; James's arguments, 74–75, 79–80, 96–98; in Peirce's logic model, 295–97; Renouvier's arguments, 71–72; Spencer's theory, 62, 73, 170–71; Ward's argument, 236. *See also* consciousness; organism-environment *entries*
interactionist position, social change, 91–94
interests, function of, 74–75, 77–78, 80, 83–84, 152–53, 320
interior medium concept, Bernard's, 68
internal-external forces. *See* natural selection hypothesis; organism-environment *entries*; variation phenomenon, causes debate
International Congress of Charities, Correction, and Philanthropy, 268–69
International Journal of Ethics, 282
intra-selection idea, Weismann's, 228–29
Introduction to Philosophy (Paulsen), 139
Introduction to Political Economy (Ely), 252–53
"Issues of Pragmaticism" (Peirce), 315

James, Denise, 286–87
James, Henry, 45, 46–47, 221
James, William: overview of role in Darwin debates, 25, 56, 100; Brazil expedition, 42–45; Campbell's reference to, 6; and cohort groupings, 10, 13; course offerings, 78, 112–13, 199, 249; criticisms of Agassiz, 42–43, 45, 47; criticisms of opponents of Darwinism, 41–42, 46, 55; and Dewey's thinking, 159, 297–98, 306, 307; education, 24n4, 40–41, 45–46, 75n41; on educational psychology, 304–5; on Fiske's book discussion, 73; on organism-environment relationship, 67–69, 80–81, 83–85; and Peirce's doubt-belief model, 295; on plasticity, 68–69; and pragmatism's logic model of inquiry, 291, 294, 304–5, 319–21, 326–27; reviews of Darwin's *Variation*, 46, 68–69; on role of habit, 206, 238, 305; on saintly conduct, 339; scholarship

James, William (cont.)
  contributions summarized, 3; tentativeness about Darwin's theories, 46–47, 69, 102; on working hypothesis, 330; Wright's influence, 49–50
James, William (and Spencer's theories): in Automata article, 84–85; in "Brute and Human Intellect" article, 80–81, 86, 120; critic role, 57, 59, 60; in Dewey's undergraduate education, 119–20; Fiske's criticism of, 90–91; in "Great Men" article, 90; Harvard Natural History courses, 78–79, 86–87; influence of European philosopher-psychologists, 72–73, 74–75; introduction to, 66–67; Johns Hopkins lectures, 130–31; in letter to publisher, 98–99; limitations of his criticisms, 81, 94–97; on natural selection hypothesis, 199; organism-environment relationship, 172, 298; in philosophy classes, 143; in "Remarks on . . . Mind" article, 79–80; in review of Carpenter's *Principles*, 75–76; in review of Wundt's *Grundzüge*, 76–78, 81–82; on role of consciousness, 83–85, 99; social-biological evolution analogy, 87, *88*; on source of ethics, 253–54; on Spencer's significance, 60–61, 99; in variation causes debate, 86, 97, 199, 205–7, 206–7; Wright's criticism of, 81–82
Jastrow, Joseph, 135
Jevons, William Stanley, 249, 251, 259
Johns Hopkins University, 11, 130–36, 155, 252, 292
Johnson, Charles Spurgeon, 13
Johnston, James Scott, 291–92

Kallen, Horace, 13, 302
Kant, Immanuel, 143–44, 159, 175–76, 181–82, 183
Kantor, Jacob Robert, 13
Kelley, Florence, 263, 265
Kidd, Benjamin, 150, 200, 230, 231, 232, 235–37, 244
Kilpatrick, William, 11
Kindergarten Magazine, 265
King, Clarence, 212–13, 219, 221, 224
Kitcher, Philip, 289, 336
Klein, Alexander, 97–98, 162
Kloppenberg, James, 254

Knies, Karl, 252
Koopman, Colin, 254
Krause, Ernst, 220
Kropotkin, Pyotr Alekseyevich, 150, 234, 241
Kuklick, Bruce, 4

labor-capitalism conflict, Pullman strike, 239–41
Labour and Life of the People (Booth), 261
Ladd-Franklin, Christine, 10, 18, 101
Lafferty, Theodore, 16
Laguna, Grace de, 4–5, 17, 323–24
Laguna, Theodore de, 4–5, 17, 323–24
laissez faire, arguments about, 230, 232, 236, 252, 260–62
Lamarckian evolution: Hegel's disagreement with, 161; in natural history courses, 198–99; Peirce's arguments, 212, 219, 221, 222–25; Spencer's statement, 201. *See also* Neo-Lamarckianism; Weismann, August
Langer, Susanne, 14–15, 16
Lankester, Edwin Ray, 202, 204
Laplace, Pierre-Simon, 144
Lathrop, Julia, 264–65, 269
Lavine, Thelma, 15, 16
Lawrence Scientific School, founding, 26
LeConte, Joseph, 105, 106, 114, 115, 116, 197–98, 207, 230–31
lectern socialists, 259–63
Lee, Harold Newton, 16
Leipzig. *See* Germany
Leonardo da Vinci, 249–50
Lessons in Botany (Gray), 103
Levi, Isaac, 16, 290
Levins, Richard, 169, 192–93
Lewes, G. H., 162–63
Lewis, C. I., 13, 16, 336
Lewontin, Richard, 169, 192–93
Lexell's Comet, 65
Liebig, Justus, 27
life functions, Dewey's argument, 318–19. *See also* functional psychology, Dewey's
Life of Agassiz, The (Wilder), 135
Lloyd, Alfred Henry, 10, 298–99, 300, 303
Locke, Alain, 13, 17, 285
Loeb, Jacques, 150
Logic (Wundt), 127, 137–38
Logical Investigations (Morris), 131
logic classes, 127, 132, 249

logic model of inquiry, pragmatism's: overview, 22, 290–91, 326–31; criticisms, 290, 323–24; Dewey's research and theorizing, 291–93, 297–304, 305–7, 312–13, 324, 327; with experimentation focus, 313–23; Johnston's biology argument, 291–93; Peirce's arguments, 293–94, 295–97, 308–11, 313, 314–16, 324–26, 327–29; relevance of James's theorizing, 294, 304–5, 319–21, 326–27; scholarship opportunities, 336–37; significance, 290–91
"Logic of Science, The" (Peirce), 40
Longfellow, Henry Wadsworth, 26
Longino, Helen, 338
Lotze, Rudolf Hermann, 128–29, 134, 139–40
Louisiana, Coast Survey expedition, 36–38, 39
love, in Peirce's *Monist* article, 211, 218–25
Lowell lectures, Peirce's, 40, 224, 311
Lurie, Edward, 27
Lyell, Charles, 42

Malisoff, William, 13–14
Mander, William, 164
Manual of Geology (Dana), 101, 104–5, 110, 113–14, 198
Manual of Zoology (Nicholson), 104
Margolis, Joseph, 4, 16
Martin, Henry Newell, 105–6, 133–34, 299
Martineau, James, 113
Marx, Karl, 232
Mary Holmes Seminary, 107
materialism: in Du Bois mental science class, 125; in Metaphysical Club discussions at Johns Hopkins University, 134–35; White's argument, 129. *See also* organism-environment *entries*
mathematics-based arguments, honeycomb structure, 32–34
Mayo-Smith, Richmond, 154, 270
McDermott, John, 16
McGranahan, Lucas, 91
McMurrich, James Playfair, 155
Mead, George Herbert: and cohort groupings, 9, 10, 11–12, 13, 16; graduate-level studies, 137–41, 184; Hull House lecture, 150; on learning process, 265–66; on Moore's significance, 321; on moral progress, 282; in organism-environment dialectic, 169–70, 192; research with Dewey, 298–99, 300, 301–2; scholarship contributions summarized, 3; on social reform, 273–74; Spencer's importance, 59; teaching positions and topics, 60, 151–52, 166–68, 249, 299–301; undergraduate education, 104–6, 109–10, *111*, 114, 122–24, 144; on working hypothesis, 331
Meadville Theological School, 34–35
mechanistic vs. teleological explanations. *See* evolution *entries*
medium of individual concept, Caird's, 175–76
Menand, Louis, 9, 10, 43–44
mental life. *See* consciousness; intelligence; psychology
mental science courses. *See* psychology courses
"Metaphysic" (Caird), 174–75
Metaphysical Club, Cambridge community, 20, 23–26, 55–57, 98–100. *See also specific members, e.g.*, Fiske, John; Peirce, Charles Sanders; Wright, Chauncey
Metaphysical Club, Johns Hopkins University, 134–35
Metaphysics (Lotze), 128
metaphysics courses, 127, 128, 137–38
meteorology, Wright's discussions, 65
methodeutic, in logic science, 294, 309–10, 313
Methods of Social Reform (Jevons), 251, 259
Methods of Study lectures, Agassiz's, 40–41
Michigan, University of. *See* University of Michigan
Microcosm magazine, 123
Mikrokosmus (Lotze), 140
Mill, John Stuart, 72, 81, 82
Miller, Charles Ransom, 208
Miller, David Louis, 15, 16
Mills, Charles Wright, 15, 16
Mind, 164–65, 174, 316
Mind and Body (Bain), 119
mind-body relationship. *See* organism-environment *entries*; psychology *entries*
Minute Logic (Peirce), 293, 325–26
Misak, Cheryl, 160
Mitchell, Clara Isabel, 265
Mitchell, Lucy Sprague, 13
Mitchell, Sandra, 338
Mivart, St. George, 50–55, 56, 124, 197, 248

Monist articles: Dewey's, 241, 247; LeConte's, 230–31; Morgan's, 238

Monist articles, Peirce's: about his existential graphs, 316; "Architecture of Theories," 212–14, 215–16, 218–19, 224, 324; "Design and Chance," 207; "Evolutionary Love," 211, 218–25; "Issues of Pragmaticism," 315; reception of, 216–17; request for, 211–12; "What Pragmaticism Is," 314, 315

Montague, William Pepperell (father), 24

Montague, William Pepperell (son), 290, 336

Montgomery, Edmund, 217, 220

Moore, Addison Webster, 11–12, 282–83, 306, 321–23, 331

Moore, James, 24

Moral Order and Progress (Alexander), 177–78

"Moral Philosopher and the Moral Life, The" (James), 253

moral progress: overviews, 282–83, 288–89; Addams's arguments, 277–79; Alexander's arguments, 177–78; Crummell's arguments, 190; Dewey's arguments, 178–79, 233, 240–42, 252, 279–81; experimentation linkage, 253–54; James's arguments, 253–54, 339; Moore's argument, 282; Morgan's argument, 282; Peirce's arguments, 327–29; Royce's arguments, 143–45; as theme in pragmatism's logic model, 282–83, 288–89, 327–29. *See also* social reform, experimental approach

Morgan, Conwy Lloyd, 238

Morgenbesser, Sidney, 16

Morley, Margaret Warren, 149–50, 149149

Morris, Charles W., 16, 17

Morris, George Sylvester, 130, 131–33, 163, 192, 292, 298

Moses, Wilson, 191

Movements of Thought course, University of Michigan, 161, 166–68

Munk, Hermann, 300

Murphey, Murray G., 16

Murray, John Clark, 146

Museum of Comparative Zoology, 39

mutual aid theory, Spencer's, 233–34, 241, 253

NAACP (National Association for the Advancement of Colored People), 1–2, 156

Nagel, Ernest, 15

Nation magazine, 219, 308, 309

National Negro Conference, 246–47, 285

"Nation-Making" (Allen), 88, 89–90

"natural history" approach, inquiry: Dewey's research and theorizing, 297–304, 305–7; James's arguments, 298, 304–5; Peirce's arguments, 293–94, 295–97, 308–13; as theme in pragmatism's logic model, 326–27

natural history courses: Agassiz's Harvard classes, 27–29; Chase's undergraduate studies, 110; curriculum requirements, 102–3. *See also* biology courses; geology courses; zoology courses

naturalistic vs. idealistic pragmatism, as false conflict, 160–61. *See also* idealism

natural selection hypothesis: criticism themes, 196–99; honeycomb structure arguments, 30–34; in James's social change argument, 90, 97; Mivart's opposition arguments, 50–54; Peirce's arguments, 218, 296–97, 310–11, 325–26; in Ritchie's Hegel-Darwin essay, 166; Wright's arguments, 47, 51–53. *See also* organism-environment *entries*; Spencer-Weismann debate; struggle/striving theme; variation phenomenon, causes debate

Nature journal, 200, 202, 204

Naturphilosophie (Hegel), 162, 165

nebular hypothesis, 121

"Need of New Ideas and New Aims for a New Era, The" (Crummell), 190

Neo-Darwinism, term origins, 202–3. *See also* variation phenomenon, causes debate; Weismann, August

Neo-Lamarckianism, 203, 204–5, 211–14

neuter insects example, in Weismann's argument, 227

Newer Ideals of Peace (Addams), 252

New School for Social Research, 13

Newton, Isaac, 208–9

New York Times, 208–9

New York World, 36

Nicholson, Henry Alleyne, 104, 198

nonideal theory, Anderson's, 288

North American Review, 30, 46

Norton, Charles Eliot, 46, 49

Norton, Grace, 49, 82

Numbers, Ronald, 25

Oberlin College, Mead's education, 104, 105–6, 109–10, *111*, 114, 122–24
odor example, Dewey's argument, 316–17
"On Archaesthetism" (Cope), 214
On the Genesis of Species (Mivart), 50–54
"On the Hypothesis That Animals Are Automata" (Huxley), 118–19
On the Justification of the Darwinian Theory (Weismann), 199
Open Court, 211–12, 231, 238
"Order of Nature, The" (Peirce), 296
organism-environment mismatch. *See* social reform, organism-environment perspectives
organism-environment relationship: Abbot's arguments, 70–71; cohort perspectives compared, 16–19; Comte's argument, 81; Darwin's statements, 68–69, 86; Dewey's arguments, 236–37, 242–45, 303–4, 307, 318–19; Dewey's interest and teaching, 119–20, 146–49; in Du Bois's teaching, 154; Fiske's arguments, 73–74; James's arguments, 67–69, 80–81, 83–85, 298, 307; Lloyd's argument, 302; in Mead's teaching, 152–53; Pillon's argument, 72; Renouvier's arguments, 71–72; scholarship opportunities, 334–35, 334–37; Spencer's theory, 62, 71, 73, 82–83, 84, 86–87, 125, 232–33; in Spencer-Weismann debate, 226–29; as theme in pragmatism logic model, 291, 326–28; Ward's argument, 236; Wright's arguments, 82–83. *See also* James, William (and Spencer's theories); social change debate; variation phenomenon, causes debate
organism-environment relationship, dialectical approach: overview, 21, 161, 168–70, 180, 192–95; Alexander's arguments, 177–78; Caird's arguments, 173, 174–77, 179; Dewey's arguments, 178–80, 194, 195, 235, 242–43; Green's arguments, 171–73; Haldane's arguments, 173–74; Spencer's foundational theorizing, 170–71
originator of the fittest, Cope's, 85, 95
Origin of Species (Darwin): AAAS discussions about, 30; Bowen's criticisms, 30–32; Fiske's endorsement of, 34; James's remarks about criticisms, 41–42; in Meadville Theological School library, 35;

B. Peirce's criticisms, 38–39; C. Peirce's remarks about, 36, 39–40; publication of, 24, 29; scientific acceptance, 24–25, 102; Wright's arguments for, 29–30, 32–34; Wundt's arguments about, 127. *See also* Darwin, Charles; evolution theory entries
Origin of the Fittest (Cope), 214
Osborn, Henry Fairfield, 2, 203, 209–11, 229–30
Otto, Max, 13, 16
Outlines of a Critical Theory of Ethics (Dewey), 178
Outlines of Cosmic Philosophy (Fiske), 73–74, 81, 92, 94
Outlines of Lectures on the History of Philosophy (Elmendorf), 124
Outlines of Psychology (Sully), 146
Owen, Robert, 234
Oxford Hegelians: on Hegel-Darwin linkage, 161, 162–66; in organism-environment dialectic, 169–79, 180

Paley, William, 113
panmixia concept, Weismann's, 231
Park, Robert, 10, 11
Pasteur example, in Peirce's arguments, 224
Paulsen, Friedrich, 139–40
Peabody, Andrew Preston, 117
Peabody, Francis Greenwood, 11, 18–19, 24, 73, 143, 255–56, 269
Peabody Institute, 136–37
Pearson, Karl, 328
"Pedagogical Experiment, A" (Dewey), 265–66
Peirce, Benjamin (father of Charles), 26, 27, 30, 38–39
Peirce, Benjamin Mills (brother of Charles), 37–38, 39, 209
Peirce, Charles Sanders: overview of role in evolution debates, 25, 56, 100, 157–58; chance arguments, 207–8; Coast Survey expedition, 36–38, 39; cohort members, 10–11; on consciousness, 215–17, 295, 317; criticism of Spencer's theories, 67; criticisms of Weismann, 219, 220–21, 245; definition of factor, 196; doubt-belief model, 295–97, 310, 326; education, 29, 39; on experimental approach, 248, 249, 250–51, 330; on Hegel-Darwin linkage, 166; on impact of *Origin of*

359

Peirce, Charles Sanders (*cont.*)
  Species, 5, 39–40; at Johns Hopkins University, 132, 134–35; and logic model arguments, 291, 293–94; in Metaphysical Club, 23, 24; on "natural history" approach, 293–94, 295–97, 308–13; on natural selection hypothesis, 296–97, 310–11; in *NY Times*'s Spencer debate, 208–9, 210–11; protoplasm properties interest, 6, 7; reasonableness arguments, 313–15, 325, 326, 328–29; scholarship contributions summarized, 3; on Spencer's significance, 59, 62; in variation causes debate, 207–8, 212–13, 218–25, 325–26; on Wright's influence, 49, 50. See also *Monist* articles, Peirce's
Peirce, Harriet Melusina (earlier Fay), 29
Peirce, James Mills, 39
Peirce, Sarah, 38
perception, in mind-environment debate, 75–77, 79
Perkins, George Henry, 106–7, 118
Phenomenology of Spirit (Hegel), 183–84, 186
Philadelphia, Seventh Ward survey project, 266–67
Philadelphia Negro, The (Du Bois), 267, 268–69, 270–71, 284
Philosophical Review, 184
philosophy courses: at Clark University, 155; at Columbia University, 321; in Germany, 128–29, 138–41, 184–85; at Harvard, 73–74, 143–45, 180–84, 199, 205, 255–56; at Johns Hopkins University, 131, 132–33, 321; at Oberlin College, 122; at University of Chicago, 166, 179–80, 266; at University of Michigan, 60, 147, 151–52, 166–68, 234–35, 292, 298
Philosophy of Evolution (Spencer), 61, 63, 67
Philosophy of History (Hegel), 185, 186
Philosophy of Science Association, 338
Philosophy of Science journal, 14
phyllotaxis article, Wright's, 52
Physics and Politics (Bagehot), 88
physiological psychology, 78, 118–19, 299–300
physiology courses, 105–6, 129, 133–34, 149
Physiology of Man (Flint), 119
Piatt, Donald Ayres, 16
Pillon, François, 72
Pinkard, Terry, 162

plasticity: Cope's argument, 214; James's arguments, 68–69, 238; King's argument, 213; Morgan's argument, 238
pleasure, Spencer's theory, 75
political economy courses, 141–42, 256–57, 258–62
political experimentation, discussion history, 250–53
"Politics" (Treitschke), 186–87
Popular Science Monthly, 115, 116–17, 121, 129, 200, 211, 295
Positive Philosophy lectures, Fiske's, 73–74, 92–93
positivism, in second cohort, 157
Potter, Alonzo, 43
Poulton, Edward Bagnall, 166, 202, 209
Practicable Socialism (Barnett and Barnett), 255
"Pragmatic Movement of Contemporary Thought, The" (Dewey), 321
Pragmatism and Its Critics (Moore), 321–22
pragmatism-biology relationship, overview: cohort methodology, 8–16; literature review, 4–5, 6; origins, 2–4; scholarship opportunities, 7–8, 16–20, 334–37; significance of inquiry model, 290–91. See also specific topics, e.g., evolution entries; logic model of inquiry, pragmatism's; Peirce, Charles Sanders
Pragmatism (James), 319, 320
"Primeval Ghost-World" (Fiske), 116
primitive people theme, in social evolution arguments, 285–87
Principles of Biology (Spencer), 45–46, 62, 63, 69, 70, 81, 209
Principles of Mental Physiology (Carpenter), 75
Principles of Morality (Spencer), 63
Principles of Philosophy (Peirce), 330
Principles of Physiological Psychology (Wundt), 76–77
Principles of Psychology (James), 96, 98–99, 152, 205–6, 298, 306
Principles of Psychology (Spencer), 35, 63, 78, 94, 119, 292
Principles of Science, The (Jevons), 249
Principles of Sociology (Spencer), 93–94, 147, 253, 292
Principles of Zoology (Agassiz and Gould), 27–29
production-preservation debate. See variation phenomenon, causes debate

"Progress: Its Law and Cause" (Spencer), 63
Progressive Journal of Education, 319
progressivism, Wright's arguments, 64–66. *See also* evolutionary philosophy, Spencer's
protoplasm properties: Peirce's interest, 6, 7; White's lecture, 129–30. *See also* germ-plasm
Psychic Factors of Civilization, The (Ward), 230
Psychological Bulletin, 307
Psychological Review, 229–30
psychology: Dewey's research and theorizing, 145, 298–304, 305–7, 312–13; James's theorizing, 304–5, 306; Peirce's arguments, 310–11; Royce's argument, 145. *See also* consciousness; intelligence; James, William (and Spencer's theories)
psychology courses: at Fisk College, 124, 125; in Germany, 138; at Harvard, 78, 86–87, 249; at Johns Hopkins, 133–34; at University of Chicago, 156, 281, 303, 304; at University of Michigan, 146–49, 152–53, 298, 299–301, 302; at University of Vermont, 118–19
public management arguments, 259–60, 272
Pullman labor strike, 239–41
Putnam, Henry Ware, 24
Putnam, Hilary, 4, 16
Putnam, James Jackson, 79

Quatrefages, Armand de, 68
Quine, W. V. O., 15, 16, 336

Races of the Old World (Brace), 45–46
race themes: in eugenics arguments, 284–85; in social evolution arguments, 155, 185–86, 187, 189–92, 275–77; in social reform discussions, 266–68, 269–71; in social science research proposals, 288–89
racism, Shaler's statements, 111–12
Rae, John, 261
Railroad Transportation (Hadley), 259
Ratner, Joseph, 15
reaction-time experiments, Wundt's, 76–77
Reade, William Winwood, 55
reasonableness, in Peirce's logic theory, 313–15, 325, 326, 328–29
reciprocal causes, in organism-environment dialectic, 169–70, 172, 173–76, 177–78, 193

redintegration concept, Hodgson's, 74–75
"Reflex Arc Concept, The" (Dewey), 180, 242
Rehnisch, Johann Eduard, 128
Reign of Law, The (Duke of Argyll), 197
"Rejoinder to Professor Weismann, A" (Spencer), 227
Religion and Science (Seydel), 138–39
Religious Aspect of Philosophy, The (Royce), 143–44
religious perspectives. *See* theological perspectives *entries*
"Remarks on Spencer's Definition of Mind as Correspondence" (James), 79–80
Remsen, Ira, 134
Renan, Ernest, 167
Renouvier, Charles, 71–73, 80, 81
Report on the Progress of Anthropology (Quatrefages), 68
Rescher, Nicholas, 16
rhetoric course, Oberlin College, 122
Ritchie, David George, 162, 164, 166, 192
Rockford College, Addams's education, 104–5, 107–8, 120–21
Romanes, George John, 202–3, 207
Rorty, Richard, 4, 16, 160, 290, 337
Royce, Josiah: and cohort groupings, 9, 11, 13; graduate studies, 127–29, 130–31; in organism-environment dialectic, 161, 192; on Peirce's arguments, 217; Spencer's importance to, 59; and Spencer's theories, 115–16, 253; teaching positions and topics, 143–46, 181–83, 184, 249; and theology-evolution issue, 114–15; undergraduate education, 103–4, 106, 114, 117
Rudner, Richard, 16
rugose coral, Holmes's thesis, 108, *109*
Russell, Bertrand, 17
Russell, Francis Calvin, 212
Russell, John Edward, 317

Saint-Simon, Henri de, 71–72
Santayana, George, 9, 10, 183–84
Saturday Club, 26–27
Schäfer, Axel, 254
Schelling, F. W. J., 173, 194
Schmoller, Gustav Friedrich, 141, 142, 251, 261–62, 269
Schneider, Herbert, 13
School of Applied Ethics, 258

science classes, University of Michigan, 110. *See also* natural history courses; psychology courses
Science journal, 209
science models, cohort comparisons, 16–19
Science of the Mind, The (Bascom), 125
Scientific Association, Rockford College, 120–21
"Scientific Bases of Anarchy, The" (Kropotkin), 234
sculpture metaphor, in mind-environment debate, 85
second cohort members, overview, 10–12
Secret of Hegel, The (Stirling), 165
Sedgwick, William Thompson, 134
Seigfried, Charlene Haddock, 254
self-adaptation concept, Caird's, 176–77
self-control, in Peirce's logic theory, 315
Selfish Gene, The (Dawkins), 6
Sellars, Roy Wood, 4
Sellars, Wilfrid, 15, 16
sensation, in mind-environment debate, 75–76, 77, 79, 82
settlement houses, experiment opportunities, 255, 258, 264, 268–69. *See also* Hull House
Seydel, Georg Karl Rudolf, 138–39
Shaler, Nathaniel Southgate, 111–12, 114, 198–99
ship-steering metaphor, moral progress, 282
Sidgwick, Henry, 163
Sigwart, Christoph, 310
Simons, May Wood, 150
Singer, Edgar, Jr., 11, 12, 14, 16
Skagestad, Peter, 6
Skorburg, Joshua, 265
Small, Albion Woodbury, 264
smell example, Dewey's argument, 316–17
Smith, John Edwin, 16
Social and Ethical Interpretations in Mental Development (Baldwin), 244
social change debate: Allen's arguments, 87–88, 89–90, 91; Bagehot's arguments, 88–89; Fiske's arguments, 90–91, 92–93, 95; James's arguments, 86–87, 90–91, 94–97; Spencer's remarks, 86–87, 93–94. *See also* social evolution
social evolution: Dewey's arguments, 1–2, 242–47, 279–82; Du Bois's arguments, 153–54, 275–76; in Du Bois's teaching, 153–54; eugenics arguments, 283–87; evaluation problem, 338–40; Guyot's argument, 35; Hadley's arguments, 153–54; Hull House lectures, 150; implications from Spencer-Weismann debate, 229–39; race themes, 155, 185–86, 187, 189–92, 246; striving/struggle theme, 187–88; in Turner's teaching, 155; Wright's argument, 35. *See also* moral progress
Social Evolution (Kidd), 230, 231, 235–36
social heredity. *See* social evolution
socialism, 231–35, 259–63
Social Philosophy and Religion of Comte, The (Caird), 173
social reform, experimental approach: overview, 268, 287–88; advocacy arguments, 250–52, 255–56, 258, 267–69, 281; Dewey's school laboratory, 265–66; Du Bois's black population survey projects, 266–67; in economics courses, 258–62; Hull House survey project, 263–65; in philosophy courses, 255–56; role of statistical methodology, 256–57
social reform, organism-environment perspectives: overviews, 268–69, 275; Addams's arguments, 273, 274–75, 277–79; Dewey's arguments, 271–72, 274–75; Du Bois's arguments, 269–71, 276–77; Mead's arguments, 273–74
"Social Reforms as Objects of University Study" (Peabody), 256
Social Statics (Spencer), 233
social Weismannism, 229–39. *See also* social evolution
Society for Psychical Research, 135
Sociobiology (Wilson), 6
sociology, Spencer's, 86–87, 90–91, 92, 93–96
sociology classes, Du Bois's teaching, 153, 154
"Some Stages of Logical Thought" (Dewey), 309
Song of Life, A (Morley), 149
"Special Report on Negro Domestic Service" (Eaton), 267
speculative grammar, 294
Spencer, Charles Worthen, 264
Spencer, Herbert: criticisms of Agassiz's theorizing, 46, 49; Darwin reputation compared, 67; and Dewey's undergraduate education, 119–20; and Du Bois's undergraduate education, 124, 125; economic life description, 252; *Encyclopaedia Britannica* article about, 211; ethics

system, 232–34, 235, 253, 281, 282–83; Fiske's defender role, 57, 59, 62–64, 73–74, 90–91, 100; Huxley's article about, 211; influence of, 57–59, 61, 332–33, 335–36; Pillon's argument about, 72; in Renouvier's review of Saint-Simon, 71; on social change factors, 86–87, 93–94; on visibility of inheritance debate, 202; Wagner's argument for, 141–42. *See also* evolutionary philosophy, Spencer's; James, William (and Spencer's theories); Spencer-Weismann debate

Spencer, Herbert, publications of: *The Data of Ethics*, 143, 233, 234; *Essays*, 219; *Factors of Organic Evolution*, 201, 202, 206, 219; *First Principles*, 62, 63, 64, 66–67, 69, 70–71, 125, 144, 147; "The Inadequacy of Natural Selection," 226; *Philosophy of Evolution*, 61, 63, 67; *Principles of Biology*, 45–46, 62, 63, 69, 70, 81, 209; *Principles of Morality*, 63; *Principles of Psychology*, 35, 63, 78, 94, 119, 292; *Principles of Sociology*, 93–94, 147, 253, 292; "Progress: Its Law and Cause," 63; *Social Statics*, 233; *Study of Sociology*, 115; *System of Philosophy*, 61, 63, 67

Spencer-Weismann debate: arguments in, 226–29, 244; and Dewey's functional psychology, 242–44; implications for social evolution, 229–39; responses to, 226–31, 237

Spinoza, Baruch, 144

Spirit of Modern Philosophy, The (Royce), 161, 181, 183

spiritualism talks, Johns Hopkins University, 134–35

spontaneous activity concept, Bain's, 82

spontaneous variation arguments. *See* variation phenomenon, causes debate

stability of species, in natural selection arguments, 51–52

Starr, Ellen Gates, 255

state management arguments, 259–60, 272

statistical methodology, 256–57

Statistics and Sociology (Mayo-Smith), 154

Stearns, Oliver, 34–35

Stephen, Leslie, 329

Stirling, James Hutchinson, 165–66

"Strivings of the Negro People" (Du Bois), 185

structure similarities, in natural selection arguments, 51, 52–53

struggle of parts hypothesis, Weismann's, 228

struggle/striving theme: Cooper's argument, 189; Crummell's argument, 190; in Darwin's theory, 188; Dewey's arguments, 236–37, 239–42, 303; Du Bois's arguments, 191–92; Guizot's argument, 189; Hegel's argument, 186, 188; Herder's argument, 188; Huxley's argument, 231; Kidd's arguments, 231–32, 235–36; Kropotkin's argument, 234; Le Conte's arguments, 230–31; Peirce's arguments, 218, 294, 295–97; Treitschke's argument, 187; Ward's argument, 230. *See also* moral progress; Spencer-Weismann debate

Stuart, Henry Waldgrave, 11

Studies in Logical Theory (Dewey), 293, 305, 308, 327

Studies in Logic (Peirce, ed.), 296–97

Studies in the Theory of Descent (Weismann), 199

Study of Ethics (Dewey), 330–31

Study of Religion (Martineau), 113

Study of Sociology (Spencer), 115

"Study of the Negro Problems, The" (Du Bois), 267–68, 269–70

subject-object relation. *See* organism-environment relationship, dialectical approach

"Subtle Problems of Charity, The" (Addams), 277–78

Sully, James, 146

Suppes, Patrick, 338

System of Logic (Mill), 82

System of Philosophy (Spencer), 61, 63, 67

System of Philosophy (Wundt), 137

System of Positive Politics (Comte), 175–76

Taft, Julia Jessie, 13, 17

"Talented Tenth, The" (Du Bois), 284–85

Talks to Teachers (James), 304

Tanner, Amy Eliza, 10, 11, 12

Taussig, William, 258

Taylor, James Branch, 126

Taylor, Paul C., 288

Tefft, Benjamin Franklin, 147

Tenney, Sanborn, 101, 103–4

textbooks, natural history courses, 27–29, 101, 103–5, 113–14

textile mill statistics, 257

theological perspectives: Abbot's, 24–25, 36; Agassiz's, 26, 27, 29, 43–44, 48; Fiske's, 24–25, 35–36, 74; Gray's, 30, 103; James's, 42–43, 83; at Meadville Theological School, 34–35; B. Peirce's, 38; Spencer's, 46, 124; Wright's, 47, 66

theological perspectives, in 19th century education: ethics course, 113; graduate-level studies, 138, 139–41; natural history classes, 103, 104, 110–11, 113–14; periodical articles, 116–17, 123–24; in philosophy and religion classes, 112–13, 122, 124; psychology class, 125; student responses, 114–16, 122–23, 126–27

"Theory of Emotion" (Dewey), 242

Theory of Method (Wundt), 127

"Theory of Probable Inference, A" (Peirce), 296–97

third cohort members, overview, 13–14

Thompson, D'Arcy, 58

Time and Space (Hodgson), 74

Toynbee Hall, London, 255, 260–61

transcendentalism, in Abbot's critique of Spencer, 70

tree example, Wright's progressivism critique, 63–64

Treitschke, Heinrich von, 186–87

Trendelenburg, Friedrich Adolf, 131, 139

Tufts, Irene, 168

Tufts, James Hayden, 11, 12, 166, 285, 298–99

Turner, Charles Henry, 155–56

Tylor, Edward Burnett, 18, 116

Tyndall, John, 48–49

Ueberweg, Friedrich, 131

Uniform Hours of Labor (Wright), 256–57

United States Coast Survey, 27, 36–38, 39

unity argument, Addams's, 240–41

University of California, 102–5, 106, 114–17, 143

University of Chicago: biology/zoology classes, 150; and cohort groupings, 11, 12, 13, 16; Dewey's arrival, 239; ethics courses, 242–43, 271, 274, 322; experimental psychology research, 265–66, 307; logic classes, 249; philosophy courses, 166, 179–80, 266; psychology courses, 156, 281, 303, 304; zoology/biology courses, 156

University of Cincinnati, 155

University of Michigan: Chase's undergraduate studies, 110; and cohort groupings, 10; Dewey's teaching, 146–49, 249, 253; ethics courses, 298; experimental psychology research, 299–302, 306; Holmes's graduate studies, 107, 108; logic classes, 249; philosophy courses, 60, 147, 151–52, 166–68, 234–35, 292, 298; psychology courses, 146–49, 152–53, 298, 299–301, 302

University of Oxford, 162

University of Pennsylvania, 14, 16, 266

University of Vermont, Dewey's education, 103–5, 106–7, 117–20, 292

University of Wisconsin, 13, 16

University Record, 152, 301

university settlements. *See* Hull House; settlement houses, experiment opportunities

Up from Slavery (Washington), 277

"Value of Saintliness, The" (James), 339

Variation of Animals and Plants under Domestication (Darwin), 46, 68–69, 86, 250

variation phenomenon, causes debate: Cope's arguments, 204, 214; Darwin's argument, 86; Dewey's arguments, 236–37, 244, 322, 324; importance of Weismann's theory, 199–200, 201–2, 205–6; James's arguments, 86, 97, 205–7; King's argument, 212–13; Lankester's arguments, 204; limitations of natural selection hypothesis, 196–99; Montgomery's argument, 220; Moore's discussion, 322–23; Osborn's arguments, 204–5, 209–10, 237; Peirce's arguments, 207–8, 212–13, 218–25, 325–26; Romanes's arguments, 202–3; Spencer's arguments, 86, 95–96, 201, 202, 206–7, 226–27; Wallace's arguments, 203–4; Weismann's arguments, 199–200, 231, 238. *See also* organism-environment entries

Vassar College, 101, 107

Verein für Sozialpolitik, 252, 259–60

Vermont, University of, Dewey's education, 103–5, 106–7, 292

Verrill, Addison Emery, 106

Vestiges of the Natural History of Creation (Chambers), 26, 35

vitalism answers, in Abbot's critique of Spencer, 71

Wagner, Adolph, 141–42, 259, 260, 261–62
Walker, Francis Amasa, 153, 260
Wallace, Alfred Russel, 35, 49, 89, 95, 136–37, 197, 203–4
Wang Yangming, 11
Ward, Lester Frank, 230, 236, 240
Ward, Thomas Wren, 69–70
Warner, Joseph Bangs, 24, 73
Washington, Booker Taliaferro, 277
Washington Post, 58
Waters, C. Kenneth, 338
Wealth and Progress (Gunton), 257
Weismann, August: overview, 21–22; at BAAS meeting, 202; Dewey's arguments, 245, 246; Du Bois's arguments, 245–46; introduction of heredity theory, 199–200; James's introduction to, 205–6; Kidd's arguments, 235–36; on Kidd's theorizing, 232; Krause's criticism, 220; Montgomery's criticism, 220; Osborn's comments about, 209, 210; Peirce's arguments against, 219, 220–21, 245; social evolution implications, 229–39; Ward's criticism, 230. *See also* heredity theory, Weismann's; Spencer-Weismann debate
Weiss, Paul, 15, 16
Welchman, Jennifer, 178
"What Pragmaticism Is" (Peirce), 314, 315
Wheeler, William Morton, 156
Whewell, William, 222
White, Carol Wayne, 189
White, Frances Emily, 129–30
White, Morton, 15, 159
Whitehead, Alfred North, 16
Whitman, Charles Otis, 156
Whitman, Sarah Wyman, 307
Wiener, Philip Paul, 4, 15
Wilberforce University, 190, 263
Wilder, Burt, 135
Williamson, Joel, 185, 187
Wilson, Edward Osborne, 6
Wimsatt, William Church, 6
Winchell, Alexander, 108, 110
Women and Economics (Gilman), 19
Women's Medical College, Pennsylvania, 129–30
Wood, Alphonso, 102
Woodbridge, Frederick, 316

Woods, Robert Archey, 264, 268–69
Woodward, James, 338
Woolley, Celia Parker, 149
working conditions, statistical investigations, 257
working hypothesis, in pragmatism's logic model, 329–31. *See also* experimental approach
"Working Hypothesis in Social Reform, The" (Mead), 273
Wright, Albert Allen, 109–10, 198
Wright, Carroll Davidson, 256–57, 258, 263, 267
Wright, Chauncey: overview of role in Darwin debates, 25–26, 54–56, 100; in Agassiz's natural history class, 27; criticism of Agassiz, 47; criticisms of Mivart's philosophizing, 53–54, 248; criticisms of Spencer's theories, 57, 59, 62, 64, 66, 100; criticisms of Wallace's arguments, 35, 49; defense of natural selection idea, 47; honeycomb structure arguments, 32–34; influence of, 49–50; on James's critique of Spencer, 81–82; in James's review of Wundt's *Principles*, 81–82; Metaphysical Club colleagues, 24; on *Origin of Species*, 29–30; Peirce's letter about *Origin of Species*, 36; progressivism critique, 64–66; review of Mivart's *Genesis of Species*, 50, 51–55; stability of species arguments, 51–52; teaching position, 29; on working hypothesis, 330
writing classes, Harvard, 143, 249
Wundt, Wilhelm, 74, 75–77, 78, 127, 133, 134, 137–38, 278, 279
Wyman, Jeffries, 41, 42

Youmans, Edward Livingston, 115, 116–17, 211
Youmans, William Jay, 211
Young, Ella Flagg, 10

Zamir, Shamoon, 183, 184
Zammito, John, 188
zoology courses: Oberlin College, 109–10, *111*, 198; Rockford College, 120–21; University of California, 103–4, 106; University of Chicago, 156; University of Vermont, 106–7, 198; Vassar College, 101, 107

Printed in the USA
CPSIA information can be obtained
at www.ICGtesting.com
LVHW020750200324
774984LV00003B/342